Lecture Notes in Mathematics

Edited by A. Dold and B. Eckmann

1344

J. Král J. Lukeš
I. Netuka J. Veselý (Eds.)

Potential Theory Surveys and Problems

Proceedings of a Conference held in Prague, July 19–24, 1987

Springer-Verlag
Berlin Heidelberg New York London Paris Tokyo

Editors

Josef Král
Mathematical Institute of the Czechoslovak Academy of Sciences
Žitná 25, 11567 Prague 1, ČSSR

Jaroslav Lukeš
Ivan Netuka
Jiří Veselý
Faculty of Mathematics and Physics, Charles University
Sokolovská 83, 18600 Prague 8, ČSSR

Mathematics Subject Classification (1980): 05C99, 20H10, 30A05, 30C85, 31-XX, 35Bxx, 35C15, 35G15, 35Jxx, 35Kxx, 41A30, 42B20, 45A05, 45B05, 45E05, 45L10, 46A55, 46E30, 46E35, 47Bxx, 47D05, 49Dxx, 53C20, 54D35, 60Jxx, 65Nxx

ISBN 3-540-50210-6 Springer-Verlag Berlin Heidelberg New York
ISBN 0-387-50210-6 Springer-Verlag New York Berlin Heidelberg

Printed in Germany

Printing and binding: Druckhaus Beltz, Hemsbach/Bergstr.
2146/3140-543210

TO THE MEMORY OF

PROFESSOR MARCEL B R E L O T (1903-1987)

PREFACE

Within the tradition of meetings devoted to potential theory, a conference on potential theory took place in Prague on 19 - 24 July 1987. There were 116 participants and 37 accompanying persons from 26 countries. Some participants were partially supported by IMU.

The Conference was organized by the Faculty of Mathematics and Physics, Charles University, with the collaboration of the Institute of Mathematics, Czechoslovak Academy of Sciences, the Department of mathematics, Czech University of Technology, the Union of Czechoslovak Mathematicians and Physicists, and the Czechoslovak Scientific and Technical Society. The Conference was held under the auspices of the Rector of Charles University, Professor Z. Češka, corresponding member of the Czechoslovak Academy of Sciences.

The Conference was concerned with various aspects of potential theory, including the applications of potential theory in other areas

Thirteen one-hour survey lectures were delivered in the course of the Conference and 11 survey papers based on these lectures are included in the first part of the present volume. Totally 69 scientific communications from different branches of potential theory were presented during the Conference. The majority of them will appear in the Proceedings of the Conference to be published by Plenum Publishing Company.

On the occasion of the Conference, a collection of problems from potential theory was compiled and is included, with additional commentaries, in the second part of this volume.

Potential theory is nowadays a vast mathematical field, which in many respects truly reflects the state of the art in mathematics and mathematical analysis in particular. From the very beginning, potential theory whose roots are in physics, has developed into an independent domain with significant applications in the theory of functions, and in differential equations including their numerical treatment, which has also disclosed unexpected connections with other branches of mathematics like probability theory, and which employs various methods of modern mathematics including topology and axiomatics.

The present volume may thus be of interest to mathematicians specializing in the above mentioned fields and also to everybody interested in the present state of potential theory as a whole.

Josef Král Ivan Netuka

Jaroslav Lukeš Jiří Veselý

Praha, October 1987

TABLE OF CONTENTS

POSITIVE HARMONIC FUNCTIONS AND HYPERBOLICITY

Alano Ancona

Campus d'Orsay, Bat.425, Université Paris 11, Orsay , France

Consider a space X equipped with a Potential theory $\mathcal{P}$, and a geometry defined by some metric d on X ; assume further that $\mathcal{P}$ and d are compatible in some reasonable sense. The first purpose of this paper is to describe , in a very simple framework, some results of the following sort : if the metric d has certain properties -with a strong hyperbolic flavor-, then it is possible to describe in simple geometrical terms the Martin boundary of $\mathcal{P}$; the key point is a form of the so-called boundary Harnack inequalities (theorem 3.3 below). These inequalities seems to be strongly related to some kind of hyperbolicity of d. Another striking consequence of the hyperbolicity of d is that the needed compatibility between $\mathcal{P}$ and d is quite weak.

Results of this kind were derived in [2] in the framework of a Potential theory on a Riemannian manifold X defined by some second order elliptic operator over X. [2] was essentially motivated by a paper of M. Anderson and R. Schoen [3] who treated the case of the Laplace-Beltrami operator over X, when X is simply connected and with sectional curvatures lying in between two negative constants. The approach in [2] is based on very simple geometrical properties and extends naturally to other settings of Potential theory.

In this exposition, we shall deal with the case of discrete potential theory over a discrete space X . In this case, the theory is almost completely elementary, and its ideas and methods transparent. The basic results are established in sections 2 ,3 and 5 ; these sections are essentially mere adaptation of [2]. As an application, we recover and extend a result of C. Series [16] on the Martin boundary of Fuchsian groups .In the last sections, we show that the theory can be smoothly applied to the hyperbolic spaces of M.Gromov [9], so that in many cases the Martin boundary of a random walk P on a hyperbolic graph X is identical to the geometric boundary of X. For such a pair (X,P) , one gets natural analogues of the Poisson kernel and the Fatou theorem for the unit ball of $\mathbb{R}^n$. In fact, many results of the classical Potential theory on the unit ball of $\mathbb{R}^n$ have counterparts in this setting; as an illustration of this, we prove an analogue of the Beurling minimum principle for positive harmonic functions on the unit disc ([4], [13], [5]),and indicate some extensions.

I. The setting

Throughout this exposition we shall consider a countable graph , i.e a countable set X equipped with a reflexive, and symmetric relation $\Gamma\subset X\times X$, and we shall assume that X is connected .This means that for every pair a , b in X, there is a chain of points $x_0,x_1,\dots,x_m$ $(m\geq 0)$ in X with $x_0=a$, $x_m=b$, and $(x_i,x_{i+1})\in\Gamma$. If we let d(a,b) to be the smallest integer m, we obtain the canonical metric d on X. We shall also assume that for every $x\in X$ $|\{y;d(x,y)\leq 1\}|\leq c$ for some constant $c=c(X)$.

Geodesics, rays .A sequence $\{x_j\}_{0\leq j\leq m}$, with $d(x_j,x_{j+1})=1$ for $0\leq j<m$, is a geodesic segment if $m=d(x_0,x_m)$, (then $d(x_k,x_l)=|k-l|$, for $0\leq k,l\leq m$). One defines similarly an infinite (one sided or two-sided) geodesic in X. (Note that with this definition a geodesic is always minimizing). An infinite geodesic $x_0,x_1,\dots,x_m,\dots$ will also be called a ray.

The transition function P. A finite function $P: X\times X\to\mathbb{R}_+$ is given on $X\times X$. P defines a Potential theory on X in the usual way, a function $f:X\to\mathbb{R}_+$ being P-excessive (notation $f\in\mathcal{S}_P$) iff $f(x)\leq Pf(x)=\sum_{y\in X} P(x,y)\, f(y)$, for $x\in X$. The Green function of P is the kernel G defined by $G(x,y)=G_y(x)=\sum_{n\geq 0} P^n(x,y)$, $x,y\in X$. (The product RS of two kernels R and S being defined as the kernel $K(x,y)= \sum_{z\in X}R(x,z)S(z,y)$). Every Green potential $G\varphi$, $\varphi\geq 0$ on X, is P-excessive and if $G\varphi$ is finite, then the only non-negative function u on X which is P-harmonic (i.e $u=Pu$), and such that $u\leq G\varphi$ is $u\equiv 0$. Conversely, each finite (and ≥ 0) P-excessive function f is in a unique way the sum of a Green potential $G(\varphi)$ and a non negative P-harmonic function h. (see [12]). If P is submarkovian ($P1\leq 1$), this potential theory is related to the Markov chain on X such that (in usual notations) $P_x(X_1=y)=P(x,y)$.

Assumptions 1.1: We shall assume in the sequel that P is admissible on X , i.e that the following relations between P and d hold:

(i) $\exists c_0>0, \nu\in\mathbb{N}^*$ such that: $x,y\in X$ and $d(x,y)\leq 1 \Rightarrow \sum_{0\leq j\leq\nu} P^j(x,y)\geq c_0$.

(ii) $\exists m_1\in\mathbb{N}^*$, such that: $x,y\in X$ and $P(x,y)>0 \Rightarrow d(x,y)\leq m_1$.

Clearly the adjoint kernel $P^*(x,y)= P(y,x)$, and the perturbed kernels $P+tI$, $t\geq 0$ are admissible. Also our assumptions on X imply the existence of an admissible, symmetric and markovian P ($P1=1$). We shall use the following three principles:

Harnack inequalities: if $f\in\mathcal{S}_P$, if $x,y\in X$ with $d(x,y)\leq 1$, then $f(x)\geq (c_0/\nu)\, f(y)$. In particular, G is >0, and either $G(x,y)\equiv\infty$ or $G(x,y)<\infty$ for all $x,y\in X$.

P-Balayage, duality formula: The reduit of a real function φ on a set $A\subset X$, is the excessive function $R^A_\varphi=\inf\{\sigma\in\mathcal{S}_P;\ \sigma\geq\varphi \text{ on } A\}$. The reduit with respect to P^* (or the co-reduit) is denoted ${}^*R^A_\varphi$; note that the Green function of P^* is the adjoint G^* of G. If μ is a positive measure on X such that $G^*(\mu)<\infty$, the balayée (or swept out) measure of μ on the subset $A\subset X$ is the unique measure $\lambda=\mu^A$ such that $G^*(\lambda)={}^*R^A_{G^*\mu}$ (a measure being seen as a function on X). Clearly, μ^A is supported by A and for each ≥ 0 measure ν on A $\int_X G(\nu)\, d\lambda = \int_X G(\nu)\, d\mu$. If $\mu=\delta_x$ (Dirac measure), $x\notin A$, $\lambda=\delta_x^A$ is the harmonic measure of x in $X\setminus A$. Note that δ_x^A is supported by $\{z\notin A;\ d(z,\partial A)\leq m_1\}$, and that $s(x)\geq\int s\, d\delta_x^A$, for every $s\in\mathcal{S}_P$. Here and in the sequel ∂A is defined as the set $\{x\notin A; d(x,A)=1\}$).

If λ and μ are positive measures on X, with $G^*_\mu<\infty$, and if A is a subset of X, we have : $\int R^A_{G(\lambda)}\, d\mu = \int R^A_{G\lambda}\, d\mu^A=\int G(\lambda)\, d\mu^A=\int G^*(\mu^A)\, d\lambda = \int {}^*R^A_{G^*\mu}\, d\lambda$. Whence, the duality formula $\int R^A_{G(\lambda)}\, d\mu = \int {}^*R^A_{G^*(\mu)}\, d\lambda$.

Minimum principle: Let $s\in\mathcal{S}_P$, $A\subset X$, and let μ be a ≥ 0 measure on $X\setminus A$; if $s\geq G\mu$ on $\{z\in X\setminus A; d(z,\partial A)\leq m_1\}$ then $s\geq G\mu$ on A. By the Harnack inequalities, it follows that $s\geq G\mu$ on ∂A implies $s\geq c\, G\mu$ on A, for some $c=c(P)>0$.

Discrete groups: Finally, we mention the special case where X arises from a group G with a given finite symmetric generating subset S, $e\in S$. If we let $\Gamma=\{(x,y);y^{-1}x\in S\}$ we obtain a left invariant graph structure. For each finitely supported (positive) measure μ on G, such that supp(μ) generates G as a semi-group, the transition function $P_\mu(x,y)=\mu(\{y^{-1}x\})$ is a left-invariant admissible P on X ; such a μ is called admissible. If $\mu(G)\leq 1$, P_μ is related to the "right" random walk on G defined by μ.

II. Some Preliminary (and elementary) estimates

The resolvant of P is the following family of kernels : $V_\lambda=\{\lambda I-(P-I)\}^{-1}$ or $V_\lambda(x,y)=\sum_{n\geq 0}(1+\lambda)^{-n-1}P^n(x,y)$.

Here, we shall be concerned with the negative values of λ, and so will set: $G^\varepsilon(x,y)=\sum_{n\geq 0}(1-\varepsilon)^{-n-1}P^n(x,y)$ for ε in $[0,1[$. Note the following facts:

(i) $G^t=G^r+(t-r)\,G^t\circ G^r$ if $0\leq r\leq t$ (resolvant equation) (ii) G^t is the Green kernel of the admissible transition function $P+tI$, so that we have for each t in $[0,1[$ a notion of excessive function for the level t. Clearly $\mathcal{S}_{P+tI}\subset\mathcal{S}_P$.

The following condition turns out to be crucial for our purpose:

Condition (*) $\exists\varepsilon>0$, such that $G^{2\varepsilon}$ is finite.

An equivalent property is to require the existence of a finite and positive function s on X, which is P+tI excessive for some (small) t>0. When P is symmetric, and submarkovian condition (*) is equivalent to the requirement that the norm of P as an operator on $\ell^2(X)$ is strictly less than 1. If moreover P is markovian, then condition (*) turns out to be a (coercivity) condition on the graph (see section 4) .We also mention here that in the case of a graph X arising from a finitely generated group, (*) is satisfied by a given admissible symmetric probability measure μ on G, or equivalently by all admissible probability measures on G , if and only if G is non amenable.(see 4.4).

Observe that if condition (*) holds, it holds also for P+tI , t>0 and small.

Proposition 2.1

If $\varepsilon>0$, and if $G^{2\varepsilon}$ is finite, then $1\leq G(x,x)\leq(2\varepsilon)^{-1}$, for $x\in X$

Proof. $G(x,x)\geq 1$ follows from the definition of G. Since $G^{2\varepsilon}=G+2\varepsilon\, GG^{2\varepsilon}$, $GG^{2\varepsilon}$ is finite and $G^{2\varepsilon}(x,x) -2\varepsilon \sum_{y\in X} G(x,y)\, G^{2\varepsilon}(y,x) =G(x,x)$. Thus, $\sum_{y\in X}(\delta_x(y)-2\varepsilon G(x,y))G^{2\varepsilon}(y,x)$ is ≥ 0 (here $\delta_x(y)=1$ if x=y and 0 otherwise). Since each term in this serie is ≤ 0 for $y\neq x$, the term with y=x must be ≥ 0. Whence the inequality.

Lemma 2.2

Suppose that G^ε is finite, $\varepsilon>0$. For every $x\in X$, the balayées μ_x and μ_x^ε of δ_x on $X/\{x\}$ with respect to P and P+εI, satisfy: $\mu_x\leq(1-\varepsilon)\,\mu_x^\varepsilon$

Proof. We just have to use the following explicit formula expressing μ_x in terms of P: $\mu_x= (1-P(x,x))^{-1} [\sum_{y\neq x} P(x,y)\, \delta_y]$. (The reader may directly check that the measure λ on the right side satisfies: $G^*(\lambda)(y)=G^*_x(y)$ for $y\neq x$). This and the similar formula for μ_x^ε reduces the proof to the inequality $(1-\varepsilon-P(x,x)) \leq (1-\varepsilon)(1-P(x,x))$.

From this lemma, a more general property may be derived. Recall that m_1 is the constant in (ii) of definition 1.1.

Proposition 2.3

Suppose that $G^\varepsilon<\infty$. Let $x\in X$, $n\in\mathbb{N}$, and let μ_x, μ_x^ε be the balayées of δ_x on the set

$F_n=\{z\in X; d(x,z)\geq 1+m_1 n\}$, with respect to P and $\varepsilon I+P$ respectively. Then, $\mu_x \leq (1-\varepsilon)^n \mu_x^\varepsilon$.

Proof. Use induction on $n\geq 0$. Assume $n\geq 1$, and let λ and λ' be the balayées of δ_x on F_{n-1}, with respect to P and $\varepsilon I+P$ respectively. λ is supported by $X\setminus F_n$, and μ_x may be deduced from λ in the following way: let $F_n=\{a_1,a_2,\ldots,a_p,\ldots\}$, each point of F_n being labelled infinitely often, and define inductively λ_p by $\lambda_0=\lambda$, λ_{p+1}=the P-balayée of λ_p on $X\setminus\{a_p\}$, for $p\geq 0$. Clearly, as p tends to ∞, λ_p tends to μ_x; there is a similar construction of a sequence λ'_p with respect to $\varepsilon I+P$; since at the pth stage only the parts $\lambda_p\{a_p\}\,\delta_{a_p}$, and $\lambda'_p\{a_p\}\,\delta_{a_p}$ of λ_p and λ'_p are modified, it is easily seen, using lemma 2.2, that $\lambda_p \leq (1-\varepsilon)^n \lambda'_p$ if p is so large that $\{a_1,a_2,\ldots,a_p\}=F_n$.

Corollary 2.4

There is a constant $\alpha=\alpha(P,\varepsilon)>0$, such that $G(y,x)\leq e^{-\alpha d(x,y)}\, G^\varepsilon(y,x)$, for all $x,y\in X$.

Proof. With the notation of proposition 2.3, $G(x,y)=\int G(z,y)\, d\mu_x(z)$ if $d(x,y)>nm_1$ and there is a similar formula for G^ε. Since $G\leq G^\varepsilon$, the inequality follows from proposition 2.3.

Remark 2.5. If P is symmetric, then G^t is bounded for $t<2\varepsilon$. In fact, by the resolvant equation, $G^t(x,y)=\sum_{n\geq 0} (t-s)^n\, (G^s)^{n+1}(x,y)$ for $s<t$, and by the symmetry of G, $(G^s)^{2n+1}(x,y) \geq \{G^s(x,y)\}^{n+1}$. Hence if $G^t<\infty$, we must have $G^s(x,y) < (t-s)^{-2}$. The above corollary shows that G^t has even a uniform exponential decay at infinity, for $t<\varepsilon$.

III. Φ-chains and the Harnack inequality at infinity.

The next definition introduces the geometrical objects which forces some Harnack inequalities at infinity, provided (*) is satisfied. (see [2], 514-515).

Definition 3.1 Let $\Phi:[0,\infty[\to]0,\infty[$ be an increasing function with $\lim_{t\to\infty}\Phi(t)=\infty$. A Φ-chain in X is a decreasing sequence of regions $U_0\supset U_1\supset\ldots\supset U_m$ in X, together with points $x_0,x_1,\ldots x_m$ in X such that:

(i) if $j=0,1,\ldots,m$ $\quad d(x_j,\partial U_j)\leq 1/\Phi(0)$, and if $j<m$: $\quad \Phi(0)\leq d(x_j,x_{j+1})\leq 1/\Phi(0)$

(ii) if $j=0,1,\ldots,m-1$, and if $z\in\partial U_j$ $\quad d(z,\partial U_{j+1}) \geq \Phi(d(z,x_j))$

A sequence $x_0,x_1,\ldots x_m$ in X will be called a Φ-chain if there exists corresponding regions $U_0,\ldots,U_m$ such that the previous conditions are satisfied. One defines similarly infinite Φ-chains, and Φ-chains in a Riemannian manifold.

3.2.Examples : **a)** Let H_n be the half-space $x_n>0$ of $\mathbb{R}^n$ equipped with the hyperbolic metric $ds^2 = x_n^{-2}(\sum dx_i^2)$, and let $\Phi(r)=\max\{r-2,1\}$. We obtain an infinite Φ-chain in H_n, by letting $x_j=(0,..,0,e^{-j})$, $U_j=\{x\in H_n;\ \|x\|<e^{-j}\}$, $j=0,1,\ ,..,m,...$(Using dilations, we have to check (ii) only for $j=0$, and observe then that $d(z,U_1)\geq d(z,\{\xi\in\mathbb{R}^n;\xi_n>e^{-1}\})$). It follows, on using self-isometries of H_n , that for every unit speed geodesic γ in H_n, $\{\gamma(j)\}_{j\geq 0}$ is a Φ-chain in H_n. The same property holds in B, the unit ball of $\mathbb{R}^n$ equipped with the hyperbolic metric $ds=2(1-|x|^2)^{-1}\,|dx|$, since B is isometric to H_n . **b)** A similar example with a graph X can be deduced as follows: let $X\subset B$ be such that: (i) $\forall x\in B$, $d_h(x,X)\leq c_1$,and (ii) $\forall x,y\in X$, $d_h(x,y)\geq c_2$, where d_h is the hyperbolic metric on B. Equip X with the graph structure Γ, such that $x\Gamma y \Leftrightarrow d_h(x,y)\leq 3c_1$, for $x,y\in X$. Then, X is a connected graph, whose metric $d=d_X$ is uniformly equivalent to d_h. If γ is a geodesic in B, and if $x_j\in B$ is such that $d(x_j\,,\gamma(3c_1 j\,))\leq c_1$,it follows from the above that $\{x_j\}_{j\geq 0}$ is a Φ_1-chain in X for some Φ_1 in the form $\Phi_1= a_1\Phi$, $a_1>0$.

We come now to the key estimate of G

Theorem 3.3

Assume (*). If $x_1,...x_m$ is a Φ-chain, and if $1<k<m$, then

$$G(x_1,x_m) \leq c\, G(x_1,x_k)\, G(x_k,x_m) \quad \text{with } c=c(\varepsilon,\Phi,c_0,\nu,m_1)<\infty$$

(Here c is essentially independent of m or k).

3.4. Remarks. Observe that the regions U_k of definition 3.1 do not appear in the conclusion of the theorem. Note also that the reverse inequality with the constant c $=2\varepsilon$ is elementary ($x\to G(x,x_m)/G(x_k,x_m)$ is by proposition 2.1 larger than the potential $(2\varepsilon)^{-1}G(.,x_k)$ at x_k ,and thus everywhere on X). So that the theorem says that $G(x_1,x_k)\,G(x_k,x_m)$ and $G(x_1,x_m)$ are really equivalent in size.

If P is submarkovian, there is a simple probabilistic interpretation of theorem 3.3 : the P-Markov chain, starting from x_1 and conditionned to hit x_m will first hit x_k with a probability larger than a constant $c(P)>0$, at least if $m-k$ is large. (See Appendix C).

The proof of theorem 3.3 breaks into two steps .The method goes back to [2'], where a combination of a technic of Carleson and a remark of Brelot were used. The first step is the following proposition:

Proposition 3.4

Let $U_1,\dots,U_m$, $x_1,\dots x_m$ be a Φ-chain. Then, for every $z\in U_m$, we have

$$G(x_1,z)\leq c\, G(x_1,x_m)\, G^{\varepsilon}(x_m,z) \quad .(c=c(\varepsilon,P)).$$

proof of proposition 3.4 : Clearly by the Harnack inequality, there is, for each m, a best constant $c_m=c(m,P,\varepsilon)<\infty$ such that the above estimate holds. We shall compare c_m with c_{m-1} $(m\geq 2)$. First observe that if R is a given positive number, it follows from the definition of a Φ-chain, that there exists a $\rho=\rho(\Phi,R)>0$ such that if $z\in U_m$ and $d(z,x_m)\geq\rho$, then $B(z,R)\subset U_{m-1}$.

By the definition of c_{m-1} (and Harnack): $G(x_1,z)\leq c_{m-1}\, c\, G(x_1,x_m)\, G^{\varepsilon}(x_m,z)$, for all $z\in U_{m-1}$ with $c=c(P,\Phi)$ .Integrating this against μ_z ,the P*-balayée of δ_z on $X\backslash B(z,R)$, with z as above, we have:

$$G(x_1,z)=\int G(x_1,y)\, d\mu_z(y)\leq c_{m-1}\; c\int G(x_1,x_m)\, G^{\varepsilon}(x_m,y)\, d\mu_z(y)$$

If μ^{ε}_z is the similar $\varepsilon I+P$-balayée, we know from proposition 2.3, that we may choose R (depending only on P ,ε and c) such that $\mu_z\leq(1/c)\,\mu^{\varepsilon}_z$. With this choice of R, (and a corresponding choice of ρ), we get :

$$G(x_1,z)\leq c_{m-1}\, G(x_1,x_m)\, G^{\varepsilon}(x_m,z)$$

for $z\in U_m$, with $d(z,x_m)\geq\rho=\rho(\Phi,P,\varepsilon)$. For $z\in U_m$, $d(z,x_m)\leq\rho$, we have by the Harnack inequalities: $G(x_1,z)\leq c'\, G(x_1,x_m)\, G^{\varepsilon}(x_m,z)$, $c'=c'(\Phi,\varepsilon,P)$. Finally, $c_m\leq\max(c',c_{m-1})$, so that $c_m\leq\max(c',c_1)$.

proof of theorem 3.3: By the first step, (for P*), we have for $z\in U_k$:

$$G^*_{x_1}(z)\leq c\, G(x_1,x_k)\; u(z)$$

where u is the co-reduit of $G^{\varepsilon}(x_k,.)$ on the region U_k , or what amounts the same, on the region $A_k=\{z\in X;\ d(z,U_k)\leq m_1+1\}$ (by Harnack). $G^{\varepsilon}(x_k,.)$ is a P*-potential: $G^{\varepsilon}(x_k,.)=G^*\lambda$, where λ is a positive measure on X. Now, by the duality formula : ${}^*R^{A_k}{}_{G^*\lambda}(x_m)=\int R^{A_k}{}_{G(.,x_m)}\, d\lambda=u(x_m)$.

Again, by proposition 3.4 , (for the Φ-chain $V_j=X\backslash U_j$, with the distinguished points x_j, $k\leq j\leq m$) $G(z,x_m)\leq c\, G^{\varepsilon}(z,x_k)\, G(x_k,x_m)$,for $z\in V_k$, and also, by the Harnack inequality , for $z\in A_k$ (with another c).Hence, <u>for $z\in X$,</u>

$$R^{A_k}{}_{G(.,x_m)}(z)\leq c\, G(x_k,x_m)\;\; G^{\varepsilon}(z,x_k).$$

Finally, using the resolvant equation , and proposition 2.1 (for $P+(3\varepsilon/2)I$):

$$\int G^{\varepsilon}(z,x_k)\, d\lambda(z)=\int G(z,x_k)\, d\lambda(z)+\varepsilon\int G(z,y)G^{\varepsilon}(y,x_k)\, d\lambda(z)\, dy=$$

$$= G^{\varepsilon}(x_k,x_k)+\varepsilon\,(G^{\varepsilon}\circ G^{\varepsilon})(x_k,x_k)\leq G^{\varepsilon}(x_k,x_k)+2\, G^{3\varepsilon/2}(x_k,x_k)\leq c=c(\varepsilon,P).$$

Puting together these inequalities, we get $G(x_1,x_m) \leq c\, G(x_1,x_k)\, G(x_k,x_m)$.

IV. Remarks on property (*)

We shall see now that condition (*) is also a geometrical property if we restrict ourselves to symmetric random walks on X (i.e P is markovian and symmetric). We first assume P, symmetric and submarkovian ($P1 \leq 1$). (P is also admissible).

It is well known that P defines then an operator on $\ell^2(X)$ with norm $\|P\|_2 \leq 1$. Condition (*) holds if and only if $\|P\|_2 < 1$. This last condition is obviously sufficient and its necessity can be proved as follows: by assumption there is a positive and finite $(1+\varepsilon)P$-superharmonic funtion s on X. Consider for each finite $A \subset X$ the number $\lambda_A = \sup\{ |(Pf,f)|;\ f \in \ell^2(A),\ \|f\|_2 = 1\}$.(we let $f=0$ on $X \setminus A$ for $f \in \ell^2(A)$). Clearly, λ_A increases with A, $0 < \lambda_A \leq 1$, and $\lambda_A \to \|P\|$, when $A \to X$. Also, (since $|(P(f),f)| \leq (P(|f|),|f|)$) there is a non-negative function $f \in \ell^2(A)$, $f \not\equiv 0$, such that $Pf = \lambda_A f$ on A. f (extended by 0 outside A) is clearly subharmonic for the kernel $Q = \lambda_A^{-1} P$. If λ_A^{-1} is smaller than $(1+\varepsilon)$, then $s - \alpha f$ is Q-superharmonic on X, for each $\alpha > 0$, and one can choose α such that $s - \alpha f$ is ≥ 0, and vanishes at one point of A. This implies, by the Harnack principle, that $s = \alpha f$, a contradiction. Hence $\lambda_A^{-1} > 1+\varepsilon$, and $\|P\|_2 \leq (1+\varepsilon)^{-1}$.

Remark 4.1. If $\|P\|_2 < 1$, then for each $a \in X$, the Green kernel G_a is in $\ell^2(X)$, with a uniform bound on $\|G_a\|_2$. In fact, since $\|P\|_2 < 1$, G induces a bounded operator on $\ell^2(X)$.

The canonical Dirichlet space of X is the Hilbert space H of all real functions $u: X \to \mathbb{R}$ such that $\|u\|_H^2 = \sum_{x \in X} |u(x)|^2 + \sum_{(x,y) \in \Gamma} |u(x)-u(y)|^2 < \infty$. In our setting, H is nothing but $\ell^2(X)$ with a new norm. The Dirichlet form of an admissible symmetric submarkovian kernel P on X is the following bilinear form on H:

$$\alpha_P(f,g) \equiv (f-P(f),g) = \sum_{x \in X} \alpha(x)\, f(x)\, g(x) + (1/2) \sum_{x,y \in X} P(x,y)\, (f(x)-f(y))\, (g(x)-g(y))$$

where α is the function $1-P1$. (To check the second equality one may first restrict to $H_0 = \{f \in H;\ \mathrm{supp}(f) \text{ finite}\}$). The next statement immediately follows from the above:

Corollary 4.2

If P is symmetric and submarkovian, condition (*) holds if and only if α_P is coercive: i.e if and only if $\alpha_P(f,f) \geq \varepsilon \|f\|_2$, for all $f \in H$ and some positive ε.

Note that the part $\sum_{x,y \in X} P(x,y)\, (f(x)-f(y))^2$ of $\alpha_P(f,f)$ does not depends very much on P: there is a $C \geq 1$, such that, for all $f \in H$,

$$C^{-1} \sum_{(x,y)\in\Gamma} |f(x)-f(y)|^2 \leq \sum_{x,y\in X} P(x,y)\,(f(x)-f(y))^2 \leq C \sum_{(x,y)\in\Gamma} |f(x)-f(y)|^2$$

It follows that among symmetric markovian kernels P, condition (*) is independent of P. This is similar to the fact that the admissible symmetric markovian kernels on X are all transient or all recurrent.(see [19])

Definition 4.3

X is called coercive, if the following Poincaré-Sobolev inequality holds on X:

$$\sum_{(x,y)\in\Gamma} |u(x)-u(y)|^2 \geq \varepsilon\, \|u\|_2^2$$

for all $u: X \to \mathbb{R}$, with finite support, and some positive ε.

It is well-known that the coercivity of X is equivalent to the isoperimetric inequality (4) below.

Proposition 4.4

The following properties are equivalent: (1) X is coercive . (2) for all symmetric admissible and submarkovian P on X , $\|P\|_2 < 1$. (3) there is a symmetric, admissible and markovian P on X satisfying condition (*) . (4) There is a constant $c = c(X) > 0$ such that , for each finite $A \subset X$, $|A| \leq c\, |\partial A|$.

Proof. (1)⇔(2)⇔(3) is contained in the above discussion. Clearly (4) follows from (1), by taking $u = \mathbb{1}_A$. To go the other way, note that (4) means that for each indicator function $u = \mathbb{1}_A$, $A \subset X$, A finite, $\sum_{(x,y)\in\Gamma} |u(x)-u(y)| \geq c \sum_{x\in X} |u(x)|$ (4') (with another $c>0$).

Now, any finitely supported $u: X \to \mathbb{R}_+$ can be written $u = \sum_{1\leq j\leq m} a_j \mathbb{1}_{A_j} = \sum_{1\leq j\leq m} u_j$ with $a_j \geq 0$, and A_j in the form $A_j = \{f \geq t_j\}$. Because $(u_j(x)-u_j(y))\,(u_k(x)-u_k(y)) \geq 0$, we obtain by adding the inequalities (4') with $u = u_j$, $1 \leq j \leq m$, the same inequality for u. Thus (4') holds for all finitely supported $u: X \to \mathbb{R}$. Applyng now (4') to u^2 , we have:

$$c \sum_{x\in X} |u(x)|^2 \leq \sum_{(x,y)\in\Gamma} |u(x)^2-u(y)^2| \leq \{\sum_{(x,y)\in\Gamma} |u(x)-u(y)|^2\}^{1/2} \{\sum_{(x,y)\in\Gamma} |u(x)+u(y)|^2\}^{1/2}$$

Since , $\sum_{(x,y)\in\Gamma} |u(x)+u(y)|^2 \leq 4\mu \sum_{x\in X} |u(x)|^2$, with $\mu = \sup_{x\in X} |B(x,1)|$, we obtain the coercivity of X.

Example 4.5 .The graph X in 3.2.b is coercive: If u is a finitely supported function on X then, using standard extension technics, we may find a C^1 function $f: B \to \mathbb{R}$, with compact support, and such that for each $x \in X$, $f = u(x)$ on the hyperbolic ball $B(x, c_2/4)$ in B, and $\|df(z)\|_h \leq c \sum_{d(x,y)\leq 3} |u(x)-u(y)|$ for $d_h(x,z) \leq c_1$, with $c = c(c_1, c_2)$. Use now the Poincaré inequality $\int f(z)^2\, d\sigma_h(z) \leq c_B \int_B \|df(z)\|_h^2\, d\sigma_h(z)$ to conclude. (Coming back to H_n , it is easy to see that the preceding inequality is equivalent to the more familiar Hardy-type inequality: $\int_{x_n>0} \varphi(x)^2 x_n^{-n} dx \leq c \int_{x_n>0} |\nabla\varphi(x)|^2 x_n^{2-n} dx$, for all C^1 and compactly supported functions φ on the half-space $\{x_n>0\}$).

There is a useful extension of (2) in 4.4 .

Proposition 4.6

If X is coercive, every admissible kernel P on X such that P and P* are submarkovian satisfies (*), and admits a Green function G such that $G(x,y) \leq C \exp(-\alpha\, d(x,y))$, for some positive constants C , α .

Proof. The begining of the argument is inspired by [19] section 4. Since P and P^* are submarkovian, P is a contraction of $\ell^2(X)$. Let Q be an admissible and symmetric markovian kernel on X and consider the kernel $R=e^{P-I}=(1/e)\ e^P$. R and R* are markovian and R dominates Q , i.e $Q \geq c R$ for $c > 0$. Hence, $R=(1/c)Q+(1-1/c)Q'$, for a kernel Q' such that Q' and Q'* are submarkovian. In particular ,Q' is a contraction of $\ell^2(X)$. Since X is coercive , Q is a strict contraction of $\ell^2(X)$:$\|Q\|_2<1$; from $\|Q'\|_2\leq 1$, we obtain that $\|R\|_2<1$.

Now, the spectral radius r_P of P (acting on $\ell_{\mathbb{C}}^2(X)$) is an element of the spectrum of P (see [15] , p.323) and $\exp(r_P-1)$ belongs to the spectrum of R. Hence, $r_P<1$. It follows that for some small $\varepsilon>0$, $G^\varepsilon=\sum_{n\geq 0} (1-\varepsilon)^{-n-1} P^n$ is bounded on $\ell_{\mathbb{C}}^2(X)$. In particular, $G^\varepsilon(x,y)$ is bounded on X×X, and the final assertion follows from 2.4.

We mention here that if P is markovian, admissible, and satisfies (*), and if X is coercive, it does not follow in general that the Green function G_a (a∈X) vanishes at infinity .(see 5.6.b).

Let us now give a closer look to the case of a graph X arising from a group G with a given finite symmetric generating set S. It is clear that the coercivity of X is independent of the choice of S, so that we could speak of coercive groups. However, it was proved by Kesten [10] , [10'] , that the operator P_μ satisfies $\|P_\mu\|_2<1$ for every symmetric admissible probability measure μ on G, if and only if G is not amenable, and that on the other hand, if G is amenable then $\|P_\mu\|_2=1$ for this class of μ . Hence, G is "coercive" if and only if G is non amenable . This property follows also from proposition 4.4 , since property (4) there is clearly equivalent to the Fölner criteria for non amenability [8] . Using proposition 4.6, it is also seen that for a non amenable G, the spectral radius of the convolution operator P_μ on $\ell^2(G)$ is strictly less than 1 for every admissible probability measure μ on G. (This was proved by Day [6] in an even more general form). In particular, for every admissible submarkovian measure μ on G, the kernel P_μ satisfies condition (*) and the Green kernel of μ vanishes at infinity. Finally, we mention that Derrienic and Guivarc'h have extended these results to locally compact groups [7].

V. Applications to the Martin boundary and the Dirichlet problem

Fix a reference point O in X. Recall that the Martin compactification of X w.r to P is defined as the unique (up to equivalence) compactification $\hat{X}$ of X, such that an unbounded sequence $\{x_j\}$ in X converges to some point in $\partial\hat{X}$, if and only if the limit $\lim_{j\to\infty} G(x,x_j)/G(O,x_j)=K(x)$ exists for all $x\in X$. If ζ is the limit point, K depends only on ζ and is denoted K_ζ. ζ is said to be minimal, if every positive P-harmonic function smaller than K_ζ is proportional to K_ζ. Recall the following two key-properties of ∂X. For every positive P-harmonic function u on X, there is a unique positive measure μ on $\partial\hat{X}$, concentrated on the minimal part of $\partial\hat{X}$, and such that $u(x)= \int K_\zeta(x)\, d\mu(\zeta)$, for $x\in X$; moreover, for each $s\in\mathcal{S}_P$, s/u admits a fine (and finite) limit at μ-almost every $\zeta\in\partial\hat{X}$. (see [17]).

Theorem 5.1

If (*) holds, every Φ-chain $\{x_j\}_{j\geq 1}$ in X converges in $\hat{X}$ to some minimal point $\zeta\in\partial\hat{X}$.

For a proof, we refer the reader to the analogue theorem of [2].

Theorem 5.2

Suppose that $\tilde{X}$ is a compactification of X such that the following geometric assumptions hold (for some function $\Phi:[0,\infty[\to]0,\infty[$ with $\lim_{t\to\infty}\Phi(t)=+\infty$):

(G.A) $\forall\zeta\in\partial X$, $\exists\{V_j\}$ a decreasing basis of neighborhoods of ζ and points $a_k\in V_{2k}\setminus V_{2k+1}$ such that every $x\in\partial V_{2k}\cap X$ and $y\in\partial V_{2k+1}\cap X$ can be joined by a Φ-chain through a_k.

Then, if condition (*) is fulfilled by P, the P-Martin compactification of X coincides with $\tilde{X}$.

Proof. With the notation of the theorem, we have from theorem 3.3,

$$G(y,x) \leq c\, G(y,a_k)\, G(a_k,x)\ ,\ c=c(P,\Phi)>0,$$

for $x\in V_{2k+1}$, $y\in X\setminus V_{2k}$ (using the maximum principle for both P and P*). This is also true if $x\in A_k=\{z;d(z,\partial V_{2k+1})\leq 1+m_1\}$ .Integrating with respect to x, we see that for every potential p supported by A_k, $c^{-1}p(a_k)\ G(y,a_k) \leq p(y) \leq c\ p(a_k)\ G(y,a_k)$, for $y\in X\setminus V_{2k+1}$ .Let H_ζ be the cone of positive P-harmonic functions u on X such that $u=O(G_a)$ at each $\zeta'\in\partial X$, $\zeta'\neq\zeta$, (where a is an arbitrary point in X). If $u\in H_\zeta$, the reduit u' of u on A_k is a potential (u' is dominated by a multiple of G_a) supported by A_k , and u'=u on $X\setminus V_{2k+1}$. Hence,for every $u\in H_\zeta$, $y\in X\setminus V_{2k}$:

$$c^{-1}\ G(y,a_k)u(a_k) \leq u(y) \leq c\ G(y,a_k)u(a_k)$$

In particular,if $u,v\in H_\zeta$, we see that on $X\setminus V_{2k+1}$, u/v is (uniformly) equivalent to its value at a_k, or what amounts the same,at any other point in $X\setminus V_{2k+1}$. Hence u/v is bounded, and by a standard argument H_ζ is at most one dimensional. On the other hand,the preceding inequalities show that any cluster value of $f_k=G(.,a_k)/G(0,a_k)$ is an element of H_ζ.

Remarks 5.3. 1. An obvious modification of the above shows the following boundary Harnack principles: Let u,v be P-excessive functions on X . If $u=O(G_0)$ on $X\setminus V_{2k+1}$, and is P-harmonic on $X\setminus V_{2k+1}$, then $u(y) \leq c\ [u(a_k)/v(a_k)]\ v(y)$, $y\in X\setminus V_{2k}$. If $u=O(G_0)$ on $V_{2k}\cap X$ and is P-harmonic on $V_{2k}\cap X$, there a similar inequality for $y\in V_{2k+1}\cap X$.

2. The proof shows that K_ζ the Martin kernel at ζ (with normalization at 0) is $O(G_0)$ at each $\zeta'\in\partial\tilde{X}$, $\zeta'\neq\zeta$ and that each $\zeta\in\partial X$ is minimal. Also, It is easy to see that $(\zeta,x)\to K_\zeta(x)$ is continuous on $\partial X\times X$.

Corollary 5.4

Assume further that $P1=1$, and that P is symmetric (or the less restrictive hypothesis that $G_0=o(1)$ at infinity) . Then, for each $f\in C(\partial\tilde{X})$ there exists a unique $u_f\in C(\tilde{X})$, which is P-harmonic on X, and coincides with f on $\partial\tilde{X}$.

Consider the Martin Kernel K ; by the Martin theory ,there is a probability measure σ on ∂X with $1=\int K_\zeta(x)\ d\sigma(\zeta)$,$x\in X$ .Put $u_f(x)= \int K_\zeta(x)\ f(\zeta)\ d\sigma(\zeta)$,and check , exactly as for the Poisson kernel of a ball in $\mathbb{R}^n$,that $\lim_{x\to\zeta} u_f(x)=f(\zeta)$ for each $\zeta\in\partial X$. (Note that if $G_0 = o(1)$ at infinity, then by the boundary Harnack principle $K_\zeta(x)$ tends to 0 at ζ', for $\zeta'\neq\zeta$, $\zeta',\zeta\in\partial\tilde{X}$) .

Remark 5.5. We also mention the following Fatou type property: Let u,v be positive P-harmonic on X, and let μ,ν be the positive measures on ∂X such that $u=\int K_\zeta\ d\mu(\zeta)$, $v = \int K_\zeta\ d\nu(\zeta)$. Then, for ν-amost all $\zeta\in\partial\tilde{X}$, the ratio u/v converges along the sequence $\{a_k\}$ (related to ζ) to $f(\zeta)=d\mu/d\nu$.(See the method outlined in [2], p.512).

5.6. Examples. a. Assume that X is a discrete approximation of the hyperbolic ball B as in example 3.2.b . Then, using the Φ-chains exhibited in 3.2.b, it is easy to see that the obvious compactification $\tilde{X}=X\cup S_{n-1}$ satisfies (G.A). Since the graph X is coercive ,we see that for every admissible P such that P and P* are submarkovian, the P-Martin boundary is S_{n-1}, and an analogue of the Fatou theorem with non-tangential convergence holds .If P is symmetric,and P1=1 the Dirichlet problem is solvable for the boundary S_{n-1} of X.

b. Let X be the following discrete approximation of H_2 : $X=\{ (k\, 2^n, 2^n) ; k,n\in\mathbb{Z} \}$, and $\xi\Gamma\xi'$ iff $d_h(\xi,\xi')\leq \mathrm{Log}(2)$. Let also $P(\xi,\xi')=\varepsilon$ if $\xi\Gamma\xi'$, $\xi'\neq\xi$ and $\xi'_2\leq\xi_2$, $P(\xi,\xi')=1-3\varepsilon$ if k is even and ξ' is the point immediately above ξ, $P(\xi,\xi')=(1-3\varepsilon)/2$ if k is odd, and ξ' is one of the two nearest points to ξ with $\xi'_2=2\xi_2$, $P(\xi,\xi')=0$ in all other cases. P is then an admissible markovian kernel on X. It easily checked that for fixed $\varepsilon<1/4$, and for $\beta>0$ and small $s(x,y)= y^{-\beta}$ is excessive for $P+tI$, $t>0$ and small. Thus, P satisfies (*) and the P-Martin boundary is ∂X the boundary in $\mathbb{R}^n\cup\{\infty\}$. The P-harmonic function 1 is represented by a probability measure σ on ∂X. Since $s\in\mathcal{S}_P$, s admits a finite fine limit at σ-almost every $\zeta\in\partial X$. This forces σ to be the Dirac measure at $\zeta_0=\infty$; thus $K_{\zeta_0}=1$. Since K_{ζ_0} and G_a $(a\in X)$ have the same behaviour at $\zeta\in\partial X$, $\zeta\neq\zeta_0$, we see that G_a does not vanish at infinity.

c. Let X be a subset of B equipped with a (connected) graph structure whose metric d satisfies : $c^{-1}\, d_h(x,y) \leq d(x,y) \leq c\, d_h(x,y)$, for all $x,y\in X$ and some $c>0$. In particular,for $x,y\in X$, $x\neq y$, $d_h(x,y)\geq c^{-1}$. Then, the closure $\bar{X}$ of X in $\bar{B}$ satisfies (G.A) (but in general X is not coercive). This is easily proved by using the following well-known property of hyperbolic geometry.

Lemma 5.7

Let $x_0,x_1,\ldots,x_m$ be a sequence in B, such that $c\,|i-j| \leq d_h(x_i,x_j) \leq c^{-1}\,|i-j|$,$c>0$, and let S be the (hyperbolic) geodesic segment x_0x_m; then, $d(x_i,S) \leq c'$, for a constant $c'=c'(c)$.

A proof of this lemma is provided in appendix A.

To illustrate example 5.6.c ,consider a Fuchsian group Γ acting on the unit disc D, (n=2) ,with limit set $K=\overline{\{\gamma(o);\gamma\in\Gamma\}}\cap S_1$ and assume that Γ is finitely generated , without parabolic elements,and that $|K|\geq3$ (i.e Γ is non-elementary). We may assume that $\gamma\in\Gamma$ and $\gamma(0)=0 \Rightarrow \gamma=\mathrm{Id}$,and hence may represent Γ by the set $X=\{\gamma(0);\gamma\in\Gamma\}$.

After choosing a generating symmetric subset S of Γ,there is a natural graph structure on Γ which we transport on X. By assumptions, Γ admits a fundamental domain V of the usual type and without cusps ,and one can easily prove ,using V, that the distance d on X is uniformly equivalent to d_h . Moreover, it is known that Γ is non-amenable [16]. Hence,for each probability measure μ on X with finite support S generating Γ as half-group,the Martin boundary of the μ-right random walk on Γ can be identified with the limit set of Γ, $L=\overline{X}\setminus X$, and again the Dirichlet problem for this boundary is uniquely solvable. This identification of the P-Martin boundary with the limit set L is a theorem of C.Series [16].

VI. Hyperbolic graphs

In [9] , M.Gromov introduces several definitions of hyperbolicity for a metric space and shows that these definitions are equivalent (under suitable conditions). He also defines a natural (geometric) boundary for these spaces. We shall see here that if X is a hyperbolic graph, then the above discussion can be used in order to identify the geometric boundary of X with the P-Martin boundary, for any admissible P on X satisfying (*). Also, when X arises from a group G with a given finite set of generators and which is not a finite extension of {0} or $\mathbb{Z}$, then condition (*) automatically holds for every **submarkovian** and left-invariant P.

Let us first recall the Gromov definition of hyperbolicity for our graph X. Let $O\in X$, and let ,for $x,y\in X$, $(x,y)_O=(1/2)\{d(O,x)+d(O,y)-d(x,y)\}$.

Definition 6.1 [9].

X is said to δ-hyperbolic , δ being a >0 number, if for all x,y,z,O in X:

$$(x,z)_O \geqslant \min\{(x,y)_O,(y,z)_O\} - \delta$$

Remarks 6.2 1. If the above inequality is satisfied for one $O\in X$,and all $x,y,z\in X$, then it is easy to check that X is 2δ-hyperbolic.(see [9])

2. If X is a tree (X contains no non trivial loops), X is δ-hyperbolic with δ=0. The graph X in example 5.6.c is δ-hyperbolic for some δ>0. (see appendix B).

The next proposition defines the geometric boundary of a hyperbolic graph .

Proposition 6.3 [9]

Suppose that X is δ-hyperbolic. Up to equivalence, there is a unique compactification $\tilde{X}$ of X such that: a sequence $\{x_i\}$ in X is converges in $\tilde{X}$ to some point $\zeta\in\partial\tilde{X}$, if and only if $\lim_{k,j\to\infty}(x_j,x_k)_0 = \infty$ (for some fixed $O\in X$).

Let us see now that in a δ-hyperbolic graph, Φ-chains arise naturally in such a way that the compactification $\tilde{X}$ satisfies assumptions GA in theorem 5.2. We first gather the required elementary properties of X. In the sequel X is assumed to be δ-hyperbolic.

Lemma 6.4

If $O,x,y\in X$ and if σ is a geodesic segment joining x to y, then

$$(x,y)_0 \leq d(O,\sigma) \leq 2\,((x,y)_0 + \delta)$$

The first inequality follows from the triangle inequality: consider $z\in\sigma$, with $d(O,z)=d(O,\sigma)$, and write $d(O,x)+d(O,y)-d(x,y) \leq 2d(O,z)+d(x,z)+d(z,y)-d(x,y)=2\,d(O,z)$.

Let z be the point on σ which is nearest to y and such that $d(x,z)\leq d(O,x)$; then, if $z\neq y$, $d(z,y)=d(x,y)-d(x,z) \leq d(O,x)+d(O,y)-d(x,z) = d(O,y)$., and $d(y,z)\leq d(O,y)$ in every case. By the δ-hyperbolicity, we have $(x,y)_0 \geq \min((x,z)_0,(z,y)_0) - \delta$. If $(x,z)_0\leq(x,y)_0+\delta$, we see that $d(O,z)\leq d(O,x)+d(O,z)-d(x,z) \leq 2(x,z)_0$ and get the result. Otherwise, we replace x by y.

Corollary 6.5

Let $O,x,z\in X$ and $C\geq 0$. If $(x,z)_0 \geq d(O,x)-C$, then $d(x,\sigma)\leq 2(C+\delta)$, for every geodesic segment σ between O and z.

Proof. By assumption, $d(O,x)+d(O,z)-d(z,x) \geq 2\,d(O,x) - 2C$. Whence, $(O,z)_x\leq C$, and the corollary follows from lemma 6.4.

Corollary 6.6

Let AB and CD be two segments in X. Assume that $O\in AB$, and that $\min[d(O,A),d(O,B)] \geq \max(d(A,C),d(B,D))+4\delta$. We then have: $d(O,CD)\leq 6\delta$.

Proof. We have $2(A,C)_0 = d(O,A)+d(O,C)-d(A,C) \geq 2[d(O,A)-d(A,C)] \geq 8\delta$. Similarly, $(B,D)_0 \geq 4\delta$. By the hyperbolicity of X, $0=(A,B)_0 \geq \min[\,(A,C)_0, (C,D)_0,(D,B)_0\,]-2\delta$ so that we must have $(C,D)_0\leq 2\delta$. Thus, by Lemma 6.4, $d(O,CD)\leq 2(2\delta+\delta)$.

Corollary 6.7

Two rays $\gamma_1,\gamma_2 : \mathbb{N}\to X$ such that $\liminf_{k\to\infty} d(\gamma_1(k),\gamma_2(\mathbb{N}))<\infty$ must also satisfy :

$\limsup_{k\to\infty} d(\gamma_1(k),\gamma_2(\mathbb{N})) \leqslant 6\delta$.

We shall now assume as we may that δ is an integer >3.

Proposition 6.8

(a) Each geodesic ray $\gamma:\mathbb{N}\to X$, converges to a point $\zeta\in\partial\tilde{X}$. If we let $U_j=\{x\in X; (x,\gamma(4j\delta))_0 \geqslant 4j\delta-2\delta\}$, $(0=\gamma(o))$, and V_j=closure in $\tilde{X}$ of U_j, then V_j is a decreasing fundamental system of neighborhoods of ζ in $\tilde{X}$.

(b) Conversely, each point of $\partial\tilde{X}$ is the end point of at least one geodesic ray γ: $\mathbb{N}\to X$, starting from a given point $0\in X$.

(c) Two rays $\gamma_1,\gamma_2:\mathbb{N}\to X$ admit the same end point on ∂X if and only if $\liminf_{k\to\infty} d(\gamma_1(k),\gamma_2(\mathbb{N}))<+\infty$ (i.e iff $\limsup_{k\to\infty} d(\gamma_1(k),\gamma_2(\mathbb{N})) \leqslant 6\delta$).

Proof.(a) The first property is straightforward since $(\gamma(k),\gamma(p))_0=2\min(k,p)\to\infty$ when k and p increase to $+\infty$.

One immediately checks,using definition 6.1,that $U_{j+1}\subset U_j$. On the other hand if $z_j\to\zeta$, then, for fixed j, $(z_k,\gamma(4j\delta))_0 \geqslant \min\{(z_k,\gamma(4k\delta))_0, 4j\delta\}-\delta \geqslant (4j-1)\delta$ if k is large. Hence $z_k\in U_j$ for large k, and V_j is a neighborhood of ζ. Moreover,if $y_j\in V_j$, $(y_j,\gamma(4j\delta))_0\to\infty$, so that $y_j\to\zeta$,and $\{V_j\}$ is a basis of neighborhoods of ζ.

b) Fix $0\in X$,and $y_j\in X$, $y_j\to\zeta\in\partial\tilde{X}$. Let $\gamma_j:\{0,1,\dots,m_j\}\to X$ be a segment joining 0 to y_j. We may assume ,after taking a subsequence that γ_j converges to some ray γ: $\mathbb{N}\to X$, and we want to see that $\gamma(p)\to\zeta$,when $p\to\infty$.For $R\in\mathbb{N}$ fixed,we have

$$(\gamma_j(R),y_i)_0 \geqslant \min\{(\gamma_j(R),y_j)_0, (y_j,y_i)_0\}-\delta = \min\{R, (y_j,y_i)_0\}-\delta.$$

Hence for large i and j , $(\gamma_j(R),y_i)_0 \geqslant R-\delta$,and a fortiori $(\gamma(R),y_i)_0 \geqslant R-2\delta$ for large i. Again , by the hyperbolicity of X, this implies $(\gamma(R'),y_i)_0 \geqslant R-3\delta$, for $R'>R$ and large i. This means that $\lim_{R'\to\infty}\gamma(R')=\zeta$.

c) If γ_1 and γ_2 are rays of X ending at the same $\zeta\in\partial\tilde{X}$,we have for a given p and for large m $(\gamma_1(p),\gamma_2(m))_{0_1} \geqslant p-2\delta$ (using an intermediate $\gamma_1(p')$,p' large). Here 0_1 is the origin of γ_1.Hence, $(\gamma_1(p),\gamma_2(m))_{0_2} \geqslant p-d(0_1,0_2)-2\delta \geqslant d(0_2,\gamma_1(p))-2d(0_1,0_2)-2\delta$ .Now by Corollary 6.5, $d(\gamma_1(p),\gamma_2(\mathbb{N})) \leqslant 4d(0_1,0_2)+6\delta$.

Theorem 6.9

Suppose that γ: $\{0,1,2,\dots,p\}\to X$ is a geodesic segment in X .Let $x_0=\gamma(o)$, $x_1=\gamma(4\delta)$, ... , $x_m=\gamma(4m\delta)$, $(4m\delta\leqslant p)$, $0=x_0$ and $U_j=\{z\in X; (x,x_j)_0 > 4j\delta-2\delta\}$. Then, the regions U_j and the corresponding points x_i form a Φ-chain for a function Φ depending only on δ.

Proof. We note (x,y) instead of $(x,y)_0$. From the δ-hyperbolicity, it is clear that $(x,x_{j+1}) > 4(j+1)\delta-2\delta$ implies $(x,x_j)\geq 4j\delta-\delta$, whence $U_{j+1}\subset U_j$. Note that $x_j\in U_j$, and $\gamma(4j\delta-2\delta)\notin U_j$, so that $d(x_j,\partial U_j)\leq 2\delta$. If $z'\in U_{j+1}$ and $z\in\partial U_j$ we have:

$$(z',x_j) \geq \min[(z',x_{j+1}),(x_{j+1},x_j)]-\delta \geq \min[4(j+1)\delta-2\delta,4j\delta]-\delta \geq 4j\delta-\delta.,$$

and $4j\delta-2\delta=d(0,x_j)-2\delta \geq (z,x_j) \geq \min[(z',z),(z',x_j)]-\delta > \min[(z',z),4j\delta-\delta]-\delta$

Hence: $(z',z)\leq d(0,x_j)-\delta$. This shows that if $z\in\partial U_j$, $z'\in\partial U_{j+1}$:$(z',z)\leq d(0,x_j)-\delta'$, with $\delta'=\delta-1$. Now,for such z,z',

$(z',x_{j+1})\geq d(0,x_{j+1})-2\delta-\frac{1}{2}$, or $d(0,z') \geq d(0,x_{j+1})+d(x_{j+1},z')-4\delta-1$

Similarly, $d(0,z) \geq d(0,x_j)+d(x_j,z)-4\delta-1$. From $(z',z)\leq d(0,x_j)-\delta'$ and these inequalities , we get:

$$[\, d(0,x_{j+1})+d(x_{j+1},z')-4\delta +d(0,x_j)+d(x_j,z)-4\delta-d(z,z')] \leq 2\, d(0,x_j)-2\delta'+2$$

and, $$d(z,z') \geq -2\delta-4+ d(x_j,z)+ d(x_{j+1},z') \geq -2\delta-4 + d(x_j,z)$$

Finally,observe that $4\delta=d(x_j,x_{j+1}) \leq d(x_j,z)+d(x_{j+1},z')+d(z,z')\leq 2[d(z,z')+\delta+2]$, so that $d(z,z')\geq \sup[\, d(x_j,z)-2(\delta+2)\, ,\, \delta-2\,] =\Phi_\delta(d(x_j,z))$.

Corollary 6.10

The assumptions (G.A) in theorem 5.2 are satisfied by the compactification $\tilde{X}$ of X.

Proof. Let γ be a ray joining 0 to $\zeta\in\partial X$, let $x,x'\in\gamma$,x nearer to 0, and such that $d(x,x')=4\delta$. Let $U=\{y;(x,y)>d(0,x)-2\delta\}$, $U'=\{y;(x',y)>d(0,x')-2\delta\}$,$z\in\partial U$,$z'\in\partial U'$. Pick the point a on xx', with $d(x,a)=\delta$ and evaluate $(z,z')_a$:

From the proof above, $d(z,z') \geq -2\delta-4+ d(x,z)+ d(x',z')\geq -6\delta-4+d(a,z)+d(a,z')$ which means that $(z,z')_a\leq 3\delta+2$ and by Lemma 6.4 $d(a,\sigma)\leq 9\delta$, for any segment $\sigma=zz'$. By the previous proposition,this means that we can join z to z' by a Φ_δ-chain whose distance to a is bounded by a fixed constant.

Since a hyperbolic group is not amenable if and only if $|\partial G|\geq 3$ (or iff G is not a finite extension of {0} or $\mathbb{Z}$),([9]), we have the following corollary.

Corollary 6.11.

If G is a (finitely generated) hyperbolic group which is not a finite extension of {0} or $\mathbb{Z}$, then for each admissible submarkovian measure μ on G, the μ-Martin compactification is naturally homeomorphic to $\tilde{G}$. If moreover μ is markovian, then for each $f\in C(\partial\tilde{G})$, there is a unique continuous extension f on $\tilde{G}$, which is μ-harmonic on G.

VII. Beurling's minimum principle and hyperbolicity.

Let B be the unit ball in $\mathbb{R}^n$, and let $\zeta\in\partial B$. Denote by K_ζ the Poisson kernel at ζ, $K_\zeta(x)= (1/s_n)(1-|x|^2)/|x-\zeta|^n$. Following Beurling, a sequence $\{z_j\}$ in B is said to be equivalent to ζ if every positive harmonic function h in B such that $h(x_j)\geq K_\zeta(x_j)$ for all j, satisfies $h \geq K_\zeta$ in B. It is plain that this concept is closely related to the notion of minimal thinness at ζ. Recall that $A\subset B$ is minimally thin at ζ, iff $R^A_{K_\zeta} \not\equiv K_\zeta$ or equivalently iff $R^A_{K_\zeta}$ is a potential on B [17]. Of course, $A_0=\{x_j; j\geq 1\}$ being polar is always minimally thin at ζ; however, if the set $A=\bigcup_{j\geq 1} B(x_j;(1-|x_j|)/2)$ (euclidean balls) is unthin, then by the Harnack inequalities $\{x_j\}$ is equivalent to ζ, and it is not hard to prove the converse ([20'] , and 7.3 below).

Beurling [4] proved the following theorem for n=2 ; the general case was established independently by Maz'ja [13] and Dahlberg [5] ,using properties of the growth of subharmonic functions. Later, Sjögren [18] gave another proof relying on his notion of weak L^1 sets.

Theorem 7.1

Let $\{x_j\}$ be a sequence in B. If $\{x_j\}$ is separated (i.e $|x_i-x_j|\geq c\delta(x_i)=c(1-|x_i|)$ for $i\neq j$ and some c>0); then $\{x_j\}$ is equivalent to $\zeta\in\partial B$ if and only if $\sum_{1\leq j<\infty}\delta(x_j)^n|\zeta-x_j|^{-n} =+\infty$

We shall see here that this statement extends naturaly to our hyperbolic framework, which moreover leads to a relatively simple proof of this theorem . The two ingredients of this proof are the boundary Harnack inequalities and an energy identity which ,together with condition (*), enable us to avoid the growth technics and to extend Beurling's theorem to situations where the size of the Poisson kernel is not known.

In the next theorem, X is again a δ-hyperbolic graph, and P an admissible kernel on X satisfying (*). We fix ζ on the boundary $\partial\tilde{X}$ and a ray γ joining 0 to ζ. Let a_j be the point on γ with $d(0,a_j)=4j\delta$ and let $V_j=\{x\in X; (x,a_j)_0 > d(0,a_j)-2\delta \}$.

Theorem 7.2

Assume that P is symmetric and submarkovian,and let $A\subset X$, $A_j= (V_j\setminus V_{j+1})\cap A$, $j\geq 1$. The set A is unthin at $\zeta\in\partial X$ if and only if $\sum_{1\leq j<\infty}\sum_{z\in A_j} G(a_j,z)^2 =\infty$.

Proof. A) We shall first prove a natural analogue of the Wiener criterion for minimal thinness (See Lelong-Ferrand [11] for a half-space, Aikawa [1], Yiping [20] for "irregular" domains in $\mathbb{R}^p$) : A is thin at ζ if and only if $\sum_j R^{A_j}{}_{K_\zeta}(0)<\infty$, (or if and only if $p=\sum_j R^{A_j}{}_{K_\zeta}$ is a potential on X).

If p is a potential, then, since $p \geqslant K_\zeta$ on A, $R^{A}{}_{K_\zeta}$ is a potential and A is thin at ζ. To show the converse, we may assume that A is thin at ζ, and that $A_{pj+k}=\varnothing$, for all $j \in \mathbb{N}$, $0<k<p$, and some fixed integer $p \geqslant 2$ to be chosen later . Let $B_j = \cup_{k<pj} A_k$, $C_j= \cup_{k>pj} A_k$, $u_j= R^{B_j}{}_{K_\zeta}$, and $v_j= R^{C_j}{}_{K_\zeta}$. It is obvious that $\lim_{j\to\infty} v_j(0)=0$. By the boundary Harnack principle 5.3.1, the ratio v_j/K_ζ is on $X \setminus V_{pj+1}$ of the order of its value at 0. Hence, if j is large enough we must have $v_j \leqslant (1/3) K_\zeta$ on A_{pj}.

Also, for similar reasons, $u_j(x)$ has on V_{pj} the same order of magnitude than $u_j(a_{p(j-1)})\, G(a_{p(j-1)},x) \sim u_j(a_{p(j-1)})\, G(a_{p(j-1)},a_{pj})\, G(a_{pj},x)$. Thus, for some constant $c>0$ (independent of p), $u_j(x) \leqslant c\, K_\zeta(a_{p(j-1)})\, G(a_{p(j-1)},a_{pj})\, G(a_{pj},x)$, $x \in V_{pj}$. Using again 5.3.1, $K_\zeta(a_{p(j-1)}) \sim K_\zeta(a_{pj})\, G(a_{pj},a_{p(j-1)})$ (uniformly in p). Hence, $u_j(x) \leqslant c\, (G(a_{p(j-1)},a_{pj}))^2\, K_\zeta(x)$ on V_{pj}. By 2.5 there is a p independent of j such that $G(a_{p(j-1)},a_{pj})$ is smaller than any given positive constant. Hence, for $p \geqslant k_0$ (independent of p and j) we have $u_j \leqslant (1/3) K_\zeta$ on A_{pj}.

Choosing $p=\max(k_0,2)$, we see that u_j+v_j is less than $(2/3) K_\zeta$ on A_{jp}, for all j after delating a finite number of A_k. The potential $\pi= R^{A}{}_{K_\zeta}$ is a sum $\sum_j \pi_j$ of potentials, π_j supported (as a potential) by A_{jp} ; we must have $\pi_j \geqslant (1/3) K_\zeta$ on A_{jp}. Hence $\sum_j R^{A_{jp}}{}_{K_\zeta}(0) \leqslant 3 \sum_j \pi_j(0) \leqslant 3$.

B) The rest of the proof consists in evaluating $R^{A_j}{}_{K_\zeta}(0)$. Using the boundary Harnack principle, we see that :(i) $R^{A_j}{}_{K_\zeta}(0)$ is (uniformly) equivalent in size to $R^{A_j}{}_{K_\zeta}(a_j) \times G(0,a_j)$, (ii) $K_\zeta(z) \sim K_\zeta(a_j)\, G(a_j,z)$, for $z \in X \setminus V_{j+1}$ and (iii) in particular with $z=0$ we get $K_\zeta(a_j) \sim [1/G(0,a_j)]$ .Hence, $R^{A_j}{}_{K_\zeta}(0) \sim R^{A_j}{}_{g_j}(a_j)$, with $g_j=G(a_j,.)$.

We let $u_j=R^{A_j}{}_{g_j}$ and we use the following identity: $\alpha_P(u_j,u_j)=u_j(a_j)$, where α_P is the Dirichlet form related to P ; to check this identity , note first that by 4.1 g_j and u_j belongs to the Dirichlet space $H=\ell^2(X)$ of X and observe that $\alpha_P(u_j,u_j) = \langle (I-P)u_j,u_j \rangle = \langle (I-P)u_j,g_j \rangle$ since $(I-P)u_j=0$ outside A_j; hence $\alpha_P(u_j,u_j)=\langle u_j,(I-P)g_j \rangle=u_j(a_j)$ because $(I-P)g_j=\mathbb{1}_{\{a_j\}}$.

Combining this identity with the coercivity of α_P:

$$R^{A_j}{}_{g_j}(a_j) \geqslant c\, \|u_j\|_2^2 \geqslant c \sum_{z\in A_j} g_j(z)^2 \qquad (c=c(P)>0)$$

We conclude that if A is thin at ζ, then $\sum_{1\leq j<\infty} \sum_{z\in A_j} g_j(z)^2 < \infty$. On the other hand $u_j \leq \sum_{z\in A_j} u_z$ where $u_z = g_j(z)\ G(z,.)/G(z,z)$ (by the domination principle [12]). Whence, with another constant c, $u_j(a_j) \leq c \sum_{z\in A_j} g_j(z)^2$. This finishes the proof.

As for the relation between thinness and Beurling's notion of equivalence, we have

Proposition 7.3

Assume that for each $z\in X$, there exists a ray γ starting at 0, and such that $d(z,\gamma)\leq C$, for some constant $C = c(X)$. Assume further that P is symmetric and submarkovian. Then ,a set $A\subset X$ is equivalent to ζ if and only if A is unthin at ζ.

Proof. Suppose that A is thin at ζ; then, there is a measure $\mu=\sum_{z\in A} \mu_z \varepsilon_z$ with $G\mu=K_\zeta$ on A. Now associate to each $z\in A$ a point z' on ∂X which is the end of a ray γ as above. By the boundary Harnack principle, it is seen that $K_{z'}(0)/K_{z'}(z) = 1/K_{z'}(z) \sim G(z,0)$. If we let $u=\sum \mu_z\ G(z,0)\ K_{z'}$,then by the domination principle $u \geq c\ G_\mu$ and $u(0) \asymp G_\mu(0)$. Hence,after delating a finite number of points in A, we have obtained a positive P-harmonic function u on X with $u\geq K_\zeta$ on A and $u(0)<1=K_\zeta(0)$. Since, if A is not equivalent to ζ, A is obviously thin at ζ , the proposition follows.

Let us mention the analogue of theorem 7.1 in the framework of the potential theory over a Lipschitz domain Ω in $\mathbb{R}^n$,with respect to an elliptic operator L in the form $L = \sum_{1\leq i,j\leq n} \partial_i(a_{ij}\partial_j .) - c$, with measurable coefficients such that (with $\delta(x)=d(x,\partial\Omega)$) , $a_{ij}=a_{ji}$, $\nu^{-1} |\xi|^2 \leq \sum a_{ij}(x)\ \xi_i\xi_j \leq \nu\ |\xi|^2$, $0 \leq c(x) \leq \nu\ \delta(x)^{-2}$ for $x\in\Omega$, $\xi\in\mathbb{R}^n$, and $\nu=\nu(L)>0$.

Let $\zeta=0\in\partial\Omega$, and assume that $\Omega\cap B(0,r_0)=\{\ x\in B(0,r_0);\ x_n>f(x_1,\ldots,x_{n-1})\ \}$ for some Lipschitz function f on $\mathbb{R}^{n-1}$ (with $f(0)=0$), and $r_0>0$. Let $A_p = (0,\ldots,0,r_0 2^{-p})$, and $V_p=\{x\in\Omega;\ r_0\ 2^{-p-2}\leq |x-\zeta|=|x| < r_0 2^{-p-1}\}$, $p\geq 3$. Denote by G the Green function of L.

Theorem 7.4

If $\{x_j\}$ is a separated sequence in Ω (i.e $d(z_j,z_k) \geq c\ \delta(z_j)$ for $j\neq k$ and some constant $c > 0$) . Then A is equivalent to ζ (w.r to L) if and only if

$$\sum_{p\geq 3} \sum_{z_j\in V_p} G(A_p,z_j)^2\ (2^{-p}\ \delta(z_j))^{n-2} = \infty$$

or, equivalently,if and only if, $\sum_{p\geq 3} \sum_{z_j\in V_p} [G(A_1,z_j)/G(A_1,A_p)]^2\ (\delta(z_j)/|z_j-\zeta|)^{n-2} = \infty$ ■.

Here,the ingredients of the proof (similar to the proof of theorem 7.2) are again the boundary Harnack estimates, the Wiener-type criterion for minimal thinness ([1], [20] for $L=\Delta$), the identity $\int_\Omega [\sum a_{ij}\, u_i(x)\, u_j(x) + c(x)\, u(x)^2]\, dx = u(a)$, where u denotes the reduit of the L-Green function G_a with pole at a on a set $A \subset \Omega$, and , instead of (*), the well-known Hardy type inequality: $\int \varphi^2(x)\, \delta^{-2}(x)\, dx \leq c_\Omega \int |\nabla\varphi|^2\, dx$, for all test functions φ on Ω.

Appendix A. Proof of lemma 5.7.

We present an argument taken from [14] , p. 74. Let M be a simply connected Riemannian manifold with sectional curvatures ≤ -1, (the reader may as well assume M to be the hyperbolic ball B) and let γ be a c-quasi geodesic in M (i.e $c \geq 1$,and γ is a finite sequence $\{x_1,\ldots,x_m\}$ in M such that $c^{-1}\, |i-j| \leq d(x_i,x_j) \leq c\, |i-j|$, for $1 \leq i,j \leq m$).

Denote by σ the geodesic arc x_1x_m , by U_R the neighborhood $\{z \in M;\ d(z,\sigma) \leq R\}$ of σ and by p the projection p: $M \to \sigma$; we use the fact that an arc τ contained in $M\backslash U_R$ projects on an arc $\tau' \subset \sigma$ whose length $|\tau'|$ satisfies $|\tau'| \leq C\, e^{-\alpha R}\, |\tau|$, for some >0 constants C and α independent of R. If there is an $x_j \notin U_R$, there is an arc $\tau \subset M\backslash U_R$, with end points a,b on ∂U_R, and such that $|\tau| \leq c\,(d(a,b)+4)$. Now, $d(a,b) \leq 2R + |\tau'| \leq 2R + C\, e^{-\alpha R}\, |\tau|$; it follows that

$$|\tau| \leq c\, (\, 2R + C\, e^{-\alpha R}\, |\tau| + 4)$$

Hence, if we fix R such that $cC\, e^{-\alpha R} \leq 1/2$, we have $|\tau| \leq 2c\, (\, 2R + 4\,)$ and this shows that $x_j \in U_\rho$, if $\rho = R + 2c(2R+4)$.

Appendix B. a) We first prove the δ-hyperbolicity of B or more generally of any simply connected Riemannian manifold M with sectional curvatures ≤ -1. Recall that if $O,x,y \in M$ are such that $\measuredangle(Ox,Oy) \geq \pi/2$ then $d(O,xy) \leq C\ (=1/2\, \mathrm{Log}(1+\sqrt{2})\,)$. Using this property, it is easily seen that, for $O,x,y \in M$, $d(O,xy) - 4C \leq (x,y)_o \leq d(O,xy)$, (xy is the geodesic segment between x and y). If H is the point on xy such that $d(O,H) = d(O,xy)$, and if $z \in M$, we have

$$(z,x)_o \leq d(O,H) + (x,z)_H \quad , \quad (z,x)_o \leq d(O,H) + (y,z)_H \quad \text{and} \quad (x,y)_o \geq d(O,H) - 4C\,.$$

Hence $\min\{(x,z)_o,(z,y)_o\} - (x,y)_o \leq \min\{(x,z)_H,(z,y)_H\} + 4C \leq 5C$. (Using that $\measuredangle(Hx,Hz) = \pi/2$ if $H \neq x$ and $H \neq y$).

b) Let now X be a connected graph, $X \subset M$, with a metric d_X uniformly equivalent to $d = d_M$, $c^{-1}\, d(x,y) \leq d_X(x,y) \leq c\, d(x,y)$. Using the result of appendix A, we see that the product in X $(x,y)_0$ satisfies $d_X(O,H) - C' \leq (x,y)_0 \leq d_X(O,H) + C'$, where H is a point on the X-segment xy chosen within a fixed distance to the projection in M of O on xy. We then conclude with the same argument as above (using appendix A again).

Appendix C. The probability in question is $\pi=(u(x_1)\ V(x_k))/V(x_1)$, where V is the equilibrium potential of $\{x_m\}$ in X and u the equilibrium potential of $\{x_k\}$ in $X'=X\setminus\{x_m\}$. Let g be the P-Green function over X' .Since $u(x)=g(x,x_k)/g(x_k,x_k)$,it will be enough by 2.1 to see that $g(x_1,x_k)\ V(x_k) \geqslant c\ V(x_1)$,c>0. Now, we have $g(x_1,x_k) = g_{x_k}(x_1)-G(x_1,x_k)-G(x_m,x_k)\ V(x_1)$, and by 2.4 ,

$$G(x_m,x_k)\ V(x_k) \leqslant G(x_m,x_k)\ G(x_k,x_m) \leqslant \exp(-2\alpha d(x_k,x_m))\ (G^{\varepsilon}\circ G^{\varepsilon})(x_m,x_m)$$
$$\leqslant (2/\varepsilon)\exp(-2\alpha d(x_k,x_m))\ G^{3\varepsilon/2}(x_m,x_m) \leqslant (4/\varepsilon^2)\exp(-2\alpha d(x_k,x_m))$$

(We use the resolvant equation, and again 2.1). Hence ,if m-k is larger than some constant (depending on P and Φ), we are led to the estimate: $G(x_1,x_k)\ V(x_k) \geqslant c\ V(x_1)$. However, by 2.1, $2\varepsilon\ G(x,x_m) \leqslant V(x) \leqslant G(x,x_m)$, and the result follows from 3.3 .

References

[1] H. Aikawa, On the thinness in a Lipschitz domain, Analysis 5, 1985, 347-382.

[2] A. Ancona, Negatively curved manifolds, elliptic operators, and the Martin boun--dary , Annals of Maths,125,1987, 495-536.

[2'] A.Ancona, Principe de Harnack à la frontière et théorème de Fatou pour un opéra--teur elliptique dans un domaine Lipschitzien, Ann. Inst. Fourier, XVIII, 4, 1978.

[3] M.Anderson, R.Schoen, Positive harmonic functions on complete manifolds of ne-gative curvature, Annals of Math.,121,1985,429-461.

[4] A.Beurling, A minimum principle for positive harmonic functions, Ann. Acad. Sc. Fenn. Ser. AI 372 ,1965, 3-7.

[5] B.Dahlberg, A minimum principle for positive harmonic functions, Proc. London Math. Soc.,33,1976, 238-250.

[6] N.M. Day , Convolution, mean and spectra, Illinois J. of Math. ,8,1964,100-111.

[7] Y. Deriennic, Y. Guivarc'h, Théorème de renouvellement pour les groupes non mo--yennables , C. R. Acad. Sc. Paris, t. 277, 1973, 613-615

[8] F.P.Greanleaf, Invariant means on topological groups, Van Nostrand.

[9] M.Gromov, Hyperbolic groups, Preprint ,I.H.E.S.

[10] H. Kesten, Full Banach mean values on countable groups,Math.Scand.,7,1959,146-156

[10'] H. Kesten, Symmetric random walks on groups,Tr. Am Math. Soc , 92, 1959, 336-354.

[11] J. Lelong-Ferrand, Etude au voisinage d'un point frontière des fonctions surharmoniques positives dans un demi-espace, Ann.Sc.Ec.Norm.Sup.,66,1949,125-159

[12] P.A. Meyer, Probabilités et Potentiel, chapitre 9 ,Hermann, Paris,1966.

[13] V.G. Maz'ja, On Beurling's theorem for the minimum principle for positive harmonic functions,(in Russian), Zapiski Naucnyh Seminarov LOMI, 30, 1972, 76-90 .

[14] P. Pansu, **Quasi-isométries des variétés à courbure négative, Thèse,** Univ. Paris 7, 1987

[15] H. H. Schaefer , Banach lattices and positive operators, Springer-Verlag, 1974 .

[16] C. Series, Martin boundaries of random walks on Fuchsian groups, Israel J. Math.,44, n°3,1983, 221-242.

[17] D. Sibony, Théorème de limites fines et problème de Dirichlet , Ann. Inst. Fourier , 18, 2, 1968, 121-134 .

[18] P. Sjögren, Une propriété des fonctions harmoniques positives d'après Dahlberg, Sem. Théorie du Potentiel 6, Springer L.N 563,1975.,275-282.

[19] N. T. Varopoulos, Brownian motion and random walks on manifolds, Ann. Inst. Fourier, 34,2, 1984,243-269.

[20] Z.Yiping, Comparaison entre l'effilement interne et l'effilement minimal, C.R. Acad. Sci. Paris, 304,1987,5-8 .

[20'] Z.Yiping, Comparaison entre l'effilement interne et l'effilement minimal, Thèse, Orsay,1987.

ORDER AND CONVEXITY IN POTENTIAL THEORY

N. Boboc and Gh. Bucur
University of Bucharest, Faculty of Mathematics
Str. Academiei nr. 14, 70109-Bucharest, Romania
and
Department of Mathematics, INCREST, Bd. Pacii
nr. 220, 79622-Bucharest, Romania

0. INTRODUCTION

An important role in the study of the set of positive superharmonic functions on a harmonic space is played by the property of Riesz decomposition with respect to the natural order discovered by G. Mokobodzki and D. Sibony in 1968 ([13]). This property is also satisfied in the convex cone of all excessive functions with respect to a general submarkovian resolvent on a measurable space. Later, G. Mokobodzki in 1970 ([12]) developed an abstract theory of superharmonic functions (Cones of potentials) which use essentially this property.

Another important property of positive superharmonic functions on a harmonic space is that any family $(s_i)_{i\in I}$ of such functions, has an infimum $\bigwedge_{i\in I} s_i$ (with respect to the natural order) and we have $\bigwedge_i s_i+s=\bigwedge_i(s_i+s)$. This property is also verified for the cone of all excessive functions with respect to a basic submarkovian resolvent.

At last we remember that in the theory of harmonic spaces a special role is played by the special class of all finite continuous potentials which are harmonic outside a compact set. Any such positive superharmonic function p is essentially characterized by the following property: for any strictly positive superharmonic function u and any increasing family $(s_i)_{i\in I}$ of superharmonic functions converging to p there exists $i\in I$ such that $p\leq s_i+u$. A similar property is meeting in the theory of excessive functions with respect to a basic submarkovian resolvent. Moreover in this convex cone there exists a countable family of such kind of excessive functions which is increasingly dense in order from below in the set of all excessive functions.

The concepts of H-cone, H-cone of functions, standard H-cone and standard H-cone of functions represent an attempt to organise in a mathematical structure the above properties. The beginning can be considered in 1970, [2]. Another important step of the development of the theory is made in 1975 [3]. The monograph "Order and Convexity in Potential Theory: H-cones" appeared in 1981 in Lecture Notes in Mathematics No. 853 contains many results concerning the whole theory.

For the last results from §3 one can see the papers [5c] and [5d]. In connection with the notion of Green potential developed in §5 we send to [5f]. Other results concerning the various versions of potentials may be found in a recent paper of authors, "Potentials in standard H-cones of functions" (to appear in Rev.Roum.Math.Pures et Appl.10,1988).

1. H-CONES AND DUAL OF H-CONES

In a natural way we may introduce the notion of ordered convex cone which is nothing else than a subcone S of positive elements of an ordered vector space E. The order relation on S induced by E is called natural order.

DEFINITION. An ordered convex cone S is termed H-cone if the following properties hold:

1) For any increasing and dominated family F in S, there exists $\bigvee F$ and we have

$$s + \bigvee F = \bigvee (s+F)$$

for any $s \in S$.

2) For any family F in S there exists $\bigwedge F$ and we have

$$s + \bigwedge F = \bigwedge (s+F)$$

for any $s \in S$.

3) S satisfies the Riesz decomposition property: for any $s, t_1, t_2 \in S$ such that $s \leq t_1 + t_2$ there exist $s_1, s_2 \in S$, such that

$$s = s_1 + s_2 \ , \quad s_1 \leq t_1 \ , \quad s_2 \leq t_2 \ .$$

If we denote by [S] the ordered vector space generated by S then [S] is a complete vector lattice with respect to the specific order (i.e. the order given by the convex cone S).

DEFINITION. A map $\mu : S \to \overline{R}_+$ is called H-integral on S if it is additive, increasing and continuous in order (natural order) from below and such that the set $\{s \in S \mid \mu(s) < \infty\}$ is increasingly dense in order from below in S.

The set S* of all H-integrals on S endowed with the usual algebraic operations and pointwise order relation is also an H-cone which is called the dual of the cone S.

In a natural way we may regard any element s of S as an H-integral $\tilde{s}$ on S* (i.e. an element of S**). The following properties hold

1) $\widetilde{\bigvee F} = \bigvee \{\tilde{s} \mid s \in F\}$

2) $\widetilde{\bigwedge F} = \bigwedge\{\tilde{s} \mid \tilde{s} \in F\}$

for any family F in S.

PROBLEMS. If S* separates the elements of S then S may be considered as a convex subcone of S**. The following unsolved (up to now) problems rise here:

a) Is the cone S solid in S** with respect to the natural order?

b) Is the cone S solid in S** with respect to the specific order given by S**?

c) Is the cone S increasingly dense in order (natural order) from below in S**?

DEFINITION. A convex cone S of positive numerical, continuous and densely finite functions on a compact space X is called <u>Stonean cone</u> if S contains the positive constant functions, separates the points of X and endowed with the usual order relation is an H-cone.

We remark that taking an increasing and dominated family F in S the element $\bigvee F$ is the upper semicontinuous regularized of the function sup F. Also an H-integral μ on S such that $\mu(1)<\infty$ is nothing else than a Radon "normal" measure on X i.e. a Radon measure for which $\mu(\bigvee F)= =\mu(\sup F)$, for any increasing and dominated family F in S. Particularly the compact space X is Stonean iff $S:=\mathcal{C}^+(X)$ is a Stonean cone on X and moreover S* separates S iff X is hyperstonean.

One can show that any H-cone S which possesses a weak unit (i.e. an element u such that $s=\bigvee_n(s\wedge nu)$ for any $s\in S$) may be represented as a Stonean cone on a compact space X.

DEFINITION. A convex cone S of positive numerical functions on a set X is called <u>H-cone of functions</u> (on X) if S separates the points of X, S is min-stable, sup $F\in S$ for any increasing and dominated family F in S, $\inf(s,1)\in S$ for any $s\in S$, there exists an increasing family F in S such that $1=\sup F$ and S endowed with the pointwise order relation is an H-cone.

We remark that any element $x\in X$ may be considered as an extreme element of the convex cone S* and therefore the set of all extreme elements of S* separates S. Hence to represent an H-cone S as an H-cone of functions on a set X is not sufficient that S* separates S.

In the sequal we give some examples of H-cones of functions:

a) The set S of all positive superharmonic functions on a harmonic space X such that 1 is superharmonic and there exists a strictly positive potential on X.

b) The set S of all strongly increasing positive real functions on an ordered set X. (A positive real function s on an ordered set X is called strongly increasing if for any x and any finite subset F of minorants of x such that $(y,z\in F \Rightarrow y\nleq z$ and $z\nleq y)$ we have $\sum_{y\in F} s(y)\leq s(x)$.

The dual of this cone is isomorphic with the cone of all strongly decreasing positive real functions on the ordered set X.

It can be shown that if X is an ordered set then the convex cone of all increasing positive real functions on X is an H-cone iff for any $x\in X$ the set of all minorants of x is totally ordered.

c) If $\mathcal{V}=(V_\alpha)_{\alpha\geq 0}$ is a submarkovian resolvent on a measurable space X which is basic with respect to a finite measure on X then the ordered convex cone $\mathcal{E}(\mathcal{V})$ of all excessive functions with respect to $\mathcal{V}$ is an H-cone. In fact $\mathcal{E}(\mathcal{V})$ is an H-cone of functions on a measurable subset Y of X such that $X\setminus Y$ is $\mathcal{V}$-negligible. If the submarkovian resolvents $\mathcal{V},\mathcal{W}$ are basic and in duality with respect to a finite measure and if their initial kernels are proper then the dual of the H-cone $\mathcal{E}(\mathcal{V})$ is isomorphic with the H-cone $\mathcal{E}(\mathcal{W})$ and therefore, particularly, $\mathcal{E}(\mathcal{V})$ coincides with $(\mathcal{E}(\mathcal{V}))^{**}$.

Further if a submarkovian resolvent $\mathcal{V}$ is basic with respect to a finite measure and if its initial kernel is proper then there exists another submarkovian and basic resolvent $\mathcal{W}$ for which its initial kernel is also proper and such that $\mathcal{V}$ and $\mathcal{W}$ are in duality with respect to a finite measure.

2. STANDARD H-CONES AND STANDARD H-CONES OF FUNCTIONS

We remember that a weak unit of an H-cone S is an element $u\in S$ such that $\bigvee_n (s\wedge nu)=s$ for any $s\in S$.

DEFINITION. Let S be an H-cone and u be a weak unit of S. An element $p\in S$ is called <u>u-continuous</u> if for any $\varepsilon>0$ and any family $(s_i)_{i\in I}$ of S increasing to p there exists $i_\varepsilon\in I$ such that $p\leq s_{i_\varepsilon}+\varepsilon u$. We say that p is <u>universally continuous</u> if it is u-continuous for any weak unit u. The H-cone S is termed a <u>standard H-cone</u> if there exist at least a weak unit in S and a countable family of universally continuous elements which is increasingly dense in order from below in S. Particularly if S is a standard H-cone then the set S_o of all universally continuous elements of S is increasingly dense in order from below in S. An element of S which is dominated by an universally continuous element is called <u>universally bounded</u>. A standard H-cone which is an H-cone of functions on a set X such that $1\in S$ is termed <u>standard H-cone of functions on the set X</u>.

THEOREM. If S is a standard H-cone then S^* is also a standard H-cone, S^* separates S and S is solid (in natural order) and increasingly dense in order form below in S^{**}.

DEFINITION. Suppose that S is a standard H-cone. The coarsest topology on S^* such that the functions on S^* of the form $\mu\to\mu(s)$ where $s\in S_o$ are continuous is called <u>natural topology on S^*</u>.

THEOREM. If S is a standard H-cone then S^* is metrizable and complete with respect to the natural topology and a subset M of S^* will be relatively compact iff there exists a weak unit $u\in S$ such that $M\subset K_u:=$ $:=\{\mu\in S^* \mid \mu(u)\leq 1\}$. If u is a weak unit of S there exists a representation of S as an H-cone of functions on the set X_u of all non-zero extreme points of the compact convex set K_u and u corespond to the constant function 1 on X_u. If S is a standard H-cone of functions on a set X then S^{**} is also an H-cone of functions on X, and $f\in S^{**}$ iff f is densely finite and for any $s\in S$ we have $\inf(f,s)\in S$. Also there exists a greatest set X_1, $X_1\supset X$ such that S is an H-cone of functions on X_1 (in fact X_1 coincides with the set of all non-zero extreme points of the compact convex set $K_1=\{\mu\in S^* \mid \mu(1)\leq 1\}$. This set X_1 is called the <u>saturated of X</u>.

DEFINITION. Let S be a standard H-cone of functions on a set X and let X_1 be the saturated of X. The coarsest topology on X (resp. X_1) for which any element of S is continuous is termed the <u>fine topology on X</u> (resp. X_1).

Obviously the fine topology on X is greater then the natural topology on X. In general X is dense in X_1 in the fine topology. If $X=X_1$ we say that X is <u>saturated</u>. Also for any $F\subset S$ the element $\wedge F$ coincides with the lower semicontinuous regularized in the natural or in the fine topology of the function inf F.

DEFINITION. Let S be a standard H-cone of functions on a set X and X_1 be the saturated of X. Then for any $\mu\in S^*$ such that $\mu(1)<\infty$ there exists a unique measure m_μ on X_1 such that $\mu(s)=\int s\,dm_\mu$, for any $s\in S$. Whenever $\mu\in S^*$ is such that there exists a measure m_μ on X for which $\mu(s)=\int s\,dm_\mu$ for any $s\in S$ then μ is termed <u>H-measure on X</u> and always we identify m_μ with μ. So, the set X will be saturated if any $\mu\in S^*$ such that $\mu(1)<\infty$ is an H-measure on X. The set X is called <u>nearly saturated</u> if any $\mu\in S^*_o$ is an H-measure on X and it is called <u>semisaturated</u> if any $\mu\in S^*$ which is dominated (in natural order) by an element of S^*_o is an

H-measure on X.

THEOREM. Let S be a standard H-cone, S* its dual and u,u* be weak units in S and S* respectively. Then there exists a set X such that S and S* are represented in the same time as H-cone of functions on X such that u=1 and u*=1 on X and such that X is nearly saturated with respect to both S and S*.

In general we can't find weak units u and u* in S and S* respectively, such that the set X to be semisaturated with respect to S or S*. U. Schirmeier showed in 1983 ([14]) that there exists weak units u and u* in S and S* respectively such that the natural topologies induced on X by S and S* coincide.

THEOREM. If S is a standard H-cone of functions on a nearly saturated set X then there exists a basic submarkovian resolvent $\mathcal{V}=(V_\alpha)_{\alpha\geq 0}$ on X such that the kernel V_o is bounded and such that S** coincides with the set of all excessive functions on X with respect to the resolvent $\mathcal{V}$. The kernel V_o can be chosen such that Vf is finite, continuous for any Borel and bounded function f on X.

3. BALAYAGES ON A STANDARD H-CONE OF FUNCTIONS. AXIOM OF POLARITY. POTENTIALS, FINE POTENTIALS AND PURE POTENTIALS.

DEFINITION. If S is an H-cone, a map $B:S\rightarrow S$ is called <u>contraction</u> if it is additive, increasing, continuous in order from below and $Bs\leq s$ for any $s\in S$. A contraction is called <u>balayage</u> if it is idempotent (BB=B).

THEOREM. The set $\mathcal{B}$ of all balayages on S is a complete lattice (with respect to the order relation $B_1\leq B_2 \Leftrightarrow (B_1 s\leq B_2 s,\ (\forall)\ s\in S)$. Moreover we have

$$B\vee(\bigwedge_i B_i) = \bigwedge_i (B\vee B_i) ,$$

$$(\bigvee_i B_i)(s)=\bigvee_i (B_i s)$$

for any $B, B_i\in\mathcal{B}$ and any $s\in S$.

The set $\mathcal{C}$ of all contractions on S is a convex set and for any two balayages B_1, B_2 on S the contraction B_1B_2 is an extreme point of the convex set $\mathcal{C}$.

There exist H-cones such that the set of all extreme points of $\mathcal{C}$ coincides with the set of all contractions of the form $B_1B_2,\ldots,B_m$

where B_i are balayages. There exists also S for which there exist three balayages B_1, B_2, B_3 on S such that the contraction $B_1 \cdot B_2 \cdot B_3$ is not an extreme point of $\mathcal{C}$.

PROBLEM. It will be important to characterize the extremality in the convex set $\mathcal{C}$.

DEFINITION. Let S be an H-cone. An element $h \in S$ is called subtractible if for any $s \in S$ such that $h \leq s$ there exists $h' \in S$ for which $h+h'=s$ (i.e. h is specifically dominated by s). An element $p \in S$ is called pure potential ([5, b]) if any subtractible minorant of s is equal zero. We say that S satisfies axiom A if there exists a weak unit of S which is a pure potential.

THEOREM. Suppose that $h \in S$ is such that there exists a weak unit $u \in S$ and a decreasing sequence $(B_n)_n$ of balayages on S such that

$$\bigwedge_n B_n u = 0 \quad \text{and} \quad B_n h = h \quad \text{for any} \quad n \in N;$$

then h is subtractive. The converse assertion holds also if S satisfies axiom A.

THEOREM. Suppose that S is a standard H-cone. Then any pure potential is a sum of a series of universally bounded element of S. If S satisfies axiom A then the converse assertion is also true.

THEOREM. Suppose that S is a standard H-cone of functions on a set X. Then S satisfies axiom A iff there is no absorbant point in X (a point x_o is termed absorbant if there exists $s \in S$ such that $s(x_o)=0$ and $s(x)>0$ for any $x \in X$, $x \neq x_o$).

DEFINITION. Let S be a standard H-cone of functions on a set X. For any $A \subset X$ and any $s \in S$ we denote

$$B_s^A = \bigwedge \{ s' \in S \mid s' \geq s \text{ on } A \}$$

$$b(A) = \{ x \in X \mid B_s^A(x) = s(x), \quad (\forall) \; s \in S \}$$

A subset $A \subset X$ is called thin at a point $x \in X$ if $x \notin b(A)$ and it is called totally thin if it is thin at any point of X. The set A is called semipolar if it is a countable union of totally thin subsets of X. The set A is called polar if $B_s^A = 0$ for any $s \in S$.

In general a polar set is a semipolar one but the converse is not true. If $x \in X$ and $A \subset X$ is such that $x \notin A$ then A is thin at x iff $X \setminus A$ is a fine neighbourhood of x. If $A \subset b(A)$ then $B_s^A = s$ on A and therefore the map

$s \to B_s^A$ is a balayage on S denoted by B^A. If B is a balayage on S we say that B is represented on X if there exists $A \subset X$ such that $A \subset b(A)$ and $B=B^A$.

THEOREM. Let S be a standard H-cone of functions on a set X. Then X is nearly saturated (resp. semisaturated) iff any compact subset of $X_1 \setminus X$ is semipolar (resp. polar).

THEOREM. Let S be a standard H-cone of functions on a set X and μ be an H-measure on X. Then μ is a sum of a sequence of universally continuous (resp. universally bounded) from S* iff μ does not change any semipolar (resp. polar) subset of X. If the measure μ is carried by a polar subset of X then μ is a subtractible element of S*. The converse assertion is also true if S* satisfies axiom A.

DEFINITION. We say that a standard H-cone S satisfies <u>axiom of polarity</u>, if there exists a representation of S as standard H-cone of functions on a saturated set X such that any semipolar subset of X is polar.

We say that a standard H-cone S satisfies <u>axiom of nearly continuity</u> if any universally bounded element of S is equal to a series of universally continuous elements of S.

THEOREM. Suppose that S is a standard H-cone of functions on a semisaturated set X. Then the following assertions are equivalent:

1. S satisfies axiom of polarity.
2. Any semipolar subset of X is polar.
3. For any subset A of X the map B^A is a balayage on S.
4. S* satisfies axiom of nearly continuity.
5. Any H-measure μ on X which is finite on any $s \in S$ does not charge any semipolar subset of X.

For a general standard H-cone of functions S on a set X it is useful to know that S satisfies the following <u>Hunt's property</u>: For any $A \subset X$ and any open subset G, $G \supset A$ we have $B^G(B^A s)=B^A s$ for any $s \in S$. Obviously if S satisfies axiom of polarity then Hunt's property holds for any representation of S as H-cone of functions on a set X. Recently ([5, g]) we proved that if S is an arbitrary standard H-cone of functions on a set X then there exists a semipolar subset B of X such that for any $A \subset X \setminus B$ and any open subset G of X such that $A \subset G$ we have $B^G(B^A s)=B^A s$ for any $s \in S$. This result can not be generally improved.

<u>In the sequel of this section</u>, S <u>will be a standard H-cone of func-</u>

<u>tions on a semisaturated set X</u>. For any $\mu\in S^*$ and any $A\subset X$ we denote by μ^A the <u>balayage of</u> μ <u>on A</u> (i.e. the element of S^* given by the relation

$$\mu^A(s)=\mu(B^A s) \qquad (\forall) \quad s\in S\ ).$$

If μ is an H-measure on X then μ^A is also an H-measure on X which is carried by the fine closure of A if A is a Borel subset of X. The following theorem obtained by G. Bucur and W. Hansen([6])is a useful tool in the theory of balayages on H-cones.

THEOREM. For any A, B two Borel measurable subsets of X, $A\subset B$ and any H-measure μ on X such that μ does not charge the set $b(B)\cap(A\setminus b(A))$ we have

$$\mu^A=\mu^B|_A+(\mu^B|_{X\setminus A})^A$$

Using this result we proved that for any subset A of X and any $x\in X$ the following <u>property of Frostman</u> holds:

THEOREM. For any sequence $(x_n)_n$ in X such that $(x_n)_n$ converges to x and such that $(\varepsilon^A_{x_n})_n$ is naturally convergent to an H-measure μ on X then there exists a real number α, $0\le\alpha\le1$ such that $\mu=\alpha\varepsilon_x+(1-\alpha)\varepsilon^A_x$.

In the frame of harmonic spaces J. Lukes and J. Maly([8, b]) and W. Hansen([10]) showed that the set of all $\alpha\in[0,1]$ for which there exists a sequence $(x_n)_n$ as in the above theorem, is contained in $\{0,1\}$ or coincides with $[0,1]$. Generally this result fails but it may be proved if S satisfies the <u>natural localization property</u> (i.e. for any closed subset F of X and any $x\notin F$ the H-measure on ε^F_x is carried by the boundary of F).

DEFINITION. We say that an element $p\in S$ is a <u>potential</u> (resp. <u>fine potential</u>) on X if for any increasing sequence $(G_n)_n$ of open (resp. fine open such that $\overline{G}^f_n\subset G_{n+1}$ for any $n\in N$) subsets of X such that $\bigcup_n G_n=X$, we have $\bigwedge_n B^{X\setminus G_n}p=0$ where $\overline{G}^f_n$ means the fine closure of G_n.

If S is a general standard H-cone on a set X and if there exists a strictly positive potential on X (or equivalently any $s\in S_o$ is a potential on X) then X is semisaturated. Moreover the existence of a strictly positive fine potential on X (or equivalently any $s\in S_o$ is a fine potential on X) is equivalent with the fact that X is semisaturated and S satisfy axiom of nearly continuity (or equivalently S^* satisfies axiom of polarity). The following results characterize the elements of S in terms of supermean property with respect to a family of measures on X which generalize the well known harmonic measures from the case of the harmonic spaces.

THEOREM. Suppose that $S=S^{**}$ and that there exists a strictly positive potential on X. Then a lower semicontinuous positive function on X which is finite on a dense subset of X, belongs to S iff for any $x \in X$ and any open neighbourhood V of x there exists a fine neighbourhood W of x, $W \subset V$ such that

$$\varepsilon_x^{X \setminus W} f \leq f(x) .$$

THEOREM. Suppose that $S=S^{**}$, X is a souslinian topological space and that there exists a strictly positive fine potential on X. Then a fine lower semicontinuous positive function f on X which is finite on a dense subset of X belongs to S iff for any $x \in X$, any open neighbourhood V of x there exists a fine neighbourhood W of x, $W \subset V$ such that $\varepsilon_x^{X \setminus W} f \leq f(x)$.

This type of characterization of elements of S generalizes similar results given in 1982 by J. Lukes and J. Maly ([8, a]) and by J.T. Lions ([11]) in the case of harmonic spaces.

In connection with the notion of potential and fine potential we mention the following results:

THEOREM. Suppose that S is such that there exists a strictly positive potential on X. Then for any pure potential p and any countable covering of open subsets $(G_n)_n$ of X we have

$$\bigwedge_{(i_1, i_2, \ldots, i_n)} B^{X \setminus G_{i_1}} B^{X \setminus G_{i_2}} \ldots B^{X \setminus G_{i_n}} p = 0 .$$

THEOREM. Suppose that S satisfies axiom of nearly continuity and let p be an element of S, finite on X. Then we have:

a) p is a fine potential iff p is the sum of a series of universally continuous elements of S.

b) if X is a souslinian topological space and $p \in S$ is bounded on X then p is a fine potential iff p is a potential. Moreover if p is a bounded potential then for any increasing sequence of fine open subsets $(G_n)_n$ of X such that $\bigcup_n G_n = X$ then we have $\bigwedge_n B^{X \setminus G_n} = 0$.

THEOREM. Suppose that any universally continuous element of S is a potential on X. Then any pure potential is a potential on X. Conversely, suppose that any universally continuous element of S is a pure potential. Then any potential on X which is finite continuous is a pure potential.

4. NATURAL AND FINE LOCALIZATION IN STANDARD H-CONES OF FUNCTIONS

In this section S will be a standard H-cone of functions on a nearly saturated set X.

DEFINITION. For any fine open subset U of Y we denote by S(U) the convex cone of all positive numerical functions f on U finite on a fine dense subset of U, for which there exists a sequence $(s_n)_n$ in S, s_n finite, such that $(s_n - B^{X\setminus U} s_n)_n$ increases to f.

THEOREM. The convex cone of functions on U given by S(U), endowed with the pointwise order relation is a standard H-cone of functions on U and U is nearly saturated. If X is semisaturated then U is also semisaturated. The fine topology on U generated by S(U) coincide with the restriction to U of the fine topology of X. The natural topology on U generated by S(U) is in general greater than the restriction to U of the natural topology on X and coincides with it if U is open.

DEFINITION. We say that S satisfies <u>the natural</u> (resp. <u>fine</u>) <u>sheaf property</u> if the map $U \to S(U)$ defined on all open (resp. fine open) sets of X is a sheaf of convex cones.

If S satisfies the natural sheaf property then X is semisaturated. Moreover we have the following characterizations:

THEOREM. S satisfies the natural sheaf property iff S satisfies the natural localization property (i.e. for any closed set $F \subset X$ and any $x \notin F$ the H-measure ε_x^F is carried by the boundary of F) or equivalently for any two open sets G_1, G_2 in X such that $G_1 \cup G_2 = X$ we have

$$B^{G_1}(B^{G_2}s) = B^{G_2}(B^{G_1}s) \qquad (\forall)\ s \in S$$

and there exists a strictly positive potential on X.

We proved also that if X is semisaturated and S satisfies the natural localisation property then Hunt's property holds on X.

THEOREM. S satisfies the fine sheaf property iff any representation of S as an H-cone of functions on a nearly saturated set X satisfies the natural sheaf property.

THEOREM. S satisfies the fine sheaf property iff S satisfies the natural sheaf property (or only the natural localisation property) and S^* satisfies the axiom of polarity or equivalent any element of S which is dominated by an element of S_o is equal with a series of elements of S_o.

5. CARRIER THEORY ON A STANDARD H-CONE OF FUNCTIONS, GREEN FUNCTIONS, GREEN POTENTIALS

DEFINITION. Suppose that S is a standard H-cone of functions on a set X. The harmonic carrier of an element $s\in S$ is the set (denoted carr s) of all points $x\in\overline{X}$ ($\overline{X}$ is the natural closure of X in S^*) for which $B^{X\setminus V}s\neq s$ where V runs the set of all natural neighbourhood of x.

THEOREM. Any non-zero bounded element of S has a nonempty harmonic carrier. Moreover if X is nearly saturated then for any universally continuous element p of S, $p\neq 0$, we have $X\cap \text{carr } p\neq\emptyset$. Any element of S for which carr $h=\emptyset$ is subtractible and for any pure potential p we have carr $p\neq\emptyset$.

PROBLEM. In general we don't know if carr $s\neq\emptyset$ for any $s\in S$.

DEFINITION. Suppose that S is a standard H-cone such that $S=S^{**}$ and let u (resp u^*) be a weak unit in S (resp S^*). We consider the saturated set $X=X_u$ (resp $X^*=X^*_u$) and we denote by X_p, X^*_p the subsets of X and X^* respectively defined by

$$X_p=\{x\in X \mid \text{carr } x\neq\emptyset\} ,$$

$$X^*_p=\{\xi\in X^* \mid \text{carr } \xi\neq\emptyset\} .$$

We remark that for any $x\in X$ (resp $\xi\in X^*$) the set carr x (resp carr ξ) has only a point if it is non empty.

THEOREM. The set X_p (resp X^*_p) is a Borel subset of X (resp X^*) and the maps

$$\theta: X^*\to\overline{X}, \quad \theta^*: X\to\overline{X}^*$$

defined by $\{\theta(\xi)\}=\text{carr } \xi$, $\{\theta^*(x)\}=\text{carr } x$ are also Borel measurable. Moreover the sets

$$E=\{x\in X_p \mid \theta^*(x)\in X^*_p \text{ and } \theta(\theta^* x)=x\},$$

$$E^*=\{\xi\in X^*_p \mid \theta(\xi)\in X_p \text{ and } \theta^*(\theta\xi)=\xi\}$$

are Borel subsets of X_p and X^*_p respectively such that $X\setminus E$, $X^*\setminus E^*$ are semipolar and for any $A\subset E$ we have

$$\mu(B^A s) = (B^{\theta^*(A)}\mu)(s)$$

for any $s\in S$ and $\mu\in S^*$.

From this theorem it follows that a subset A of E is polar (resp. semipolar) iff $\theta^*(A)$ is a polar (resp. semipolar) subset of E^*.

DEFINITION. Let S, S^*, u, u^* as in the previous definition and let E and E^* as in the preceding theorem. We identify E and E^* following the maps θ and θ^*. We get in this way a set $\Omega=\Omega(u,u^*)$ which is a subset of both X and X^* and which is termed the Green set of S associated to the weak units u and u^*.

Thus, S and S^* are representable as standard H-cones of functions on a same set Ω such that Ω is nearly saturated with respect to S and S^* and such that for any $A\subset\Omega$ we have

$$\mu(B^A s) = {}^*B^A\mu(s)$$

where B^A (resp ${}^*B^A$) means the balayage on A with respect to S (resp S^*). The function defined on $\Omega\times\Omega\subset X\times X^*$ given by

$$G(x,y) := y(x)$$

is called the Green function on the Green set Ω.

If μ is a measure on Ω then the function $G\mu$ (resp ${}^*G\mu$) defined on Ω by

$$G\mu(x) = \int G(x,y)\,d\mu(y) \quad (\text{resp } {}^*G\mu(X) = \int G(y,x)\,d\mu(y))$$

is called Green potential (resp. Green co-potential) associated with μ if it is finite on a dense subset of Ω.

THEOREM. Let $\Omega=\Omega(u,u^*)$ be the Green set of S corresponding to the weak units u and u^*. The following assertions are equivalent:

1) The set of all Green potentials on Ω is a solid subcone (with respect to the natural order) of S.

2) Any universally bounded element of S is a Green potential on Ω.

3) Ω is a semisaturated set with respect to S^*.

4) Any pure potential is a Green potential on Ω.

THEOREM. Let $\Omega=\Omega(u,u^*)$ be the Green set of S corresponding to weak units u, u^* and let $p\in S$. The following assertions are equivalent:

1) p is a Green potential on Ω

2) for any $q\in S$, $q\prec p$ (q specifically smaller then p) and any co-na-

tural (or co-fine) neighbourhood V of the set $\Omega \cap \text{carr } q$ we have $B^V q=q$.

3) for any $q \in S$, $q \lesssim p$, $q \neq 0$ there exists $x \in \Omega$ such that for any co-natural neighbourhood V of x we have $B^{X \setminus V} q \neq q$.

DEFINITION. An element $p \in S$ is termed *-potential on Ω if for any increasing sequence $(D_n)_n$ of co-natural open subsets of Ω such that $\bigcup_n D_n = \Omega$ we have

$$\bigwedge_n B^{X \setminus D_n} p = 0$$

From the preceding theorem it follows that if p is a *-potential on Ω then it is a Green potential on Ω.

The following axiom (called Green axiom) represents the key to characterize the Green potentials on Ω. We say that (S,Ω) satisfies <u>Green's axiom</u> if there exists a strictly positive *-potential on Ω (or equivalently any universally continuous element of S is a *-potential on Ω).

THEOREM. If (S,Ω) satisfies Green's axiom then an element $p \in S$ is a Green potential iff it is a *-potential on Ω.

In general in the preceding considerations we cannot replace the term of *-potential by that of potential. Obviously if the co-natural topology on Ω coincides with the natural topology on Ω and there exists a strictly positive potential on Ω then a Green potential on Ω is nothing else than a potential on Ω.

The last theorems can be considered as the extensions in the frame of standard H-cones of some results obtained by B. Fuglede ([7]) for the theory of fine superharmonic functions on the harmonic spaces for which there exists a good adjoint structure.

PROBLEM. It is not known whether for any standard H-cones S there exists a weak unit u in S and a weak unit u* in S* such that the Green set $\Omega=\Omega(u,u^*)$ satisfies Green's axiom.

6. MARKOV PROCESSES ASSOCIATED WITH A STANDARD H-CONE

In this paragraph S will be a standard H-cone of functions on a saturated set X such that $S=S^{**}$. We denote by Y the closure of X in S* endowed with the natural topology. The topological space Y is a metrisable compact space.

<u>DEFINITION</u>. A Ray semigroup $P=(P_t)_{t \geq 0}$ on Y is called <u>a strong Ray semigroup</u> if for any $\alpha>0$ and any bounded Borel function f on Y the

function $\int_0^\infty e^{-\alpha t} P_t(f)dt$ is continuous on Y and the set of all finite continous, excessive functions with respect to P separates Y.

We have the following two results ([4]).

<u>THEOREM</u>. Suppose that $P=(P_t)_{t\geq 0}$ is a strong Ray semigroup on Y and let Y_o be the set of all nonbranching points of Y with respect to P. Then the ordered convexe cone $\mathcal{E}_P$ of all excessive functions with respect to P restricted to Y_o is a standard H-cone of functions on the set Y_o and Y_o is a saturated set with respect to $\mathcal{E}_P$.

An element $p\in S$ is termed a generator of S if for any $s\in S$ there exists an increasing sequence $(p_n)_n$ in S and a sequence $(\alpha_n)_n$ in $\mathbb{R}_+$ such that $\bigvee_n p_n = s$ and $p_n \prec \alpha_n p$ for any $n\in N$.

<u>THEOREM</u>. Let $p\in S$ be a generator of S of the form $p=\sum_{n=1}^{\infty} p_n$, $p_n \leq S_o$, $p_n \leq \frac{1}{2^n}$. Then there exists a strong Ray semigroup $P=(P_t)_{t\geq 0}$ on Y such that $\int_0^\infty P_t 1 dt = p$, X coincides with the set of all nonbranching points with respect to P and S is exactly the set of all excessive functions with respect to P, restricted to X.

Using the previous theorem and standard procedures in the theory of Ray processes one can construct a Markov process with values in X having the point "0" of Y as heaven . More precisely we have:

<u>THEOREM</u>. Let $(P_t)_{t\geq 0}$ be a strong Ray semigroup constructed as above. Then for any positive finite measure μ on X there exists a right continuous Markov process with values in X having $(P_t)_{t\geq 0}$ as semigroup of transitions, μ as initial low and the point "0" as heaven. Moreover any element of S is right continuous on the paths.

This theorem allow us to use the Markov process technics as a powerful tool in the study of standard H-cones of functions on a set X. Obviously, from the above considerations, the set X must be saturated or at least semisaturated.

It would be desirable to construct on the same set X Markov processes in duality associated to S and S* as above. Such a construction may be realised if S satisfies both axiom of polarity and axiom of nearly continuity (the eliptic case) or more generally if S and S* are represented as standard H-cones of functions on a Green set Ω which is semisaturated with respect to S as well to S*. To be more precisely let u^* be a weak unit in S^* and let $X^*=X^*_{u^*}$ be a saturated set on which the H-cone S^* is represented as a standard H-cone of functions.

We consider now the Green set Ω associated to S and to the weak units $u=1\in S$ and $u^*\in S^*$. Obviously (see §5) Ω is a subset of X as well of X^*.

We consider now a finite measure μ on Ω of the form $\mu=\sum_n \mu_n$ such that

$$G^{\mu_n}\in S_o \ , \quad G^{*\mu_n}\in S^*_o \qquad (\forall)\ n\in\mathbb{N}$$

and such that the element $\sum_{n=1}^{\infty} G^{\mu n}$ (resp. $\sum_{n=1}^{\infty} G^{*\mu n}$) is a generator of S (resp. S^*). Such a measure μ may be allways chosen.

We have the following result:

<u>THEOREM</u>. Let Ω be a Green set as above such that $X\setminus\Omega$, respectively $X^*\setminus\Omega$, is a Borel and polar subset of X, respectively X^*.

Let also μ be a finite measure on Ω as above. Then there exists two Markov processes on Ω

$$(\Omega,(X_t)_{t\geq 0}\ ,\ (P^x)_{x\in\Omega}),\quad (\Omega,(\hat{X}_t)_{t\geq 0},\ (P^x)_{x\in\Omega})$$

which are right continuous with respect to the natural, respectively conatural topology on Ω having the same heaven point δ, $\delta\notin\Omega$ and being in duality with respect to the measure μ i.e.

$$P^x(\int_o^\infty f_o X_t dt)=\int G(x,y)f(y)d\mu(y)\ ,\quad \hat{P}^*(\int_o^\infty f\cdot\hat{X}_t dt)=\int G(y,x)f(y)d\mu(y)$$

<u>REMARKS</u>. a) In the above theorem the assertion that the sets $X\setminus\Omega$ and $X^*\setminus\Omega$ are polar seems to be necessary for the existence of the Markov processes on Ω in duality associated with S and S^*.

Usually the set $X\setminus\Omega$ is called the <u>exit boundary</u> of Ω with respect to S and $X^*\setminus\Omega$ is called the <u>entrance boundary</u> of Ω with respect to S. There are many interesting examples of H-cones S for which both boundaries of Ω are polar (we recall, for instance, the case of the standard H-cone S of all positive supercaloric functions on a band Ω in $\mathbb{R}^{n+1}=\mathbb{R}^n\times\mathbb{R}$ given by

$$\Omega=\{(x,t)\mid \alpha<t<\beta,\ x\in\mathbb{R}^n\}\ .$$

b) If at least one of the sets $X\setminus\Omega$, $X^*\setminus\Omega$ is not polar in X, respectively X^*, then for any finite measure μ on Ω as above we may construct Markov processes

$$(X_t, (P^x)_{x \in X}), \quad (\hat{X}_t, (\hat{P}^y)_{y \in X^*})$$

on X respectively X* which is right continuous with respect to the natural topology on X respectively conatural topology on X*, and being in duality on Ω (only) with respect to μ i.e.

$$P^x(\int_0^\infty f \cdot X_t dt) = \int G(x,y) f(y) d\mu(y) \qquad (\forall) \quad x \in \Omega$$

$$\hat{P}^x(\int_0^\infty f \ \hat{X}_t dt) = \int G(y,x) f(y) d\mu(y) \qquad (\forall) \quad x \in \Omega$$

and any positive, Borel function f on Ω (on the left hand side the function f vanishes on $X \setminus \Omega$ and $X^* \setminus \Omega$ respectively).

For the last theorem one can see our paper "Standard H-cones on Markov Processes" Preprint series in Mathematics, no.51, 1979, INCREST Bucharest.

REFERENCES

1. J. Bliedtner, W.Hansen: Potential Theory, Universitext Springer-Verlag,Berlin 1986.
2. N. Boboc, A.Cornea: a) Cônes convexes ordonnés, H-cônes et adjoints. C.R.Acad.Sci.Paris 270, 596-599, 1970.
   b) Cônes convexes ordonnés, H-cônes et bi-adjoints de H-cônes. C.R.Acad.Sci.Paris 270, 1679-1682, 1970.
   c) Cônes convexes ordonnés. Representations integrals C.R.Acad.Sci.Paris 271, 880-883, 1970.
3. N. Boboc, Gh.Bucur, A. Cornea, H-cones and potential theory. Ann. Inst.Fourier 25, 71-108, 1975.
4. N. Boboc, Gh. Bucur, A. Cornea, Order and Convexity in Potential Theory: H-Cones. Lecture Notes in Mathematics, 853, Springer-Verlag, Berlin 1981.
5. N. Boboc, Gh. Bucur: a) Natural localization and natural sheaf property in Standard H-cones of functions. Preprint Series in Math. 32, 1982, INCREST Bucureşti; Rev.Roum.Math.Pures et Appl. 1, 1-21; 2, 193-219, 1985.
   b) Potentials and pure potentials in H-cones. Rev. Roum. Math. Pures et Appl.5, 529-549, 1982.
   c) Potentials and supermedean functions on fine open sets in standard H-cones. Preprint series in Math. 59, 1984, INCREST Bucureşti; Rev. Roum. Math. Pures et Appl. 9, 745-775, 1986.
   d) Fine potentials and supermedean functions on standard H-cones. Preprint series in Math. 47, 1986, INCREST Bucureşti; Rev. Roum. Math. Pures et Appl. 1987, to appear.
   e) Stonean structures in potential theory, Preprint series in Math. 52, INCREST Bucureşti; Studii şi Cercetări matematice, 5, 411-437, 1986.
   f) Green potentials on standard H-cones. Preprint series in Math. 35, 1985, INCREST Bucureşti; Rev. Roum. Math. Pures et Appl. 4, 293-320, 1987.
   g) Sur une hypothèse de C.A. Hunt. C.R.Acad. Sci. Paris, 302, 701-703, 1986.
6. Gh. Bucur, W. Hansen, Balayage, quasi-balayage and fine decomposition properties in standard H-cones of functions. Rev. Roum. Math. Pures et Appl., 19-41, 1984.
7. B. Fuglede: a) Localization in fine potential theory and uniform approximation by subharmonic functions. J. Func. Anal. 49, 57-72, 1982.
   b) Representation integrale des potentiels fins. C.R.
8. J. Lukes, J. Maly: a) Fine hyperharmonicity without axiom D.Math. Ann. 26, 299-306, 1982.
   b) On the boundary behaviour of the Perron generalized solution. Math. Ann. 257, 355-366, 1981.
9. J. Lukes, J. Maly, L. Zajicek, Fine Topological Methods in real Analysis and Potential Theory, Lecture Notes in Math. 1189, Springer Verlag, Berlin 1986.
10. W. Hansen: a) Convergence of balayage measures, Math.Ann.264, 437-446, 1983.
11. J.L. Lions, Cones of lower semicontinuous functions and a characterization of finely hyperharmonic functions. Math. Ann. 261, 293-298, 1982.
12. G.Mokobodzki, Cônes de potentiels et moyaux subordonnés, in "Potential Theory" (CIME, 1° Ciclo, Stresa 1969, Ed.Cremonese (1970), 207-248).
13. G. Mokobodzki, D. Sibony, Sur une propriété caracthérisque des cônes de potentiels. C.R.Acad. Sci. Paris 266, 215-218, 1968.
14. U. Shirmeier, Continuity properties of the carrier map. Rev. Roum. Math. Pures et Appl. 5, 431-451, 1983.

PROBABILITY METHODS IN POTENTIAL THEORY

Kai-Lai Chung*

Mathematics Department, Stanford University

Stanford, CA 94305-2125, USA

Introduction.

The Dirichlet boundary value problem for Laplace's equation is a cornerstone in classical potential theory. In the axiomatic theory the solvability of the problem for base sets is a main postulate for the harmonic space. A probabilistic treatment of the problem has been known for nearly half a century, an exposition of which may be found in my book [4], to be referred to below as "basic probability theory".

The extension of Dirichlet's problem from Laplace's to Schrödinger's equation is of quite recent origin. In §1 the principal theorems will be reviewed with comments. In the case where the function q is bounded these were proved first in [3] and partly presented in a more didactive manner in [4]. Subsequently, the general case discussed here was undertaken in [1]. This is a natural extension from the probabilistic standpoint. It requires certain technical detours and improvements, but the development follows the old lines when probabilistic, rather than operator-analytic, methods lead the way. Details of the theorems quoted in §1 will be given in a monograph under preparation by Z. Zhao and the author.

The probabilistic method yields solutions in a form which can be "tinkered with" to produce new problems and results. This potentiality is particularly pronounced in the present case owing to the simple structure of the Feynman-Kac functional. We give three examples in §2 to show that the new method accomplishes more than merely re-solving old problems, a point often lost on the traditional analysts.

In §3, some of the results in §1 are recast in the more formal and less explicit setting of semigroup and and potential theory. An alternative representation of the fundamental solution formula for the Dirichlet problem is adduced, leading to a new application. In §4, a review of the conditional gauge theorem is given together with the related analysis of the Green function for a Lipschitz domain.

In classical potential theory more general kinds of boundary value problems are considered. A particular case is known as Neumann's problem, in which the normal derivative of the solution at the boundary is prescribed. The probabilistic approach to this problem begins with the construction of a new process called the reflecting Brownian motion, and the associated local time on the boundary, under a strong smoothness assumption. Then a gauge can be defined, and a gauge theorem holds which reads just like Theorem 3 below, see [9]. When the gauge is finite, the unique solution of the problem is given by an explicit formula obtained by "plugging in" the boundary

* This work is supported in part by NSF grant DMS83–01072 and by AFOSR grant 85–0330.

data into the gauge, as in the Dirichlet case to be described below; see [20]. A more general "mixed problem" is the topic of the doctoral dissertation by V. Papanicolaou, currently in progress.

1. The gauge.

We denote by $X = \{X_t,\ t \geq 0\}$ the standard Brownian motion process in R^d, $d \geq 1$, with continuous paths; P^x and E^x are the probability and expectation for paths starting at x. A domain D is a nonempty, open and connected set in R^d, $\overline{D}$ its closure, ∂D its boundary, both in R^d; there is no point at infinity. The first exit time from D is defined to be

$$\tau_D = \inf\{t > 0 :\ X_t \notin D\} \tag{1}$$

where the inf of an empty set is $+\infty$. The class of real, extended valued Borel measurable functions on a set S is denoted by $\mathcal{B}(S)$; the subscripts "+" and "b" signify the subclasses of "positive (≥ 0)" and "bounded" functions. When $S = R^d$ it is omitted from the notation.

A special class of functions in $\mathcal{B}$ will be designated by **J** which satisfies a strengthened form of integrability condition along the paths of X, as follows: $q \in \mathbf{J}$ iff

$$\lim_{t \downarrow 0} E^x \left\{ \int_0^t |q(X_s)|ds \right\} = 0 \quad \text{uniformly in } x \in R^d \tag{2}$$

Obviously $\mathcal{B}_b \subset \mathbf{J}$. Next, $\mathbf{J}_0$ denotes the class of q such that $q1_D \in \mathbf{J}$ for every bounded open set D. It can be shown (see [1]) that $q \in \mathbf{J}_0$ if and only if the function $x \to \int_D K(x-y)|q(y)|dy$ is continuous in R^d, where $K(u) = |u|^{2-d}$ if $d \geq 3$, $K(u) = \log^+(|u|^{-1})$ if $d = 2$, $K(u) = |u|$ if $d = 1$.

As an abbreviation, we put

$$e_q(t) = \exp\left\{ \int_0^t q(X_s)ds \right\}, \qquad t \geq 0. \tag{3}$$

This is a multiplicative functional for X, called the Feynman-Kac functional. The condition (2) implies that $0 < e_q(t) < \infty$ almost surely, and there is a constant $C(t)$ such that

$$\sup_{x \in R^d} E^x\{e_q(t)\} = C(t) < \infty; \quad \lim_{t \downarrow 0} C(t) = 1. \tag{4}$$

This is due to Khas'minskii, and is essential for the development.

The situation changes radically when the constant time t above is replaced by the random (optional) time τ_D. We now introduce the central probabilistic object below, well-defined for $f \in \mathcal{B}_+(\partial D)$:

$$u_f(x) = E^x\{\tau_D < \infty;\ e_q(\tau_D)f(X(\tau_D))\}, \tag{5}$$

and the special case

$$u_1(x) = E^x\{\tau_D < \infty;\ e_q(\tau_D)\}. \tag{6}$$

The condition (2) still implies that $0 < e_q(\tau_D) < \infty$ on the set $\{\tau_D < \infty\}$; but in contrast to (4) it is no longer true that $u_1(x)$ is always finite. We begin with the following general proposition.

Theorem 1. *Let D be an arbitrary domain, and $q \in \mathbf{J}$. For each compact subset C of D there exists a constant $A = A(D, C, q)$, $1 \le A < \infty$, such that for all $f \in \mathcal{B}_+(\partial D)$ we have*

$$\sup_{x \in C} u_f(x) \le A \inf_{x \in C} u_f(x) . \tag{7}$$

As an immediate consequence, for each indicated f either $u_f \equiv +\infty$ in D or u_f is bounded in every compact subset of D; also, either $u_f \equiv 0$ in D or u_f is bounded away from 0 in every compact subset of D.

When $q \equiv 0$, each u_f is a harmonic function in D by basic probability theory, and the theorem is known under the name of "Harnack inequality". Numerous papers have been written on this item. For the purposes of this survey, the next result of more recent origin is more important. We denote by m the Lebesgue measure in R^d and by $L^1(D)$ the usual class of integrable functions on D. From here on we shall make the following assumptions on D and q:

(H) D is a domain in R^d, $d \geq 3$; or a bounded domain in R^d, $d \leq 2$;
$q \in \mathbf{J} \cap L^1(D)$.

Any modification of the assumptions later on will be specified.

Theorem 2. *Suppose $f \in \mathcal{B}_{+,b}(\partial D)$. Then either $u_f \equiv +\infty$ in D, or u_f is bounded in $\overline{D}$.*

The function u_1 in (6) is called the "gauge" for (D, q), and Theorem 2 in this case is the "gauge theorem". It can be complemented as follows.

Theorem 3. *If the gauge $u_1 \not\equiv +\infty$ in D, then it is bounded above and away from zero in R^d, and continuous in D.*

In this case we shall say that (D, q) is "gaugeable". Theorem 1 and 2 were first proved for $q \in \mathcal{B}_b$ in [2], and for a bounded D and $q \in \mathbf{J}_0$ in [25]. An extension of it to a general class of Markov processes and multiplicative functionals is given in [16]. We shall not discuss the latter generalization here.

To proceed to the Dirichlet boundary value problem we introduce some further notation. Let $G = G_D$ be the Green operator for D, defined most expediently as follows:

$$G_D\varphi(x) = E^x \left\{ \int_0^{\tau_D} \varphi(X_s) ds \right\} , \qquad \varphi \in \mathcal{B}_+(D) . \tag{8}$$

Write also

$$h_f(x) = E^x \{ f(X(\tau_D)) \} , \qquad f \in \mathcal{B}_b(\partial D) . \tag{9}$$

Then u_f reduces to h_f when $q \equiv 0$. It is basic probability theory that h_f is harmonic in D. A point $z \in \partial D$ is "regular" iff $P^z \{ \tau_D = 0 \} = 1$. The set of such points will be denoted by $(\partial D)_r$. If $(\partial D)_r = \partial D$ we say D is regular.

The importance of Theorem 2 becomes apparent from the next result which solves the Dirichlet boundary value problem for the Schrödinger equation. We denote by $C(S)$ the class of continuous functions on S.

Theorem 4. *Suppose (D,q) is gaugeable. Then for any $f \in \mathcal{B}_b(\partial D)$, we have $u_f \in C_b(D)$ and the integral equation:*

$$u_f = G_D(q u_f) + h_f . \tag{10}$$

Consequently u_f is a solution of the equation

$$\left(\frac{\Delta}{2} + q\right)\varphi = 0 \quad \text{in } D \tag{11}$$

in the weak (or distributional) sense, satisfying the following boundary condition. At each $z \in (\partial D)_r$ where f is continuous, we have

$$\lim_{\overline{D} \ni x \to z} \varphi(x) = f(z) . \tag{12}$$

Moreover, for any f on ∂D there is at most one solution of (11) satisfying (12) at all $z \in (\partial D)_r$.

A solution φ of (11) which belongs to $C(D)$ will be called "q-harmonic" in D. If it also satisfies (12) it is called a solution of the Dirichlet problem with boundary value f. When $q \equiv 0$ and $f \in C(\partial D)$ this reduces to the classical notions. An important variant of Theorem 4 is the following representation theorem.

Theorem 5. *Suppose (D,q) is gaugeable. Then any q-harmonic function u which belongs to $C_b(\overline{D})$ can be represented as the u_f in (5) with $f = u$ on ∂D.*

Note that in Theorem 5 neither the boundedness nor the regularity of D is assumed; it is the continuity of u in $\overline{D}$ that is essential. The result implies immediately the representability of u on any subdomain D_0 with $\overline{D_0} \subset D$ because u is continuous in $\overline{D_0}$, and the gaugeability of (D,q) implies that of (D_0,q). The latter assertion is not trivial because q is not always positive, but is an important property of the gauge[1].

From the standpoint of probability theory, it is the simple explicit formula u_f as defined by (5), which yields the solution in Theorem 4 and the representation in Theorem 5, that distinguishes it from other methods. While the solution of the Dirichlet boundary value problem is more or less known under stronger hypotheses, even though it is not easy to nail down the exact statement[2], a useful expression of the solution is not generally available in classical analysis. Indeed, the representation given in (5) is the stochastic version of the Poisson formula for a harmonic function when $q \equiv 0$, and D is a ball. It involves only the primary Brownian motion process, and is simple enough for manipulations. Numerous consequences flow from it. We shall give three easy examples to illustrate this essential point before further discussion.

Examples.

(i) Consider the gauge u_1 for a ball B in R^d. Without loss of generality we may suppose that B has center at the origin and radius r. It is basic probability that

$$E^x\{\tau_B\} = \frac{r^2 - |x|^2}{d}, \qquad x \in B; \tag{13}$$

so that

$$\sup_{x\in B} E^x\{\tau_B\} = \frac{r^2}{d}. \tag{14}$$

The number in (14) is known as "rigidity coefficient" in applied mathematics. Now suppose $|q| \le Q$ where Q is a constant. Then it follows by the explicit representation in (6) that

$$u_1(x) \le E^x\left\{e^{Q\tau_B}\right\} \tag{15}$$

Now Khas'minskii's lemma is applicable to the hitting time τ_D as well as a constant, and yields the following result. If $q \in \mathbf{J}_0$ and

$$\sup_{x\in D} E^x\left\{\int_0^{\tau_D} |q(X_s)|ds\right\} < \epsilon < 1 \tag{16}$$

then

$$\sup_{x\in D} E^x\left\{e_{|q|}(\tau_D)\right\} < \frac{1}{1-\epsilon} < \infty. \tag{17}$$

Applying this with $D = B$ and $q = Q$, we see from (14) and (15) that $u_1(x) < \infty$ if $r < Q^{-1/2}d^{-1/2}$; indeed u_1 is bounded. Theorem 2 is not needed here. Thus (B, q) is gaugeable under this condition and so by Theorem 4, the Dirichlet boundary value problem with any $f \in \mathrm{C}(\partial B)$ is uniquely solvable. In particular if $f \equiv 0$, there is a nonzero solution only if $r \ge Q^{-1/2}d^{-1/2}$. This solution is the eigenfunction corresponding to the eigenvalue zero. For $d = 1$, a theorem by de la Vallée-Poussin (see [2][3]) states that such a solution can exist only if $r \ge (8Q)^{-1/2}$. Thus the crude estimation above yields a result $\sqrt{8}$ times sharper than this. Incidentally, the example can be extended to any bounded domain D in R^d, using a result due to E. Lieb (see [1]) which states that the rigidity coefficient is maximized by a ball for all domains with a given volume (measure). It follows that if $|q| \le Q$, and

$$m(D) < A_d\left(\frac{d}{Q}\right)^{d/2}, \qquad A_d = \frac{2\pi^{d/2}}{d\Gamma(d/2)}$$

then (D, q) is gaugeable.

(ii) This is a result on the nodal structure of a quantum wave function, an object germane to the Schrödinger name; see [19][4]. In our terminology it may be stated as follows:

Proposition. *Let u be q-harmonic in a connected neighborhood U of x_0 and $u(x_0) = 0$. Then either $u \equiv 0$ in U or it must change signs there.*

Proof. Let $x_0 \in D$ with $\overline{D} \subset U$, where D is a domain. We are assuming $q \in \mathbf{J}_0$ which includes the celebrated Coulomb potential, see [1]. The analytic characterization of $\mathbf{J}$ mentioned above makes it easy to check that (16) is true if $m(D)$ is small enough, hence (17) follows. Namely: $(D,|q|)$, and *a fortiori* (D,q), is gaugeable provided $m(D)$ is small enough for fixed q. The case discussed in Example 1 is a very special case of this useful result. Take D to be such a domain; then since $u \in C(\overline{D})$ we have by Theorem 5 the representation:

$$u(x) = E^x\{e_q(\tau_D)u(X(\tau_D))\}, \quad x \in D. \tag{18}$$

For comparison put also

$$h(x) = E^x\{u(X(\tau_D))\} \tag{19}$$

Suppose u does not change sign in $\overline{D}$, then we may suppose $u \geq 0$ there. Since $u(x_0) = 0$ it follows from (18) and the remark made earlier that $e_q(\tau_D) > 0$ almost surely that $h(x_0) = 0$. But h is harmonic in D, hence $h \equiv 0$ in D by basic probability (or analysis). Going back to (18) we see that this implies $u \equiv 0$ in D. Since U is connected, this implies $u \equiv 0$ in U as desired. The argument above may be made somewhat plainer if balls are used for D.

(iii) This example will illustrate the use of path behavior to prove a "fine" property involving the boundary. We begin by reviewing an old result.

Theorem 6′. *If there exists a q-harmonic function in D satisfying*

$$0 < \inf_{x\in D} u(x) \leq \sup_{x\in D} u(x) < \infty, \tag{20}$$

then (D,q) is gaugeable.

Note the converse to Theorem 6′ is true; for the gauge, if not identically ∞ in D, is a q-harmonic function by Theorem 4 and satisfies (20) by Theorem 3.

For an arbitrary domain, $q \in C_b(D)$ and a strict solution u of the Schrödinger equation (11), this is proved in [3], see Corollary 2 of Theorem 2.2 there. The extension to $q \in J_0$ by the method (stochastic calculus) used there is straightforward. As a special case, let D be bounded and regular, $f \in C(\partial D)$ and $f > 0$. If there is a solution of (11) which belongs to $C_+(\overline{D})$ with $u = f$ on ∂D, then u is in fact strictly positive in $\overline{D}$ by the Proposition in Example 2, and consequently (20) is satisfied. Theorem 6′ asserts that the existence of one such solution guarantees the gaugeability of (D,q). When D is not assumed to be regular, the following result is a sharpening of Theorem 6′, valid in fact for an arbitrary domain D.

Theorem 6 *If there exists a positive and bounded q-harmonic function u in D and a constant $\delta > 0$ such that*

$$\varliminf_{D\ni y\to z} u(y) \geq \delta, \quad \text{for all } z \in (\partial D)_r \tag{21}$$

then (D,q) is gaugeable.

Proof. There exist bounded and regular domains D_n such that $\overline{D}_n \subset D_{n+1}$ and $\bigcup_{n=1}^{\infty} D_n = D$. Since $u \in C(D)$, the condition (20) is satisfied when the D here is replaced by D_n, for each n. Hence by Theorem 6′, (D_n,q) is gaugeable, and by Theorem 4 we have

$$u(x) = E^x\{e_q(\tau_{D_n})u(X(\tau_{D_n}))\}, \quad x \in D_n \tag{22}$$

Fix x and let $n \to \infty$. Then P^x - almost surely, $\tau_{D_n} \uparrow \tau_D$ and $X(\tau_D) \in (\partial D)_r$ on $\{\tau_D < \infty\}$. The latter assertion is a consequence of the basic probability theory that $\partial D \setminus (\partial D)_r$ is a polar set, hence is almost never hit by X. Hence by (21),

$$\underline{\lim}_{n\to\infty} u(X(\tau_{D_n})) \geq \underline{\lim}_{D\ni y\to(\partial D)_r} u(y) \geq \delta .$$

Since $0 \leq e_q(\tau_{D_n}) \to e_q(\tau_D)$, P^x-a.s., it follows from (22) by Fatou's lemma that

$$u(x) \geq E^x\{e_q(\tau_D)\}\,\delta \tag{23}$$

Since u is assumed to be bounded in D, the boundedness of the gauge in D follows from (23). A simple argument extends this to $\overline{D}$.

Corollary. *Under the assumption (H), the boundedness of u in the hypothesis of Theorem 6 may be omitted.*

For in this case Theorem 3 applies, and the validity of (23) for one x in D implies gaugeability.

When $q \in \mathcal{B}_b$ Theorem 6 was proved by T. Sturm [23] who has also communicated to me a proof of the general case. His method is quite different and leans heavily on functional analysis. A further extension using q-superharmonic functions can be derived by the method shown above. Theorem 6 (or 6′) gives a necessary and sufficient condition for gaugeability, as already remarked. Another formulation is as follows. Suppose we require that in the solution of the Dirichlet boundary value problem for Schrödinger's equation that $f \geq 0$ implies $u_f \geq 0$, where f is the given boundary function and u_f is the corresponding solution (as in Theorem 4), then the gaugeability is not only a sufficient but also a necessary condition. Indeed, the necessity follows already from this requirement for just one specially chosen f, such as $f \equiv 1$. The sufficiency is obvious by the explicit formula (5).

3. Semigroup and potential.

For definiteness we assume D is a bounded domain in this section, though many results can be extended to the case where $m(D) < \infty$. If $q \in \mathbf{J}_0$, then it is easy to show that q is locally integrable so that the assumption (H) is in force.

We introduce for $t \geq 0$ the operator T_t as follows, for $\varphi \in \mathcal{B}_b(D)$:

$$T_t\varphi(x) = E^x\{t < \tau_D;\ e_q(t)\varphi(X_t)\}\ , \qquad x \in D . \tag{24}$$

Thus T_0 is the identity and $\{T_t,\ t \geq 0\}$ is a semigroup on $\mathcal{B}_b(D)$ which may be called the Feynman-Kac semigroup on D. It follows from (4) that $\|T_t\| \leq C(t)$ where $\|\ \ \|$ denotes the sup-norm. This semigroup is not necessarily submarkovian but is exponentially bounded. It can be considered in the spaces $L^p(D)$, $1 \leq p \leq \infty$; in fact it maps $L^1(D)$ into $L^\infty(D)$ (see [1]). Note that our $\mathcal{B}_b(D)$ is $\mathcal{B} \cap L^\infty(D)$. Next, we define the potential operator:

$$V = \int_0^\infty T_t\, dt \tag{25}$$

When $q \equiv 0$, T_t reduces to the semigroup of the Brownian motion on D (killed off D), and V reduces to the Green potential G in (8). Thus V may be called the q-Green potential. The following link is important:

(26) $$V = G + V(q \cdot G) = G + G(q \cdot V)$$

see [8]. When q is a negative constant, (26) reduces to a case of the resolvent equation.

In terms of $\{T_t\}$ or V, the gaugeability of (D, q) may be expressed in several ways.

Theorem 7. *The following propositions are equivalent*[5].

(i) (D,q) *is gaugeable.*

(ii) $V1 \not\equiv \infty$ *in* D.

(iii) $V1$ *is bounded in* D.

(iv) $V|q|$ *is bounded in* D.

(v) There exist constants $a > 0$, $b > 0$ *such that*

$$\|T_t\| \leq a\,e^{-bt}, \qquad t \geq 0\,.$$

Let us indicate a revealing formula here. Suppose (iv) is true, then we have by Fubini's theorem:

$$Vq(x) = E^x\left\{\int_0^{\tau_D} e_q(t)q(X_t)dt\right\} = E^x\left\{e_q(\tau_D) - 1\right\} = u_1(x) - 1$$

Thus (iv) implies (i). Furthermore, we have by (26):

$$\frac{\Delta}{2}u_1 = \frac{\Delta}{2}Vq = \frac{\Delta}{2}\left(Gq + G(q \cdot V)\right) = -q - qV = -qu_1$$

verifying that u_1 is q-harmonic.

In the same way, we can show that

(27) $$u_f = V(qh_f) + h_f$$

where h_f is given in (9). This new formula is a covert expression of (5), convenient for algebraic (versus probabilistic) manipulations. For instance, we can derive (10) from (27) via (26). A novel application will be given shortly.

We are ready to solve the extension of Dirichlet problem to the inhomogeneous case (known as Poisson equation when $q \equiv 0$):

(28) $$\left(\frac{\Delta}{2} + q\right)\varphi = g \quad \text{in } D\,, \qquad \varphi = f \quad \text{on } \partial D\,.$$

Theorem 8. *Suppose* $|g| \leq c_1 + c_2|q|$ *where* c_1 *and* c_2 *are positive constants, and* $f \in C(\partial D)$. *Then the unique weak solution of (28) such that* $\varphi \in C(D)$ *and*

$$\lim_{D \ni x \to z} \varphi(x) = f(z), \qquad \text{for all } z \in (\partial D)_\gamma$$

is given by $u_f - Vg$ with u_f as in (5) or (27).

In fact, $-V$ is the inverse to $\frac{\Delta}{2}+q$ in the appropriate operator sense in several spaces, just as $-G$ is the inverse to $\frac{\Delta}{2}$.

When D is bounded and regular, it is proved in [12] that $\{T_t\}$ is a doubly-Feller semigroup. In fact, the arguments there show that T_t is a compact operator in $\mathbf{C}_0(D)$, the subclass of $\mathbf{C}(D)$ which "vanishes at ∂D". If (D,q) is gaugeable so that u_f is bounded with $f \in \mathcal{B}_b(\partial D)$, an application of the Markov property yields

$$T_t u_f(x) = E^x\left\{t < \tau_D;\ e_q(\tau_D) f(X(\tau_D))\right\}\ ,$$

and so

$$(I - T_t)\, u_f(x) = E^x\left\{\tau_D \le t;\ e_q(\tau_D) f(x(\tau_D))\right\} \tag{29}$$

where I is the identity operator. The right member in (29) may be called a "truncated gauge" and will be denoted by $u_{f,t}$ below. Unlike the gauge itself, it is always bounded in x for each t, because its aboslute value does not exceed

$$E^x\left\{e_{|q|}(t)\right\} \|f\|$$

which is bounded in x by (4), with $|q|$ replacing q there.

Since T_t is a compact operator in $\mathbf{C}_0(D)$, the Fredholm-Riesz alternative asserts that either the homogeneous equation

$$(I - T_t)\varphi = 0 \tag{30}$$

has a nonzero solution in $\mathbf{C}_0(D)$: or for any $g \in \mathbf{C}_0(D)$ the inhomogeneous equation

$$(I - T_t)\,\varphi = g \tag{31}$$

has a unique solution. In the latter case, $I - T_t$ has a bounded inverse in $\mathbf{C}_0(D)$.

Theorem 9. *For a given $t > 0$, suppose the second alternative obtains. Then for each $f \in \mathbf{C}(\partial D)$, the unique solution of (11) such that $\varphi \in \mathbf{C}(\overline{D})$ and $\varphi = f$ on ∂D is given by*

$$(I - T_t)^{-1}\, u_{f,t} \tag{32}$$

The assumption in the theorem may be translated as follows: "zero is not an eigenvalue of the Schrödinger operator on $\mathbf{C}_0(D)$". Clearly it is a necessary condition for the unique solvability of the Dirichlet problem.

Theorem 9 was proved by Ma and Zhao [22]. In the course of writing this article I found another proof which is motivated by (27) and the heuristic relation $(I - T_t)V = \int_0^t T_s ds$. Thus let us put

$$v(x) = \int_0^t T_s(qh_f)ds + (I - T_t)h_f\ . \tag{30}$$

Using the known result that $\int_0^t T_s|q|ds < \infty$ for every $t < \infty$ we can verify by the invevitable Fubini theorem that $v = u_{f,t}$. Since $\frac{\Delta}{2} + q$ is the infinitesimal generator of T_t in $C_0(D)$, it commutes with $(I - T_t)^{-1}$ and a direct calculation yields that

$$\left(\frac{\Delta}{2} + q\right)(I - T_t)^{-1}v = 0 .$$

The last step is similar to that in [20]; only the identification of the truncated gauge in (30) is new.

4. Conditional gauge.

There is a strengthening of the notion of gauge which involves the exit time τ_D but not the exit place $X(\tau_D)$. In traditional schemes such as random walk it is natural to consider both time and space and treat them conjointly. This leads to the conditional expectation below, in outmoded but serviceable notation:

(31) $$E^x\left\{e_q(\tau_D) \,\middle|\, X(\tau_D) = z\right\} .$$

There is a Borel measurable version of this which will be denoted by $u(x,z)$, $(x,z) \in D \times \partial D$. In particular, we have the representation of the gauge, to be denoted by u rather than u_1 here:

(32) $$u(x) = \int u(x,z)H(x,dz)$$

where $H(x,dz) = P^x\{X(\tau_D) \in dz\}$ is the "harmonic measure" supported of course by ∂D. If D is a bounded Lipschitz domain, then there is a "kernel function" $K(x,z)$ such that for each $z \in \partial D$, $K(\cdot,z)$ is strictly positive harmonic in D, $K(\cdot,\cdot)$ is continuous in $D \times \partial D$, and we have

(33) $$H(x,dz) = K(x,z)\sigma(dz)$$

where σ is the Lebesgue measure on the "surface" ∂D. When $q \equiv 0$ and D is a ball, K is the Poisson kernel.

Now if h is a positive harmonic function on a domain D, Doob (see [17]) has defined the h-conditioned Brownian motion process on D. When applied to the D and $K(\cdot,z)$ just mentioned, the $K(\cdot,z)$-conditioned process is nothing but the Brownian motion *stopped* at ∂D, and *conditioned* by $\{X(\tau_D) = z\}$. The associated probability and expectation will be denoted by P_z^x and E_z^x. In particular we have

(34) $$u(x,z) = E_z^x\{e_q(\tau_D)\} .$$

We can formulate a conditional gauge theorem as follows:

If $u(\cdot,\cdot) \not\equiv \infty$ in $D \times \partial D$, then it is bounded there.

In contrast to the gauge theorem, the validity of this result depends on the smoothness of ∂D. The difficulty can be pinned down by the following lemma given in [10].

Lemma. Said result *will hold if: for each $\epsilon > 0$ there exists $\delta(\epsilon) > 0$ such that for any open subset U of D with $m(U) < \delta(\epsilon)$ we have*

$$\sup_{(x,z)\in U\times\partial D} E_x^z\left\{\int_0^{\tau_U} |q(X_s)|\,ds\right\} < \epsilon. \tag{35}$$

In practice, U is a "cordon sanitaire" impinging on ∂D where all the kinks lie. The result postulated in the Lemma has been proved by Cranston, Fabes and Zhao [15] for a bounded Lipschitz domain in R^d, $d \geq 3$, and by Zhao for a Jordan domain in R^2. The basic estimate in the former case is furnished by the curious inequality below, where G is theGreen function for D, and g is that of R^d, $d \geq 3$, also called the fundamental potential, namely: $g(u) = |u|^{2-d}$ apart from a constant C_d. For all x, y and z in D we have

$$\frac{G(x,y)G(y,z)}{G(x,z)} \leq C\frac{g(x,y)g(y,z)}{g(x,z)} \tag{36}$$

where C is a constant depending on D.

The proof of (36) is analytic. It relies on the finer properties of the Lipschitzian character (see [21]) and is devoid of probabilistic interpretation, so far. It was first given for a ball by Brossard; for a more direct proof see [14]. By contrast, the reduction lemma stated above is immediately generalizable to an arbitrary bounded domain endowed with the celebrated Martin boundary, whereby the Lipschitz kernel is replaced by the Martin kernel. Indeed, it is known that the Martin boundary coincides with the Euclidean one in the Lipschitz case. The definition of the conditional gauge in (31) still makes sense with z belonging in the ramified boundary, and reduces to that in (34) under the general interpretation. But the theory is yet to be developed. To summarize:

Theorem 10. *The conditional gauge theorem is true for a bounded Lipschitz domain. Moreover $u(\cdot,\cdot)$ is strictly positive and continuous in $D \times \partial D$.*

In fact, a further extension of u to $\overline{D} \times \overline{D}$ is possible but we must omit the details here, as well as several important applications of the conditional gauge, including a boundary Harnack inequality for the Schrödinger equation. These topics will be treated in the monograph mentioned in the Introduction. Here we close with a simple application of conditional gauge which actually motivated its consideration.

Let $A \subset \partial D$ with $\sigma(A) > 0$, and put $f = 1_A$ in (5).

Corollary. *If $u_{1_A} \not\equiv \infty$ in D, then u_1 is bounded in $\overline{D}$.*

This follows from (32), (33) and the strict positivity of $u(\cdot,\cdot)$ and $K(\cdot,\cdot)$. The corollary was proved by R. J. Williams [24] for a C^2-domain. Then Falkner [18] proved the conditional gauge theorem for a class of domains including the C^2 ones. A simpler proof is given in [7]. Later Kenig showed that the class actually includes C^1-domains.

The conditional gauge for a small ball was used by Zhao [25] to solve the only difficulty in generalizing the gauge theorem from a bounded q to the class $\mathbf{J}_0$, thereby replacing an interesting time reversal argument in [1]. An essentially different method is introduced in [15] to extend the

gauge theorem to a class of Markov provesses which may have discontinuous paths. Even in that general framework a conditional gauge may be defined as in (31), which should prove worthy of investigation.

Footnotes

1. The general problem of the variation of the gauge with D is studied in [6]. It is related to the variation of the principal eigenvalue of the Schrödinger operator and provides another good example of application of the gauge.

2. It is hard to find in textbooks on partial differential equations a simple statement of the solution of Dirichlet's problem under the general conditions given here; see [13] for some details on this subject.

3. In hindsight, the method used in this paper to solve Dirichlet's problem is "reactionary" in that known analytic solvability was used to verify the probabilistic solution. Actually in R^1, besides the general methods discussed here, we can derive the differential equation by a modification of Schwarz's generalized second derivative, see [5].

4. It is instructive to see what laborious estimates are employed for lack of a proper formulation of the problem as a boundary value problem: "time must have a stop". For an amusing historical account see [11].

5. Another popular proposition in terms of the spectrum is omitted here. It is tedious to specify the details, and superfluous in view of (v) below; and it is not easily generalizable to the framework of [16]. Indeed, there we have a gauge in search of an operator.

References

1. **Aizenman, M., Simon, B.: Brownian motion and Harnack inequality for Schrödinger operators, *Comm. Appl. Math.* 35, 209–273 (1982).**
2. **Chung, K. L., S. R. S. Varadhan: Kac functional and Schrödinger equation, *Studia Math.* 68, 249–260 (1980).**
3. **Chung, K. L., Rao, K. M.: Feynman-Kac functional and the Schrödinger equation, *Seminar on Stochastic Processes* 1, 1–29 (1981) Birkhäuser, Boston.**
4. **Chung, K. L.: *Lectures from Markov Processes to Brownian Motion*, Grundlehren der Mathematishen Wissenschaften 249, Springer-Verlag, 1982.**
5. **Chung, K. L.: Brownian motion on the line, *J. of Math. Research and Exposition* 2, 87–98 (1982).**
6. **Chung, K. L., R. Durrett, Z. Zhao: Extension of domains with finite gauge, *Math. Ann.* 264, 73–79 (1983).**
7. **Chung, K. L.: Conditional gauges, *Seminar on Stochastic Processes* 3, 17–22 (1983).**
8. **Chung, K. L.: Notes on the inhomogeneous Schrödinger equation, *Seminar on Stochastic Processes* 4, 55–62 (1984).**

9. Chung, K. L. Hsu, P.: Gauge theorem for the Neumann problem, *Seminar on Stochastic Processes* **4**, 63–70 (1984).

10. Chung, K. L.: The gauge and conditional gauge theorem, *Séminaire des probabilités* XIX, 496–503 (1983/84), see also XX, 423–425 (1984/85).

11. Chung, K. L.: Probabilistic approach to boundary value problems for Schrödinger's equation, *Expo. Math.* **3**, 175–178 (1985).

12. Chung, K. L.: Doubly-Feller process with multiplicative functional, *Seminar on Stochastic Processes* **5**, 63–78 (1985).

13. Chung, K. L., Li, P., Williams, R. J.: Comparison of probability and classical methods for the Schrödinger equation, *Expo. Math.* **4**, 271–278 (1986).

14. Chung, K. L.: Green's function for a ball, *Seminar on Stochastic Processes* **6**, 1–14 (1986).

15. Chung, K. L., Rao, K. M.: General gauge theorem for multiplicative functionals, *Trans. Amer. Math. Soc.* (to appear).

16. Cranston, M., Fabes, E., Zhao, Z.: Conditional gauge and potential theory for the Schrödinger operator, *Trans. Amer. Math. Soc.* (to appear).

17. Doob, J. L.: *Classical Potential Theory and its Probabilistic Counterpart*, Grundlehren der Mathematischen Wissenschaften 262, Springer-Verlag, 1983.

18. Falkner, N.: Feynman-Kac functionals and positive solutions of $\frac{1}{2}\Delta u + qu = 0$. *Z. Wahrscheinlichkeitstheorie verw. Gebiete* **65**, 19–34 (1983).

19. Hoffman-Ostenhof, M., Hoffman-Ostenhof, T., Simon, B.: Brownian motion and a consequence of Harnack's inequality: Nodes of quantum wave functions, *Proc. Amer. Math. Soc.* **80**, 301–305 (1980).

20. Hsu, P.: Reflecting Brownian motion, boundary local times and the Neumann problem, Doctoral Dissertation, Stanford University, June 1984.

21. Jersion, D., Kenig, C.: Boundary value problems on Lipschitz domains, *M. A. A. Studies in Math.* Vol. **23**, 1–68 (1982).

22. Ma, Z., Zhao, Z.: Truncated gauge and Schrödinger operator with eigenvalues of both signs, *Seminar on Stochastic Processes* **6**, 149–154 (1986).

23. Sturm, T.: On the Dirichlet-Poisson problem for Schrödinger operators, *C. R. Math. Rep. Acad. Sci. Canada* **9**, 149–154 (1987).

24. Williams,R. J.: A Feynman-Kac gauge for solvability of the Schrödinger equation, *Advances Appl. Math.* **6**, 1–3 (1985).

25. Zhao, Z.: Conditional gauge with unbounded potential, *Z. Wahrscheinlichkeitstheorie verw. Gebiete* **65**, 13–16 (1983).

26. Zhao, Z: Uniform boundedness of conditional gauge and Schrödinger equation, *Commun. Math. Phys.* **93**, 19–31 (1984).

LAYER POTENTIAL METHODS FOR BOUNDARY VALUE PROBLEMS ON LIPSCHITZ DOMAINS

Eugene Fabes

Introduction

This survey will concentrate on the method of layer potentials for solving boundary value problems in Lipschitz domains. It will begin with a study of the classical single and double layer potentials corresponding to Laplace's equation for such domains. Emphasis will be given to those techniques, developed in the past five years, needed to solve the Dirichlet and Neumann problems in the form of one of these potentials. The data in the general case will be allowed to belong to an appropriate L^p-space on the boundary, but detailed analysis will emphasize the case of L^2-boundary data. (Section 1)

The layer potential method reduces the boundary value problem to that of an integral equation on the boundary. For Laplace's equation in domains with smooth boundaries one is able to use the classical Fredholm theory to resolve the integral equation. In the case of Lipschitz boundaries this theory no longer applies. In 1982 G. Verchota [11] found that so-called "Rellich Inequalities" could be used in the L^2-case to establish the invertibility of the integral operators. These inequalities, obtained through integration by parts, prove the equivalence of the L^2-norms over the boundary of the normal and tangential components of the gradient of a harmonic function [10]. When applied to the single layer potential the identities are used to show the invertibility on L^2 of the integral operator on the boundary associated with the Neumann problem. Because of duality relationships the result for the Neumann problem leads to the invertibility for the integral operator associated with the Dirichlet problem.

A significant advantage to the above method is that it does not rely on positivity properties of the Laplacian; namely the maximum principle and Harnack's inequality. Because of this, the L^2-theory mentioned above carries over to the Dirichlet and Neumann-type boundary value problems on Lipschitz domains for several constant coefficient elliptic systems. The survey will review recent work concerning boundary value problems associated with the system of elasticity (Section 2)

Throughout this survey D will denote a bounded domain in R^n and for simplicity we take $n \geq 3$. Points of D will generally be denoted by capital letters X or Y, and points on the boundary of D, ∂D, will be represented by the capital letters P or Q.

<u>Definition</u>: D is a Lipschitz domain means that corresponding to each point $Q \in \partial D$ there is a system of coordinates of R^n, isometric with the usual coordinate system, and a sphere $B_\delta(Q)$ with center Q and radius $\delta > 0$ such that relative to this coordinate system Q is the origin and

$$D \cap B_\delta(Q) = \{(x,t) : x \in R^{n-1}, \ t > \varphi(x)\} \cap B_\delta(Q)$$

where φ is a Lipschitz function on R^{n-1} and $\varphi(0) = 0$

<u>Remark</u>: If D is a Lipschitz domain we can find a number m and a finite number of spheres $\{B_{\delta_j}(Q_j)\}_{j=1}^{j=\ell}$, $Q_j \in \partial D$, such that

$$\partial D \subset \bigcup_{1 \leq j \leq \ell} B_{\delta_j}(Q_j)$$

$$D \cap B_{2\delta_j}(Q_j) = \{(x,t) : x \in R^{n-1}, \ t > \varphi_j(x)\} \cap B(Q_j, 2\delta_j),$$

$\varphi_j(0) = 0$ and $\|\nabla \varphi_j\|_{L^\infty(R^{n-1})} \leq m$. We will describe the pair of number ℓ and m as the Lipschitz character of D.

<u>Section 1</u>. <u>Laplace's Equation</u>

In this section we discuss the classical Dirichlet and Neumann problems for Laplace's equation in a Lipschitz domain. We seek to represent our solutions in the form of a single or double layer potential when the data is taken from $L^2(\partial D)$, the space of functions square integrable over ∂D with respect to surface measure, $d\sigma$.

We now define the classical layer potentials:

$$\mathcal{D}f(X) = \frac{1}{\omega_n} \int_{\partial D} \frac{(X-Q)\cdot N_Q}{|X-Q|^n} f(Q)\, d\sigma_Q .$$

and

$$\mathcal{S}f(x) = \frac{1}{(n-2)\omega_n} \int_{\partial D} \frac{1}{|X-Q|^{n-2}} f(Q)\, d\sigma_Q , \quad n > 2 ,$$

where ω_n is the area of the unit sphere in R^n , i.e. the area of $\{X : |X| = 1\}$, and N_Q denotes the unit inner normal to ∂D at Q . (N_Q exists for a.e. $(d\sigma)$ $Q\in\partial D$.)

In the case of smooth domains and smooth density f , it is well known that for $P\in\partial D$

$$\lim_{X\to P,\, X\in D} \mathcal{D}(f)(X) = \frac{1}{2} f(P) + Kf(P)$$

and $$\lim_{X\to P,\, x\in R^n\setminus\bar{D}} \mathcal{D}(f)(X) = \frac{1}{2} f(P) - Kf(P)$$

where $$Kf(P) = \frac{1}{\omega_n} \int_{\partial D} \frac{(P-Q)\cdot N_Q}{|P-Q|^n} f(Q)\, d\sigma_Q .$$

(See [7]) Also, in the same setting of smoothness,

$$\lim_{X\to P,\ X\in D} N_P\cdot\nabla\mathcal{S}f(X) = \frac{1}{2} f(P) - K^* f(P)$$

and $$\lim_{X\to P,\ X\in R^n\setminus\bar{D}} N_P\cdot\nabla\mathcal{S}f(X) = \frac{1}{2} f(P) + K^* f(P)$$

where K^* denotes the adjoint of the operator K .

From the above jump relation the (interior) Dirichlet and Neumann problems for Laplace's equation with appropriate boundary data g are reduced to solving respectively the integral equations

$(\frac{1}{2}I + K)f = g$ and $(\frac{1}{2}I - K^*)f = g$.

Our purpose in Section 1 is to show that this layer potential method continues to work in the case of Lipschitz domains.

The first problem that confronts us is the meaning of the operation Kf. In the case of Lipschitz domains the kernel $\frac{(P-Q)\cdot N_Q}{|P-Q|^n}$ is not locally integrable over ∂D and the operator Kf must be interpreted as a Calderon-Zygmund singular integral operator. Using the fundamental results of A.P. Calderon ([1]) and R. Coifman, A. McIntosh, Y. Meyer ([2]), the following results for K can be established:

Theorem 1.0. Set $K_\epsilon f(P) = \frac{1}{\omega_n} \int_{|P-Q|>\epsilon} \frac{(P-Q)\cdot N_Q}{|P-Q|^n} f(Q)\, d\sigma_Q$ and $\tilde{K}f(P) = \sup_{\epsilon>0} |K_\epsilon f(P)|$. Then

a) for $1<p<\infty$, $\|\tilde{K}f\|_{L^p(\partial D)} \leq c\|f\|_{L^p(\partial D)}$ where c depends only on p, n, and the Lipschitz character of D ,

b) $Kf = \lim_{\epsilon\to 0} K_\epsilon f$ exists in $L^p(\partial D)$ and pointwise almost everywhere (a.e. $d\sigma$) ,

c) parts a and b also hold for the corresponding operators K_ϵ^*, $\tilde{K}^*$, and K^* .

For a proof of the above result the reader is referred to [6]. In this survey we will assume the boundedness properties of K and K^* and concentrate on those methods developed in [11] for inverting $(\frac{1}{2}I + K)$ on $L^2(\partial D)$ and $(\frac{1}{2}I - K^*)$ on $L_o^2(\partial D) = \{f\in L^2(\partial D) : \int_{\partial D} f\, d\sigma = 0\}$. The main tool in this regard is the Rellich inequality.

Before stating and proving the Rellich inequality we need to define $\frac{\partial u}{\partial N}(P) \equiv N_P\cdot\nabla u(P)$, the interior normal derivative of u at $P\in\partial D$, and

$\nabla_\tau u(P) = \nabla u(P) - \frac{\partial u}{\partial N}(P)N_P$, the tangential component of ∇u at $P\in\partial D$

Lemma 1.1. Suppose D is a Lipschitz domain in R^n . Then there exists $C > 0$, depending only on the Lipschitz character of D such that if

$\Delta u = 0$ in D, then

$$\int_{\partial D} |\nabla_\tau u|^2 d\sigma \le C\Big[\int_{\partial D} \Big(\frac{\partial u}{\partial N}\Big)^2 d\sigma + \int_{\partial D} u^2 d\sigma \Big]$$

and

$$\int_{\partial D} \Big(\frac{\partial u}{\partial N}\Big)^2 d\sigma \le C \Big[\int_{\partial D} |\nabla_\tau u|^2 d\sigma + \int_{\partial D} u^2 d\sigma \Big] .$$

Proof Since D is a Lipschitz domain we can find a smooth vector field $h(X) = (h_1(X),\ldots,h_n(X))$ on R^n such that $\operatorname*{essinf}_{\partial D} h(Q)\cdot\nu_Q > 0$ where ν_Q = exterior unit normal to ∂D at Q. From Gauss's divergence theorem

$$\int_{\partial D} h(Q)\cdot\nu_Q \; |\nabla u(Q)|^2 d\sigma = \int_D \operatorname{div}\,(h(X)\;|\nabla u(X)|^2)dX$$

$$= \int_D (\operatorname{div} h)\;|\nabla u|^2 dX + 2\int_D h_i(X)\,\frac{\partial^2 u}{\partial X_i \partial X_j}\,\frac{\partial u}{\partial X_j}\,dX .$$

(Repeated index means summation on that index.) Since u is harmonic

$$\frac{\partial^2 u}{\partial X_i \partial X_j}\,\frac{\partial u}{\partial X_j} = \frac{\partial}{\partial X_j}\Big[\frac{\partial u}{\partial X_i}\frac{\partial u}{\partial X_j}\Big] .$$

So

$$\int_{\partial D} h(Q)\cdot\nu_Q\;|\nabla u(Q)|^2\,d\sigma = \int_D (\operatorname{div} h)\;|\nabla u|^2 - 2\int_D \frac{\partial h_i}{\partial X_j}\,\frac{\partial u}{\partial X_i}\,\frac{\partial u}{\partial X_j} \tag{1.2}$$

$$+ \int_{\partial D} h(Q)\cdot\nabla u(Q)\,\frac{\partial u}{\partial \nu}(Q)\,d\sigma .$$

(Here $\frac{\partial u}{\partial \nu}(Q) \equiv \nu_Q\cdot\nabla u(Q)$.)

Write $\nabla u(Q) = \frac{\partial u}{\partial \nu}(Q)\,\nu_Q + \nabla_\tau u(Q)$ and note that $|\nabla u(Q)|^2 =$

$\left[\frac{\partial u}{\partial \nu}(Q)\right]^2 + |\nabla_\tau u(Q)|^2$. Substituting in (1.2) we have

$$\int_{\partial D} h(Q)\cdot\nu_Q \; |\nabla_\tau u(Q)|^2 d\sigma = \int_{\partial D} h(Q)\cdot\nu_Q \left[\frac{\partial u}{\partial \nu}(Q)\right]^2 d\sigma + \int_D (\mathrm{div}\, h)|\nabla u|^2 dX -$$

$$- 2\int_D \frac{\partial h_i}{\partial X_j}\frac{\partial u}{\partial X_i}\frac{\partial u}{\partial X_j}\, dX + 2\int_{\partial} h(Q)\cdot\nabla_\tau u(Q)\,\frac{\partial u}{\partial \nu}(Q)\, d\sigma .$$

Since the essential $\operatorname*{ess\,inf}_{\partial D} h(Q)\cdot\nu_Q > 0$ and $\int_D |\nabla u|^2 = \int_{\partial D} u(Q)\frac{\partial u}{\partial \nu}(Q)\, d\sigma$ the Rellich inequalities easily follow.

<u>Lemma 1.3</u>. Suppose $u(X)$ is harmonic in $R^n\setminus\bar{D}$ and $\lim_{R\to\infty}\int_{\partial B_R(0)} |u||\nabla u| d\sigma = 0$.

Then once again

$$\int_{\partial D} |\nabla_\tau u|^2 d\sigma \le C\left[\int_{\partial D}\left[\frac{\partial u}{\partial \nu}\right]^2 d\sigma + \int_{\partial D} u^2 d\sigma\right]$$

and $$\int_{\partial D}\left[\frac{\partial u}{\partial \nu}\right]^2 d\sigma \le C\left[\int_{\partial D} |\nabla_\tau u|^2 d\sigma + \int_{\partial D} u^2 d\sigma\right]$$

where C is a constant depending only on the Lipschitz character of D .

<u>Proof</u>. We apply the proof of Lemma 1.1 over the domain $B_R(0)\setminus\bar{D}$ with R large and we choose the vector field $h(X)$ so that $h(X) = 0$ for $|X|$ sufficiently large, depending only on D . Letting $R\to\infty$ and observing from our hypothesis, that $\lim_{R\to\infty}\int_{B_R(0)\setminus\bar{D}} |\nabla u|^2 \le \int_{\partial D} |u|\left|\frac{\partial u}{\partial \nu}\right| d\sigma$, we easily obtain the desired inequalities.

<u>Remark</u>: The interior Rellich inequalities of Lemma (1.1) can be improved by dropping the term $\int_{\partial D} u^2(Q) d\sigma$. We will need the inequality

$$(1.3)' \quad \int_{\partial D}\left[\frac{\partial u}{\partial \nu}\right]^2 d\sigma \le c\int_{\partial D} |\nabla_\tau u|^2 d\sigma$$

for functions u satisfying $\Delta u = 0$ in D. This last estimate is a consequence of applying the second inequality in Lemma 1.1 to $u(X) - \frac{1}{\sigma(\partial D)} \int_{\partial D} u d\sigma$ and the Poincare inequality to $\int_{\partial D} |u(Q) - \frac{1}{\sigma(\partial D)} \int_{\partial D} u d\sigma|^2 d\sigma$.

Using the above Rellich inequalities we will outline the proof of the invertibility of the boundary integral operators.

<u>Theorem 1.4</u>. The operators $\frac{1}{2}I + K^*$ and $\frac{1}{2}I + K$ are invertible on $L^2(\partial D)$. The operator $\frac{1}{2}I - K^*$ is invertible on $L^2_o(\partial D) \equiv \{f \in L^2(\partial D) : \int_{\partial D} f d\sigma = 0\}$.

<u>Proof</u>. It is of course sufficient to show the results for $\frac{1}{2}I + K^*$ and $\frac{1}{2}I - K^*$. An interesting relationship occurs between these operators when we apply the Rellich inequalities to the harmonic function $u(X) = \mathcal{S}(f)(X)$, the single layer potential with density $f \in L^2(\partial D)$. From inside D, i.e. from (1.3)', we have

$$(1.5) \qquad \left\| \left[\frac{1}{2}I - K^*\right] f \right\|_{L^2(\partial D)} \leq C \|\nabla_T \mathcal{S} f\|_{L^2(\partial D)}$$

while from outside D, i.e. using the first inequality of Lemma 1.3,

$$\|\nabla_T \mathcal{S} f\|_{L^2(\partial D)} \leq C\left[\|(\tfrac{1}{2}I + K^*) f\|_{L^2(\partial D)} + \|\mathcal{S} f\|_{L^2(\partial D)}\right]$$

Since the single layer potential is continuous across the boundary of D, the tangential derivative $\nabla_T \mathcal{S} f$ is indeed the same in both of the above inequalities. Hence

$$(1.6) \qquad \|(\tfrac{1}{2}I - K^*) f\|_{L^2(\partial D)} \leq C\left[\|(\tfrac{1}{2}I + K^*) f\|_{L^2(\partial D)} + \|\mathcal{S} f\|_{L^2(\partial D)}\right]$$

<u>Claim 1</u> $\frac{1}{2}I + K^*$ is 1-1 (injective).

$$\text{If } (\tfrac{1}{2}I + K^*)f = 0 \text{ then } \int_{R^n\setminus\bar{D}} |\nabla \mathcal{S}f|^2 dX = 0$$

and therefore $\mathcal{S}f$ is constant in each component of $R^n\setminus\bar{D}$. In particular $\mathcal{S}f$ is constant on each component of ∂D and from (1.5), $\|(\frac{1}{2}I - K^*)f\|_{L^2(\partial D)} = 0$. Hence $f = 0$.

<u>Claim 2</u>. $\frac{1}{2}I + K^*$ has closed range.

Assume $(\frac{1}{2}I + K^*)f_j \longrightarrow g$ in $L^2(\partial D)$. If a subsequence of the f_j's converges weakly in $L^2(\partial D)$ to say f, then for this subsequence, which we again denote by $\{f_j\}$, and for any $h \in L^2(\partial D)$,

$$\int_{\partial D} gh\, d\sigma = \lim_j \int_{\partial D} (\tfrac{1}{2}I + K^*)f_j\, h\, d\sigma = \int_{\partial D} f(\tfrac{1}{2}I + K)h\, d\sigma = \int_{\partial D} (\tfrac{1}{2}I + K^*)f\, h\, d\sigma.$$

Hence $(\frac{1}{2}I + K^*)f = g$.

If no subsequence converges weakly in $L^2(\partial D)$, then we have $\|f_j\|_{L^2(\partial D)} \longrightarrow \infty$ and $(\frac{1}{2}I + K^*)f_j \longrightarrow g$ in $L^2(\partial D)$. By dividing through by $\|f_j\|_{L^2(\partial D)}$ we may assume $\|f_j\|_{L^2(\partial D)} = 1$ and $(\frac{1}{2}I + K^*)f_j \longrightarrow 0$ in $L^2(\partial D)$. Some subsequence, call it again $\{f_j\}$, converges weakly in $L^2(\partial D)$ to 0 from the above argument and the fact that $\frac{1}{2}I + K^*$ is 1-1. Since $\mathcal{S}$ is a compact operator on $L^2(\partial D)$, $\mathcal{S}(f_j) \longrightarrow 0$ in $L^2(\partial D)$. Inequality (1.6) now implies

$$\|(\tfrac{1}{2}I - K^*)f_j\|_{L^2(\partial D)} \longrightarrow 0 \quad \text{in } L^2(\partial D)$$

$$\text{But then } 1 = \|f_j\|_{L^2(\partial D)} \le \|(\tfrac{1}{2}I + K^*)f_j\|_{L^2(\partial D)} + \|\tfrac{1}{2}I - K^*)f_j\|_{L^2(\partial D)}\,.$$

$$\longrightarrow 0 \text{ as } j \to \infty\,.$$

This contradiction proves the claim that the range of $\frac{1}{2}I + K^*$ is closed.

<u>Claim 3</u>. $\frac{1}{2}I + K^*$ is onto (surjective) .

From Claim 2 it is sufficient to show that the range is dense in $L^2(\partial D)$. In fact we will indicate the proof that the restriction to ∂D of a function $g \in C^\infty(R^n)$ is in the range of $\frac{1}{2}I + K^*$.

Given a bounded Lipschitz domain, D , there is a sequence of C^∞ domains $D_j \supset D$ and homemorphisms $\Lambda_j : \partial D \to \partial D_j$ such that

a) $\sup_{Q \in \partial D} |Q - \Lambda_j(Q)| \to 0$ as $j \to \infty$

b) there are positive functions $w_j : \partial D \to R_+$ bounded away from zero and infinity, uniformly in j , such that $\int_E w_j \, d\sigma = \int_{\Lambda_j(E)} d\sigma_j$ for any measurable set $E \subset \partial D$,

c) $w_j \to 1$ pointwise a.e. $(d\sigma)$,

d) the unit normals to D_j, $N_{\Lambda_j(Q)}$, converge pointwise a.e. $(d\sigma)$ to N_Q.

(See [11, p. 581])

Let K_j^* denote the adjoint of the double layer potential over ∂D_j . One can find $f_j \in L^2(\partial D_j)$ (even smooth) such that $(\frac{1}{2}I + K_j^*) f_j = g|_{\partial D_j}$. Define on ∂D $F_j(Q) = f_j(\Lambda_j(Q)) w_j(Q)$. If a subsequence of $\{\|f_j\|_{L^2(\partial D_j)}\}$ is bounded, the same holds for a subsequence of $\{\|F_j\|_{L^2(\partial D)}\}$. If then $F_j \to F$ weakly in $L^2(\partial D)$ then for $h \in C^\infty(R^n)$,

$$\int_{\partial D_j} (\tfrac{1}{2}I + K_j^*) f_j \, h \, d\sigma_j = \int_{\partial D} F_j \, (\tfrac{1}{2}I + K)(h) \, d\sigma +$$

$$+ \int_{\partial D} F_j \, [(\tfrac{1}{2}I + K_j)(h) \circ \Lambda_j - (\tfrac{1}{2}I + K)h] \, d\sigma$$

Clearly the left side converges to $\int_{\partial D} g \, h \, d\sigma$. By weak convergence the first term on the right converges to $\int_{\partial D} (\frac{1}{2}I + K^*)(F) \cdot h \, d\sigma$ while the second converges to zero. This is because

$$\|(\tfrac{1}{2}I + K_j)(h) \circ \Lambda_j - (\tfrac{1}{2}I + K)h\|_{L^2(\partial D)} \longrightarrow 0$$

(To see this note that

$$(\tfrac{1}{2}I + K_j)(h) \circ \Lambda_j(P) - (\tfrac{1}{2}I + K)h(P) = h(P_j) - h(P) +$$

$$\frac{1}{\omega_n}\left\{\int_{\partial D_j} \frac{(Q_j-P_j)\cdot N_{Q_j}}{|Q_j-P_j|^n}[h(Q_j) - h(P_j)]d\sigma_{Q_j} -\right.$$

$$\left.\int_{\partial D} \frac{(Q-P)\cdot N_Q}{|Q-P|^n}[h(Q) - h(P)]d\sigma_Q\right\}.)$$

If $\|f_j\|_{L^2(\partial D_j)} \longrightarrow \infty$ we reduce consideration to $\|f_j\|_{L^2(\partial D_j)} = 1$, $\|(\frac{1}{2}I + K_j^*)(f_j)\|_{L^2(\partial D_j)} \longrightarrow 0$, and, arguing as above, to $F_j \longrightarrow 0$ weakly in $L^2(\partial D)$. Applying (1.6) to each D_j we will arrive to a contradiction when we establish $\|\mathcal{S}_j f_j\|_{L^2(\partial D_j)} \longrightarrow 0$ as $j \to \infty$, where

$$\mathcal{S}_j f_j(P_j) = \frac{1}{\omega_n(n-2)}\int_{\partial D} f_j(Q_j)\frac{1}{|P_j-Q_j|^{n-2}}\,d\sigma_{Q_j}$$

$= \frac{1}{\omega_n(n-2)}\int_{\partial D} F_j(Q)\frac{1}{|\Lambda_j(P)-\Lambda_j(Q)|^{n-2}}d\sigma_Q^*$. To prove this we write for each $\epsilon > 0$,

$$\mathcal{S}_j(f_j)(P_j) = \frac{1}{\omega_n(n-2)}\left[\int_{|P-Q|>\epsilon} + \int_{|P-Q|\le\epsilon} F_j(Q)\frac{1}{|\Lambda_j(P)-\Lambda_j(Q)|^{n-2}}\,d\sigma_Q\right]$$

$$\equiv A_{j,\epsilon}(P) + B_{j,\epsilon}(P) .$$

It is easy to show $\|B_{j,\epsilon}\|_{L^2(\partial D)} \le C\epsilon$ with C independent of j and for ϵ fixed that $A_{j,\epsilon} \longrightarrow 0$ in $L^2(\partial D)$.

We have completed the proof of the invertibility on $L^2(\partial D)$ of $\frac{1}{2}I + K^*$ and hence also of $\frac{1}{2}I + K$. Since very similar arguments can be used to show the invertibility of $\frac{1}{2}I - K^*$ on L^2_o, we will leave this part of Theorem 1.4 to the reader and also make reference to [11, p. 588]

Theorems 1.0 and 1.4 are the main results needed to construct solutions to the Dirichlet and Neumann problems for Laplace's equation respectively in the form of a double and single layer potential. We give the precise statements of these results.

Definition. An open nonempty circulary cone Γ_P with vertex $P \in \partial D$ is called a nontangential cone if there exist another open cone Γ'_P with vertex P and $\delta > 0$ such that

$$\bar{\Gamma}_P \cap B_\delta(P) \setminus \{P\} \subset \Gamma'_P \cap B_\delta(P) \subset D .$$

A regular family Γ of cones for D is a family of nontangential cones Γ_P, one for each point $P \in \partial D$, for which there exist $\beta > 1$ and $\delta > 0$ such that for each $P \in \partial D$

$$\Gamma_P \cap B_\delta(P) \subset \{X \in D : |X-P| < \beta \text{ dist } (X, \partial D)\} \cap B_\delta(P) .$$

For a regular family Γ of cones for D and a function u defined on D we set

$$M(u)(P) = \sup_{X \in \Gamma_P \cap B_\delta(P)} |u(X)|$$

Theorem 1.7. Suppose $g \in L^2(\partial D)$. There exists a unique function $u(X)$ such that $\Delta u(X) = 0$ in D, $M(u) \in L^2(\partial D)$, and $u(X) \to g(P)$ as $X \in \Gamma_P \to P$ for a.e. $(d\sigma)$ $P \in \partial D$. We can write

$$u(X) = \frac{1}{\omega_n} \int_{\partial D} \frac{(X-Q) \cdot N_Q}{|X-Q|^n} (\tfrac{1}{2}I + K)^{-1}(g)(Q) \, d\sigma_Q .$$

In particular $\|M(u)\|_{L^2(\partial D)} \le C\|g\|_{L^2(\partial D)}$, where C depends only on β, δ, and the Lipschitz character of D.

Theorem 1.8. Suppose $g \in L^2(\partial D)$ and $\int_{\partial D} g \, d\sigma = 0$. Then modulo constants there exists a unique function $u(X)$ such that $\Delta u(X) = 0$ in D, $M(|\nabla u|) \in L^2(\partial D)$, and $N_P \cdot \nabla u(X) \to g(P)$ as $X \in \Gamma_P \to P$ for a.e.$(d\sigma)$ $P \in \partial D$. We may write

$$u(X) = \frac{1}{\omega_n(n-2)} \int_{\partial D} \frac{1}{|X-Q|^{n-2}} (\tfrac{1}{2}I - K^*)^{-1} g(Q) \, d\sigma_Q .$$

In particular $\|M(|\nabla u|)\|_{L^2(\partial D)} \leq C \|g\|_{L^2(\partial D)}$, where C depends only on β, δ, and the Lipschitz character of D.

For complete proofs of Theorems 1.7 and 1.8 the reader is referred to [6] and [11].

In Theorem 1.7 we described the smoothness near the boundary of the solution of the Dirichlet problem with L^2-data by the statement $M(u)$ belonged to $L^2(\partial D)$. There is another way of describing the smoothness in terms of classical Sobolev or Besov spaces. The <u>harmonic</u> function u with L^2-boundary values as described in Theorem 1.7 belongs to the Sobolev-Besov space $H^{1/2}(D)$, and, conversely, any <u>harmonic</u> function $u \in H^{1/2}(D)$ is the solution to the Dirichlet problem described in Theorem 1.7 for some $g \in L^2(\partial D)$. This is the content of the next two theorems.

Theorem 1.9. Let D be a bounded Lipschitz domain in R^n. Assume $\Delta u = 0$ in D and

$$\iint_{DD} \frac{|u(X)-u(Y)|^2}{|X-Y|^{n+1}} \, dX dY < \infty \quad (\text{i.e. } u \in H^{1/2}(D)) .$$

Then if $d(X) = \operatorname{dist}(X, \partial D)$ we have

$$\int_D d(X) \, |\nabla u(X)|^2 dX \leq C \iint_{DD} \frac{|u(X)-u(Y)|^2}{|X-Y|^{n+1}} \, dX dY .$$

Therefore, from the results in [3], $M(u) \in L^2(\partial D)$ and u converges nontangentially, i.e. in the sense of Theorem 1.7, at a.e.$(d\sigma)$ $P \in \partial D$.

<u>Proof</u>. Fix $X \in D$. Since u is harmonic,

$$|\nabla u(X)| \le \frac{C}{d(X)^{n+1}} \int_{B_{d(X)}(X)} |u(Y)-u(X)|dY \le$$

$$\frac{C}{d(X)^{(n+1)/2}} \int_{B_{d(X)}(X)} \frac{|u(Y)-u(X)|}{|Y-X|^{(n+1)/2}} dY$$

Using Schwartz's inequality we have

$$d(X) \; |\nabla u(X)|^2 \le C \int_D \frac{|u(Y)-u(X)|^2}{|Y-X|^{n+1}} \, dY$$

The inequality in the conclusion is now immediate.

<u>Theorem (1.10)</u>. If u is the solution of $\Delta u = 0$ in D, $u|_{\partial D} = g$, in the sense of Theorem 1.7 with $g \in L^2(\partial D)$, then

$$\iint_{DD} \frac{|u(X)-u(Y)|^2}{|X-Y|^{n+1}} \, dXdY < \infty \quad (\text{i.e. } u \in H^{1/2}(D)) .$$

<u>Proof</u>. We first consider $\displaystyle\int_{B_{\frac{1}{4}d(X)}(X)} \frac{|u(Y)-u(X)|^2}{|Y-X|^{n+1}} dY$ which is bounded by

$$C \int_{B_{\frac{1}{4}d(X)}(X)} \frac{1}{|X-Y|^{n-1}} \, \frac{1}{d(X)^n} \Big(\int_{B_{\frac{1}{2}d(X)}(X)} |\nabla u(Z)|^2 dZ \Big) dY .$$ Therefore,

$$\iint_{DB_{\frac{1}{4}d(X)}(X)} \frac{|u(X)-u(Y)|^2}{|X-Y|^{n+1}} dYdX \le C \int_D d(Z) \; |\nabla u(Z)|^2 dZ .$$ The results in [Dahl] imply this last integral is finite. Similarly

$$\iint_{DB_{\frac{1}{4}d(Y)}(Y)} \frac{|u(X)-u(Y)|^2}{|X-Y|^{n+1}} \, dXdY \le C \int_D d(Z) \; |\nabla u(Z)|^2 dZ .$$

From the above observations, to control the double integral in the theorem it is sufficient to show the finiteness of the integral taken over the set $E = \{(X,Y)\in D\times D : |X-Y| > \frac{1}{4}\max(d(X),d(Y))\}$. We may also assume that X and Y range over the same coordinate patch

$$B\cap D = \{(z,r) : r > \varphi(z),\ z\in R^{n-1},\ \varphi(0) = 0,\ |\nabla\varphi| \le m\}\cap D .$$

In this situation we will not use the harmonicity of u and will show for general u with support in $B\cap\overline{D}$ that

$$(1.11)\qquad \iint_{(B\cap D\times B\cap D)\cap E} \frac{|u(X)-u(Y)|^2}{|X-Y|^{n+1}}\, dXdY \le C\int_D d(Z)\ |\nabla u(Z)|^2 dZ .$$

Using the coordinate system of R^n describing $D\cap B$ we rewrite the integral on the left side of the above inequality as

$$(1.12)\qquad \iint_{(R^n_+\times R^n_+)\cap\tilde{E}} \frac{|v(x,t) - v(y,s)|^2}{[|x-y|^2+(t-s+\varphi(x)-\varphi(y))^2]^{(n+1)/2}}\, dx\ dt\ dy\ ds$$

where $R^n_+ = \{(z,r) : z\in R^{n-1},\ r > 0\}$, $v(z,r) = u(z,r+\varphi(z))$, and $\tilde{E} = \{(x,t,y,s)\in R^n_+\times R^n_+ : (|x-y|^2 + (t-s+\varphi(x)-\varphi(y))^2)^{1/2} > \delta\max(t,s))\}$ with δ a fixed positive number. We split the integral over the set where $|t-s| \le C\,|x-y|$ and where $|t-s| > C|x-y|$; C will depend only on n and m. (Remember $|\nabla\varphi| \le m$.) If $|t-s| \le C|x-y|$ then $(|x-y|^2 + (t-s+\varphi(x)-\varphi(y))^2)^{1/2}$ is equivalent to $|x-y|$ and we can bound this part of the integral by a constant times

$$\iint_{R^n_+\times R^n_+\ \cap\ \{|x-y|>\eta\max(t,s)} \frac{|v(x,t)-v(y,s)|^2}{[|x-y|^2+(t-s)^2]^{(n+1)/2}}\, dx\ dt\ dy\ ds ,$$

η a positive number. This last expression is bounded by

$$(1.13)\qquad C\Big[\int_0^\infty t\int_{R^{n-1}}\int_{R^{n-1}} \frac{|v(x,t)-v(y,t)|^2}{|x-y|^{n+1}}\, dx\ dy\ dt$$

$$+ \int_{R^{n-1}}\int_0^\infty\int_0^\infty \left[\frac{1}{t}\int_s^t \frac{\partial v}{\partial r}(y,r)dr\right]^2 dt\ ds\ dy\ \Big] .$$

Next observe by the Besov-Sobolev result in R^{n-1},

$\int_{R^{n-1}} \int_{R^{n-1}} \frac{|v(x,t)-v(y,t)|^2}{|x-y|^{n+1}} \, dx \, dy \le C \int_{R^{n-1}} |\nabla v(x,t)|^2 \, dx$, and by Hardy's Theorem

$$\int_0^\infty \left[\frac{1}{t} \int_s^t \frac{\partial v}{\partial r}(y,r)dr \right]^2 dt \quad \le C \int_s^\infty \frac{\partial v}{\partial r}(y,r)^2 \, dr .$$

We conclude that the integrals in 1.13 are bounded by

$C \int_{R^n_+} |\nabla v(z,r)|^2 r dz \, dr$

Finally for the case of the integral in 1.12 restricted further to the set $|t-s| > C|x-y|$ we will show that

$$\iint_{E_1} \frac{|v(x,t)-v(y,s)|^2}{[|x-y|^2+(t-s)^2]^{(n+1)/2}} \, dx \, dt \, dy \, ds \le C \int_{R^n_+} |\nabla v(z,r)|^2 r \, dz \, dr$$

where $E_1 = \{(x,t,y,s) \in R^n_+ \times R^n_+ : 0 < s < t \, , \, t-s > C\,|x-y|, \text{ and } t-s > \eta t, \ \eta \text{ a positive number}\}$.

Since the corresponding situation for $0 < t < s$ is of the same form, the proof of Theorem 1.10 will be completed once the above inequality is established. To this end we consider two integrals:

$$(1.14) \qquad \iint_{E_1} \frac{|v(x,t)-v(y,t)|^2}{[|x-y|^2+(t-s)^2]^{(n+1)/2}} \, dx \, dt \, dy \, ds \quad \text{and}$$

$$\iint_{E_1} \frac{|v(y,t)-v(y,s)|^2}{[|x-y|^2+(t-s)^2]^{(n+1)/2}} \, dx \, dt \, dy \, ds .$$

The first integral above is bounded by

$$C \int_0^\infty \frac{1}{t^n} \iint_{|x-y| \le t} |v(x,t)-v(y,t)|^2 \, dx \, dy \, dt \le C \int_{R^n_+} |\nabla v(z,t)|^2 \, t \, dz \, dt .$$

We bound the second integral in (1.14) by

$$C \int_{R^{n-1}} \int_0^\infty \int_0^\infty \frac{1}{t^2} \left[\int_s^t \left| \frac{\partial v}{\partial r}(y,r) \right| dr \right]^2 dt\, ds\, dy$$

$$= C \int_{R^{n-1}} \int_0^\infty \int_0^\infty \frac{1}{t^2} \left[\int_s^t \left| \frac{\partial v}{\partial r}(y,r) \right| \underset{0<s<r}{X}(r)\, dr \right]^2 dt\, ds\, dy \ .$$

We apply Hardy's Theorem in the t-integration and then integrate in the s-variable to bound the above by $C \int_{R^n_+} \left|\frac{\partial v}{\partial r}(y,r)\right|^2 r\, dy\, dr$. This completes Theorem 1.10.

The proof of Theorem 1.10 is based upon an unpublished proof of the same result by Max Jodeit, Jeff Lewis, and the author.

I would like to conclude the discussion of the classical layer potentials with a theorem on the regularity of the solution to the Dirichlet problem when the data has tangential derivatives.

Definition. $W^2_1(\partial D)$ will denote the space of functions $f \in L^2(\partial D)$ with the property that for any coordinate ball B satisfying $B \cap \partial D = \{(x,\varphi(x));\ x \in R^{n-1}\ ,\ |\nabla\varphi| \le m\} \cap \partial D$ and for any $\psi \in C^1_0(B)$, the function $\psi(x,\varphi(x))\, f(x,\varphi(x))$ has partial derivatives, in the sense of distributions, given by functions in $L^2(R^{n-1})$. We define a norm on $W^2_1(\partial D)$ by taking a finite partition of unity $\{\psi_j\}$ of ∂D with each $\psi_j \in C^1_0(B_j)$, $B_j \cap \partial D = \{(x,\varphi_j(x)) : x \in R^{n-1}\ ,\ |\nabla\varphi_j| \le m\} \cap \partial D$, and setting

$$\|f\|_{W^2_1(\partial D)} = \|f\|_{L^2(\partial D)} + \sum_j \|\nabla(\psi_j(x,\varphi_j(x)) f(x,\varphi_j(x))\|_{L^2(R^{n-1})} \ .$$

Theorem 1.15. Suppose $g \in W^2_1(\partial D)$. Then the solution $u(X)$ of $\Delta u(X) = 0$ in D , $u|_{\partial D} = g$, in the sense of Theorem 1.7 satisfies:

$\|M(\nabla u)\|_{L^2(\partial D)} \leq C\|g\|_{W_1^2(\partial D)}$. In fact

$$\mathcal{S} : L^2(\partial D) \xrightarrow[\text{onto}]{1-1} W_1^2(\partial D) , \quad \tfrac{1}{2}I + K : W_1^2(\partial D) \xrightarrow[\text{onto}]{1-1} W_1^2(\partial D)$$

and we may write

$$u(X) = \frac{1}{(n-2)\omega_n} \int_{\partial D} \frac{1}{|X-Q|^{n-2}} \mathcal{S}^{-1}(g)(Q)\, d\sigma_Q$$

or

$$u(X) = \frac{1}{\omega_n} \int_{\partial D} \frac{(X-Q)\cdot N_Q}{|X-Q|^n} (\tfrac{1}{2}I + K)^{-1}(g)(Q)\, d\sigma_Q .$$

Proof. The proof of the continuity of the operators in question follow essentially from the results of Coifman, McIntosh, and Meyer ([2]) and can be found in [11, p. 588] and [6, p. 176] . I will briefly discuss the invertibility of $\frac{1}{2}I + K$ on $W_1^2(\partial D)$.

Since $W_1^2(\partial D) \subset L^2(\partial D)$ clearly $\frac{1}{2}I + K$ is 1-1 or injective on $W_1^2(\partial D)$. To see that $\frac{1}{2}I + K$ has closed range in $W_1^2(\partial D)$ we follow the argument in Theorem 1.4. That argument will prove closed range provided we can show the following leads to a contradiction: $\|f_j\|_{W_1^2(\partial D)} = 1$ and $(\frac{1}{2}I + K)f_j \longrightarrow 0$ in $W_1^2(\partial D)$. To show the contradiction we consider $\mathcal{D}(f_j)(X)$, the double layer potential of f_j . Since $f_j \in W_1^2(\partial D)$ the normal derivative of $\mathcal{D}(f_j)$ at the boundary does not produce a jump when computed from the interior of D or exterior of D . Hence the first Rellich inequality in Lemma 1.3 and the second Rellich inequality in Lemma 1.1 imply

$$\|(\tfrac{1}{2}I - K)f_j\|_{W_1^2(\partial D)} \leq C\, \|(\tfrac{1}{2}I + K)f_j\|_{W_1^2(\partial D)} .$$

This inequality and the assumption $(\frac{1}{2}I + K)f_j \longrightarrow 0$ in $W_1^2(\partial D)$ give $f_j \longrightarrow 0$ in $W_1^2(\partial D)$, a contradiction.

Finally, to show the range of $\frac{1}{2}I + K$ on $W_1^2(\partial D)$ is dense in the same space we follow the corresponding proof in Theorem 1.4. Using the smooth domains D_j found in Theorem 1.4, we reduce the argument to deriving a contradiction in the situation: $\|f_j\|_{W_1^2(\partial D_j)} = 1$, $\|(\frac{1}{2}I + K_j)(f_j)\|_{W_1^2(\partial D)} \longrightarrow 0$, where K_j is the double layer potential associated with D_j. This contradiction is obtained in the exact same manner as in the above paragraph with the use of the appropriate Rellich inequalities.

Section 2. Elliptic systems

The layer potential techniques introduced in Section 1 to resolve the classical boundary value problems for Laplace's equation in a Lipschitz domain do not rely on positivity properties of the Laplacian, e.g. the maximum principle and Harnack inequalities. The main ingredients were: 1) Rellich inequalities relating L^2-norms over ∂D of tangential derivatives of a solution with L^2-norms of a chosen normal or normal-type derivative, 2) the existence of a global fundamental solution to define directly the single layer potential and to define the double layer potential by appropriately applying the normal-type derivative to the fundamental solution. In Section 2 we will use these ideas to study the Dirichlet problem for constant coefficient elliptic systems.

Unless otherwise indicated, throughout the discussion on systems we will use the convention of summation on repeated indeces. The letters i and j will denote integers ranging from 1 to n, the dimension of the underlying Euclidean space, and the letters r and s will denote integers ranging from 1 to m, the number of solutions and equations. We set

$$A = (D_{X_i} a_{ij}^{rs} D_{X_j}) .$$

A is an m x m matrix with second order operators on R^n as entries. If

$$\vec{u}(X) = \begin{bmatrix} u^1(X) \\ \vdots \\ u^m(X) \end{bmatrix} \quad \text{then} \quad (A\vec{u})_r = D_{X_i} a_{ij}^{rs} D_{X_j} u^s \ .$$

In general we assume two conditions:

1) the Legendre-Hadamard condition

$$a_{ij}^{rs} \xi_i \xi_j \eta^r \eta^s \geq \lambda \ |\xi|^2 |\eta|^2 \ , \ \lambda > 0 \ .$$

2) the symmetry condition $a_{ij}^{rs} = a_{ji}^{sr}$.

We begin our study of the Dirichlet problem for A seeking Rellich inequalities for systems. We refer to such inequalities as Necas-Rellich inequalities because of the influence of the book by Necas ([9]) in this regard.

<u>Lemma 2.1</u>. Suppose D is a Lipschitz domain in R^n . Let $\frac{\partial \vec{u}}{\partial \nu_A}(Q)$ denote the vector function defined on ∂D by $\left[\frac{\partial \vec{u}}{\partial \nu_A}\right]^r (Q) = \nu_i(Q) \ a_{ij}^{rs} D_{X_j} u^s$. ν_Q = exterior normal to ∂D at Q . Also define $\nabla\vec{u}(X) = (D_{X_j} u^r(X))$. If $h(X)$ is a smooth vector field on R^n and $A(\vec{u}) = 0$, then

$$\int_{\partial D} h(Q)\cdot\nu_Q \ a_{ij}^{rs} D_{X_i} u^r D_{X_j} u^s d\sigma = \int_D \operatorname{div} h \ a_{ij}^{rs} \ D_{X_i} u^r \ D_{X_j} u^s \ dX$$

$$- \ 2 \int_D D_{X_i} h_\ell \ a_{ij}^{rs} \ D_{X_\ell} u^r D_{X_j} u^s \ dX + 2 \int_{\partial D} \frac{\partial \vec{u}}{\partial \nu_A} \cdot \nabla\vec{u}(Q)(h(Q)) \ d\sigma$$

<u>Proof</u>. Using the symmetry condition and $A(\vec{u}) = 0$,

$$\operatorname{div} \left(h(X) \ a_{ij}^{rs} \ D_{X_i} u^r \ D_{X_j} u^s\right) = \operatorname{div} h(X) \ a_{ij}^{rs} \ D_{X_i} u^r \ D_{X_j} u^s$$

$$+ \ 2 \ h_\ell \ D_{X_i} \left(a_{ij}^{rs} \ D_{X_\ell} u^r \ D_{X_j} u^s\right) \ .$$

Integrate the above equality over D and use the divergence theorem to obtain Lemma 2.1.

<u>Corollary 2.2</u> Set $|\nabla\vec{u}|^2 = \sum_r |\nabla u^r|^2$. If $a_{ij}^{rs} D_{X_i}u^r D_{X_j}u^s \geq \delta|\nabla\vec{u}|^2$ with δ a fixed positive number, then for solutions of $A(\vec{u}) = 0$,

$$\int_{\partial D} |\nabla\vec{u}|^2 \, d\sigma \leq C \left[\int_{\partial D} \left|\frac{\partial\vec{u}}{\partial\nu_A}\right|^2 d\sigma + \int_{\partial D} |\vec{u}|^2 \, d\sigma \right]$$

where C depends only on δ and the Lipschitz character of D. In particular,

$$\int_{\partial D} |\nabla_\tau \vec{u}|^2 d\sigma \equiv \sum_r \int_{\partial D} |\nabla_\tau u^r|^2 \leq C \left[\int_{\partial D} \left|\frac{\partial\vec{u}}{\partial\nu_A}\right|^2 d\sigma + \int_{\partial D} |\vec{u}|^2 d\sigma \right].$$

<u>Proof</u>. As in Lemma 1.1 we pick a smooth vector field h such that $\operatorname*{essinf}_{\partial D} h\cdot\nu > 0$. The inequality follows as in the proof of Lemma 1.1 once we note that

$$\int_D |\nabla\vec{u}|^2 dx \leq \frac{1}{\delta} \int_D a_{ij}^{rs} D_{X_i}u^r D_{X_j}u^s \, dX = \frac{1}{\delta} \int_{\partial D} \frac{\partial\vec{u}}{\partial\nu_A} \cdot \vec{u} \, d\sigma .$$

<u>Lemma 2.3</u>. If $A\vec{u} = 0$ in D and h is a smooth vector field in R^n, then

$$\int_{\partial D} h(Q)\cdot\nu_Q \, a_{ij}^{rs} D_{X_i}u^r D_{X_j}u^s \, d\sigma = \int_{\partial D} \vec{B}_j^{rs}(Q)\cdot\nabla u^r(Q) \, D_{X_j}u^s(Q) d\sigma$$

$$- \int_D \operatorname{div} h \, a_{ij}^{rs} D_{X_i}u^r D_{X_j}u^s \, dX + 2 \int_D D_{X_i}h_\ell \, a_{ij}^{rs} D_{X_\ell}u^r D_{X_j}u^s \, dX$$

where $\vec{B}_j^{rs}(Q)$ is a bounded tangential vector, i.e. $\vec{B}_j^{rs}(Q)\cdot\nu_Q = 0$ for each r,s,j.

<u>Proof</u>. Using Lemma 2.1,

$$\int_{\partial D} h(Q)\cdot\nu_Q \, a_{ij}^{rs} D_{X_i}u^r D_{X_j}u^s \, dQ =$$

$$2\int_{\partial D}[h(Q)\cdot\nu_Q\, a_{ij}^{rs}\, D_{X_i}u^r\, D_{X_j}u^s - \frac{\partial\vec{u}}{\partial\nu_A}\cdot\nabla\vec{u}(Q)(h(Q))]d\sigma$$

$$-\int_D \operatorname{div} h\;\; a_{ij}^{rs}\, D_{X_i}u^r\, D_{X_j}u^s + 2\int_D D_{X_i}h_\ell\, a_{ij}^{rs}\, D_{X_\ell}u^r\, D_{X_j}u^s\;.$$

We rewrite the boundary integral on the right side of the above equality as:

$$2\int_{\partial D}\left[h(Q)\cdot\nu_Q\, a_{ij}^{rs}\, D_{X_i}u^r - \nu_i\, a_{ij}^{rs}\, h_\ell\, D_{X_\ell}u^r\right] D_{X_j}u^s\, d\sigma$$

$$= \int_{\partial D}(\vec{B}_j^{rs}(Q)\cdot\nabla u^r(Q))\, D_{X_j}u^s(Q)\, d\sigma\;,\text{ where for fixed } r,s,j\;,$$

$$\vec{B}_j^{rs}(Q)\cdot\nabla u^r(Q) = \left[h(Q)\cdot\nu_Q\, a_{\ell j}^{rs} - \nu_i\, a_{ij}^{r,s}\, h_\ell\right] D_{X_\ell}u^r(Q)\;.$$

It is easy to verify that $\vec{B}_j^{rs}(Q)\cdot\nu_Q = 0$.

Corollary 2.4. If $a_{ij}^{rs}\, D_{X_i}u^r\, D_{X_j}u^s \geq \delta|\nabla\vec{u}|^2$ for some $\delta > 0$ and $A(\vec{u}) = 0$, then

$$\int_{\partial D}|\nabla\vec{u}|^2 d\sigma \leq C\left[\int_{\partial D}|\nabla_T\vec{u}|^2\, d\sigma + \int_{\partial D}|\vec{u}|^2 d\sigma\right]$$

where C depends only on δ and the Lipschitz character of D . (Recall $|\nabla_T\vec{u}|^2 = \sum_r |\nabla_T u^r|^2$.)

Proof. The proof follows immediately from the hypotheses and Lemma 2.3 noting as in Corollary 2.2 that

$$\int_D|\nabla\vec{u}|^2\, dX \leq \frac{1}{\delta}\int_D a_{ij}^{rs}\, D_{X_i}u^r\, D_{X_j}u^s\, DX = \frac{1}{\delta}\int_{\partial D}\frac{\partial\vec{u}}{\partial\nu_A}\cdot\vec{u}\, d\sigma\;.$$

We combine Corollaries 2.2 and 2.4 into

Theorem 2.5 (Necas-Rellich inequalities).

Assume there exists $\delta > 0$ such that $a_{ij}^{rs} D_{X_i}u^r D_{X_j}u^s \geq \delta |\nabla\vec{u}|^2$. Then there exists $C > 0$ depending only on δ and the Lipschitz character of D such that

$$\int_{\partial D} \left|\frac{\partial\vec{u}}{\partial\nu_A}\right|^2 d\sigma \leq C \left[\int_{\partial D} |\nabla_\tau \vec{u}|^2 d\sigma + \int_{\partial D} |\vec{u}|^2 d\sigma\right]$$

and

$$\int_{\partial D} |\nabla_\tau \vec{u}|^2 \, d\sigma \leq C\left[\int_{\partial D} \left|\frac{\partial\vec{u}}{\partial\nu_A}\right|^2 d\sigma + \int_{\partial D} |\vec{u}|^2 d\sigma\right].$$

We now turn from the general theory to the particular case of linear elastostatics in R^n. We consider the operator

$$A(\vec{u}) = \mu\Delta\vec{u} + (\lambda+\mu)\, \nabla \operatorname{div} \vec{u}\ ,\ \mu > 0,\ \lambda \geq 0\ .$$

Here we can take

$$a_{ij}^{rs} = \mu\delta_{ij}\,\delta_{rs} + (\lambda+\mu)\,\delta_{ir}\delta_{js}$$

$1 \leq i,j,r,s \leq n$ and we have

$$a_{ij}^{rs}\ \xi_i\xi_j\ \eta_r\eta_s = \mu|\xi|^2|\eta|^2 + (\lambda+\mu)(\xi\cdot\eta)^2\ ,$$

$$a_{ij}^{rs} D_{X_i}u^r D_{X_j}u^s = \mu\ |\nabla\vec{u}|^2 + (\lambda+\mu)(\operatorname{div}\vec{u})^2\ ,$$

$$\frac{\partial\vec{u}}{\partial\nu_A} = \mu\,\frac{\partial\vec{u}}{\partial\nu} + (\lambda+\mu)(\operatorname{div}\vec{u})\nu\ .$$

Let $\Gamma(X) = (\Gamma_{ij}(X))$ denote the fundamental solution matrix corresponding to A, the operator of linear elastostatics ([8]). Define

$$\mathcal{S}\vec{f}(X) = \int_{\partial D} \Gamma(X-Q)\ \vec{f}(Q)\ d\sigma_Q$$

and

$$\mathcal{D}\vec{f}(X) = \int_{\partial D} \{\frac{\partial}{\partial \upsilon_A} (\Gamma(X-\cdot))(Q)\}^{tr} \vec{f}(Q)\, d\sigma_Q$$

where $\frac{\partial}{\partial \upsilon_A}$ is applied to each column vector of the matrix Γ and tr denotes the transpose matrix.

All the ingredients, especially the Necas-Rellich inequalities, are available to us to proceed as in the harmonic case to obtain

<u>Theorem (2.6)</u> (Dahlberg, Kenig, Verchota [5])

1) $\|M(\mathcal{D}\vec{f})\|_{L^2(\partial D)} \leq C\, \|\vec{f}\|_{L^2(\partial D)}$

and $\mathcal{D}(\vec{f})(X) \longrightarrow (\frac{1}{2}I + K)\vec{f}(P)$ pointwise for a.e. $(d\sigma)$ $P\in\partial D$ as $X\longrightarrow P$, $X\in\Gamma_P$, where

$$K\vec{f}(P) = \lim_{\epsilon\to 0+} \int_{|P-Q|>\epsilon} \{\frac{\partial}{\partial \upsilon_A} (\Gamma(P-\cdot)(Q))\}^{tr} \vec{f}(Q) d\sigma_Q \ .$$

2) $\|M(\nabla\mathcal{S}f)\|_{L^2(\partial D)} \leq C\|\vec{f}\|_{L^2(\partial D)}$ and

$\frac{\partial}{\partial \upsilon_A} \mathcal{S}\vec{f}(X) \longrightarrow (\frac{1}{2}I - K^*)\, \vec{f}(P)$ pointwise for a.e. $(d\sigma)$ $P\in\partial D$ as $X\longrightarrow P$, $X\in\Gamma_P$.

3) $\frac{1}{2}I + K$ is invertible on $L^2(\partial D)$.

The constants in Theorem 2.6 depend only on β, δ (see the definition of a regular family of cones for D) and the Lipschitz character of D . From Theorem 2.6 follows

<u>Theorem 2.7</u>. Set $A = \mu\Delta + (\lambda+\mu)\,\nabla\mathrm{div}$, $\mu > 0$, $\lambda \geq 0$. i) Given $\vec{g}\in L^2(\partial D)$ there exists a (unique) function $\vec{u}$ such that $A\vec{u} = 0$ in D , $M(\vec{u})\in L^2(\partial D)$, and $\vec{u}(X) \longrightarrow \vec{g}(P)$ for a.e. $(d\sigma)$ $P\in\partial D$ as $X\to P$ $X\in\Gamma_P$. We can write

$$\vec{u}(X) = \int_{\partial D} \{\frac{\partial}{\partial \upsilon_A} (\Gamma(X-\cdot))(Q)\}^{tr} (\tfrac{1}{2}I + K)^{-1} \vec{g}(Q)d\sigma_Q$$

$(\frac{\partial}{\partial \upsilon_A} = \mu \frac{\partial}{\partial \upsilon} + (\lambda+\mu)(\mathrm{div})\upsilon)$.

ii) If $\vec{g} \in W_1^2(\partial D)$ then the solution in (i) also satisfies:

$\|M(\nabla\vec{u})\|_{L^2(\partial D)} \leq C \|\vec{g}\|_{W_1^2(\partial D)}$ and we can write

$$\vec{u}(X) = \int_{\partial D} \vec{\Gamma}(X-Q)\ \vec{f}(Q)\ d\sigma_Q$$

with $\|\vec{f}\|_{L^2(\partial D)} \leq C \|\vec{g}\|_{W_1^2(\partial D)}$.

Again all constants in Theorem 2.7 depend only on β,δ, and the Lipschitz character of D .

We end this survey with a list of unresolved questions in the area of boundary value problems in Lipschitz domains for elliptic systems.

1) Can we solve the Dirichlet problem in a Lipschitz domain with L^2-boundary data for a general second order elliptic operator satisfying:

$$a_{ij}^{rs}\ \xi_i\xi_j\eta_r\eta_s \geq c|\xi|^2|\eta|^2\ ,\ c > 0 \quad\text{and}\quad a_{ij}^{rs} = a_{ji}^{jr}\ ?$$

The strange difficulty here is a lack of uniqueness in writing down a second order system. As an example there are various ways of writing the equation $\mu\Delta\vec{u} + (\lambda+\mu)$ div $\vec{u} = 0$ as

$$D_{X_i}(a_{ij}^{rs}\ D_{X_j}u^s) = 0 \quad \text{for each} \quad r\ .$$

For fixed i,j,r,s we may set

$$a_{ij}^{rs}\ (1) = \mu\delta_{ij}\ \delta_{rs} + (\lambda+\mu)\ \delta_{ir}\ \delta_{js}$$

or $a_{ij}^{rs}\ (2) = \mu\delta_{ij}\delta_{rs} + \mu\delta_{jr}\delta_{is} + \lambda\delta_{ir}\delta_{js}$.

With $k = 1,2$, we have

$$a_{ij}^{rs}(k)\xi_i\xi_j\eta_r\eta_s = \mu|\xi|^2|\eta|^2 + (\lambda+\mu)(\xi\cdot\eta)^2 .$$

However, $a_{ij}^{rs}(2)\frac{\partial u^r}{\partial X_i}\frac{\partial u^s}{\partial X_j} = \frac{\mu}{2}\left[\frac{\partial u^r}{\partial X_s} + \frac{\partial u^s}{\partial X_r}\right]^2 + \lambda(\operatorname{div}\vec{u})^2$.

So we cannot say

$$a_{ij}^{rs}(2)\frac{\partial u^r}{\partial X_i}\frac{\partial u^s}{\partial X_j} \geq \delta\sum_{r,s}\left[\frac{\partial u^r}{\partial X_s}\right]^2 , \ \delta > 0 .$$

Note: $\frac{\partial\vec{u}}{\partial\nu}_{A(2)} = \mu(\nabla\vec{u}(\nu) + (\nabla\vec{u})^{tr}(\nu)) + \lambda(\operatorname{div}\vec{u})\,\nu$

This is called the traction of $\vec{u}$ on ∂D. (See [5].)

2) <u>Regularity of the solution</u>:

Suppose $\mu\Delta\vec{u} + (\lambda+\mu)\nabla\operatorname{div}\vec{u} = 0$ in D and $\vec{u}|_{\partial D} = \vec{g}$ where $\vec{g}$ is smooth, for example $\vec{g}$ is the restriction to ∂D of a C^∞ function in R^n. Is $\vec{u}$ Holder continuous in $\bar{D}$ (or even bounded)? $\vec{u}$ is Holder continuous for $n = 3$ since there exists $\varepsilon > 0$ $\mathcal{S}$: $L^{2+\varepsilon}(\partial D) \xrightarrow[\text{onto}]{1-1} W_1^{2+\varepsilon}(\partial D)$. (See [5]). This is unknown for $n \geq 4$.

3) L^p-theory for systems:

Again consider $\mu\Delta + (\lambda+\mu)\nabla\operatorname{div} = A$.

If the Dirichlet data $\vec{g}\in L^p(\partial D)$, $2 < p < \infty$, does the solution satisfy

$$\|M_\beta(\vec{u})\|_{L^p(\partial D)} \leq C\,\|\vec{g}\|_{L^p(\partial D)} ?$$

Is $\frac{1}{2}I + K$ invertible on $L^p(\partial D)$ for $2 < p < \infty$?

In the case of Laplace's equation $\frac{1}{2}I + K$ was shown to be invertible on $L^p(\partial D)$, $2 \leq p < \infty$, and $\frac{1}{2}I + K^*$ was shown to be invertible on

$$L^p_o(\partial D) \equiv L^p(\partial D) \cap \{ \int_{\partial D} f = 0\} \ , \ 1 < p \leq 2 \ .$$

by B. Dahlberg and C. Kenig ([4]) . Their method of proof relies on positivity properties associated with the Laplace operator and hence does not extend in its present form to systems.

References

1. A.P. Calderon, Cauchy integrals on Lipschitz curves and related operators, Proc. Nat. Acad. Sci. U.S.A., 74 (1977), 1324-1327.

2. R.R. Coifman, A McIntosh, Y. Meyer, L'intégrale de Cauchy définit un opérateur borné sur L^2 pour les courbes Lipschitziennes, Annals of Math. 116 (1982), 361-387.

3. B.E.J. Dahlberg, Weighted norm inequalities for the Lusin area integral and nontangential maximal functions for functions harmonic in a Lipschitz domain, Studia Math. 67 (1980), 297-314.

4. B.E.J. Dahlberg, C.E. Kenig, Hardy spaces and the L^p Neumann problem for Laplace's equation in a Lipschitz domain, Annals of Math., 125 (1987), 437-465.

5. B.E.J. Dahlberg, C.E. Kenig, G.C. Verchota, Boundary value problems for the systems of elastostatics on a Lipschitz domain, preprint available.

6. E.B. Fabes, M. Jodeit, Jr., and M.M. Riviere, Potential techniques for boundary value problems on C^1 domains, Acta Math. 141 (1978), 165-186.

7. O.D. Kellogg, Foundations of Potential Theory, New York, Dover, 1954.

8. V.D. Kupradze, Thru Dimensional Problems of the Mathematical Theory of Elasticity and Thermoelasticity, North Holland, New York, 1979.

9. J. Nečas, Les Méthodes Directes en Théorie des Equations Elliptiques, Academia, Prague, 1967.

10. F. Rellich, Darstellung der Eigenwerte von $\Delta u+\lambda u$ durch ein Randintegral, Math. Z. 46 (1940), 635-646.

11. G.C. Verchota, Layer potentials and regularity for the Dirichlet problem for Laplace's equation in Lipschitz domains, Journal of Functional Analysis, 59 (1984), 572-611.

Fine Potential Theory

BENT FUGLEDE
Matematisk Institut, Københavns Universitet
Universitetsparken 5, DK-2100 København Ø

INTRODUCTION

The title of the present survey is short for "Potential theory based on the fine topology". This topology (on $\mathbf{R}^n$, say) was introduced by H. Cartan in 1940 as the weakest topology making all subharmonic functions continuous. The fine topology is strictly stronger than the usual (Euclidean) topology (when $n \geq 2$); and an open set in the fine topology need not have inner points in the Euclidean topology.

In the sequel we outline some of the developments in fine potential theory during the past two decades. We do not consider the fine topology at the Martin boundary or other ideal boundaries, cf. Brelot [12]. The important probabilistic aspects of fine potential theory will barely be hinted at, cf. Doob [24]. For the non-linear fine potential theory we refer to the lecture by G. Wildenhain in these Proceedings.

As applications we discuss the existence of asymptotic paths for subharmonic functions (§6), uniform approximation by harmonic or continuous superharmonic functions (§7), approximation and duality in Dirichlet space (§8), and finely holomorphic functions— an extension of one-dimensional complex analysis in the spirit of Borel's monogenic functions (§§9,10).

The classical theory of harmonic and subharmonic or superharmonic functions, defined in open subsets of $\mathbf{R}^n$, has to a large extent been carried over to various axiomatic frameworks such as harmonic spaces, more generally to balayage spaces [4] and still more generally to H-cones [7]. In the present lecture we shall mainly consider harmonic spaces. On such a space, local notions of "harmonic" and "superharmonic" functions are available. We refer the reader to the excellent survey article by Bauer [2]. One may even forget about any reference to axiomatic potential theory and confine oneself to the two basic examples I and II below, which together are fairly typical as harmonic spaces.

1. HARMONIC SPACES

Let X denote a *harmonic space* in the sense of Bauer [1], or of Constantinescu & Cornea [16] with a countable base for its topology. We also assume that there exists a potential > 0 defined on all of X. (A potential means a superharmonic function having 0 as its greatest harmonic, or equally well subharmonic minorant.)

Example I. Classical potential theory on $X = \mathbf{R}^n$, $n \geq 3$, or more generally on a Green domain $X \subset \mathbf{R}^n$, $n \geq 2$. The harmonic functions on an open subset of X are here those C^2-functions u which satisfy the Laplace equation $\Delta u = 0$. The associated notion of superharmonic functions goes back to F.Riesz in its general form. A potential p on X is the same as the Green potential (supposed $\not\equiv +\infty$)

$$p = G\mu = \int_X G(\cdot, y)\, d\mu(y)$$

of a positive measure μ on X; $G(\cdot, y)$ being the Green function on X with pole at $y \in X$. If $X = \mathbf{R}^n$, $n \geq 3$, then $G(x, y) = N(x - y)$, where

$$N(x) = \frac{1}{(n-2)\omega_n |x|^{n-2}}$$

is the Newtonian kernel on $\mathbf{R}^n$, ω_n denoting the surface of the unit sphere in $\mathbf{R}^n$. Thus N is the usual fundamental solution of the operator $-\Delta$.

Example II. Parabolic potential theory on $\mathbf{R}^{n+1}$. In this example, which goes back to Doob [22] in 1956, the Laplace operator Δ is replaced by the heat- or diffusion operator D on $\mathbf{R}^{n+1}$ (with points $(t, x) \in \mathbf{R} \times \mathbf{R}^n$):

$$D = \partial/\partial t - \Delta,$$

where Δ is taken with respect to $x \in \mathbf{R}^n$. The standard fundamental solution to D is

$$E(x,t) = \begin{cases} (4\pi t)^{-\frac{n}{2}} \exp\left(-\frac{|x|^2}{4t}\right) & , \quad t > 0, \\ 0 & , \quad t \leq 0. \end{cases}$$

The harmonic functions (called caloric functions) are the solutions to $Du = 0$, and the superharmonic functions and potentials are now quite analogous to those of Example I, writing

$$G(z, w) = E(z - w) \quad \text{for} \quad z, w \in \mathbf{R}^{n+1}.$$

In what follows we shall often refer to Axiom D, or the *axiom of domination*, introduced by Brelot [10] in 1958. It asserts that, given a locally bounded potential which is harmonic off some closed set $F \subset X$, and given a superharmonic function $s \geq 0$ on X, which majorizes p on F; then $s \geq p$ on all of X. Cartan had shown [14] that *Axiom D is satisfied in Example I.* However, it is *not* satisfied in Example II. Axiom D can be formulated in a number of rather different looking ways (see [16,§9.2]), some of which we shall meet subsequently.

2. THIN SETS AND FINE TOPOLOGY

The Cartan fine topology has its roots in the notions of *irregular* and *unstable* boundary point for the Dirichlet problem in a domain (resp. on a compact set) in $\mathbf{R}^n$—we are now in Example I. According to the famous Wiener criterion [59] from 1924, such a point x is characterized by the complement E of the domain (resp. compact set) in question being "thin" (French: effilé) at x, in the sense that (if $n \geq 3$)

$$\sum_{k=1}^{\infty} \frac{\operatorname{cap}^*(E \cap A_k)}{\operatorname{cap}^* A_k} < \infty,$$

where A_k is the ball of radius 2^{-k} centered at x , and cap* denotes outer Newtonian capacity.

This notion of thinness makes sense for quite arbitrary sets $E \subset \mathbf{R}^n$, and a systematic study thereof was initiated by Brelot [9] in 1939. He found an equivalent definition of thinness which proved more useful for the general theory: E is thin at x if either x is not a limit point of E or else if

$$s(x) < \liminf_{y \to x,\, y \in E \setminus \{x\}} s(y)$$

for some superharmonic function s defined on a neighbourhood of x.—Recall that s generally is only l.s.c. (=lower semicontinuous).

Shortly after, Cartan showed that Brelot's definition is further equivalent to the complement of $E \setminus \{x\}$ being a neighbourhood of x in what he called the *fine topology*—the weakest topology on $\mathbf{R}^n$ in which all superharmonic functions on open subsets of $\mathbf{R}^n$ are continuous.—An irregular point for a domain, or an unstable point of a compact set, is now simply a finely isolated point of the complement. (The prefix fine(ly) always refers to the Cartan fine topology. Topological notions not preceded by this prefix refer to the Euclidean topology on $\mathbf{R}^n$, or—in the case of a harmonic space X—to the initially given, locally compact topology on X).

It was clear from the beginning that the fine topology on $\mathbf{R}^n, n \geq 2$, is Hausdorff, completely regular, and Baire. But it is neither first countable nor Lindelöf, and there are no infinite compact sets in the fine topology. For some time this new topology was therefore considered difficult to handle and perhaps a bit artificial. To the probabilists, however, the fine topology became extremely natural: a Borel set U is finely open if and only if a Brownian particle, starting at a point of U , remains in U almost surely in some positive time interval (J.L. Doob [21], 1954). And in 1966, Doob [23] discovered that the fine topology is *quasi Lindelöf* in the sense that the union of any family of finely open sets differs by at most a polar set (= a set of zero outer capacity, if $n \geq 3$) from the union of some countable subfamily.

The above properties possessed by the fine topology on $\mathbf{R}^n$ (as well as its definition) carried over to the parabolic case II and even to general harmonic spaces. When it came to *connectivity properties* of the fine topology, however, the parabolic case became pathological. As noticed by C. Berg in 1968, the space $\mathbf{R}^{n+1}$ is totally disconnected in the fine topology in Example II. Naturally, this circumstance led to the question about connectivity properties of the fine topology in Example I.

It turned out shortly after that, in Example I, and more generally in any harmonic space satisfying Axiom D, the fine topology is *locally connected*, and every usual subdomain is also finely connected [28]. Moreover, in Example I, any two points of a fine domain U (= a finely open, finely connected set) can be joined by a finite polygonal line contained in U (Lyons[53]; in dimension $n = 2$ this property—or even connectedness in the Euclidean topology—is equivalent to fine connectedness of a finely open set U ([35], Gamelin & Lyons [46]). Earlier, Nguyen-Xuan-Loc and T. Watanabe [58] had established the arcwise connectedness of any fine domain, using pieces of Brownian paths.

3. Balayage

The results of the present section are standard and can mostly be found, e.g., in [1] or [16].

We denote by S the convex cone of all superharmonic functions ≥ 0 on the harmonic space X; S is in fact an H-cone. In particular, S is a lower complete lattice in the natural, pointwise order $\leq$, and the infimum $\bigwedge \mathcal{E}$ of any non-empty subset $\mathcal{E}$ of S is the greatest l.s.c. (and also the greatest *finely* l.s.c.) minorant of the pointwise infimum inf $\mathcal{E}$.

For any set $A \subset X$ the fundamental operator $B^A : S \to S$ of *balayage*, or sweeping, onto A is defined as follows. For any $u \in S$, the swept-out function of u on A is defined by

$$B^A u := \bigwedge \{ v \in S \mid v \geq u \text{ on } A \}.$$

Dually, for a positive measure μ of compact support in X, the swept-out measure μ^A is characterized by the condition

$$\int u \, d\mu^A = \int B^A u \, d\mu \quad \text{for every } u \in S.$$

In particular, a set A is *thin* at a point $x \in X$ if and only if $\varepsilon_x^A \neq \varepsilon_x$.

For a *finely open* set V and a point $x \in V$, the swept-out $\varepsilon_x^{\complement V}$ of the Dirac measure ε_x onto $\complement V$ is called the (fine) *harmonic measure* associated with (V, x). It is carried by the boundary ∂V. And $\varepsilon_x^{\complement V}$ is carried even by the *fine boundary* $\partial_f V$ of V if and only if Axiom D holds, as shown by Mme Hervé [49, §28].—In probabilistic terms, the harmonic measure $\varepsilon_x^{\complement V}$ is, in Example I, the first exit distribution from V for Brownian motion starting at x.

In a harmonic space X, a *polar* set is a set $A \subset X$ such that $B^A u = 0$ for every $u \in S$, or equivalently such that $A \subset [u = +\infty]$ for some $u \in S$. A statement is said to hold *quasi everywhere* (q.e.) if it holds except at most in some polar set.

In the rest of this section we suppose for the sake of simplicity that the constant function 1 is superharmonic, i.e., $1 \in S$. (In Examples I and II it is even a harmonic function.) The set function

$$A \to B^A 1$$

with values in S serves as a kind of capacity. A function $f : X \to [-\infty, \infty]$ is said to be *quasi l.s.c.* (quasi lower semicontinuous) if

$$\bigwedge\{B^G 1 \mid G \text{ open in } X, \quad f_{|\complement G} \text{ is l.s.c.}\} = 0,$$

that is, if f is l.s.c. on the complements of open sets G of "arbitrarily small capacity".

Every quasi l.s.c. function is finely l.s.c. q.e. Conversely we have the following fundamental theorem due to Brelot [11], cf. also [16, §8.4, §9.2].

THEOREM 1. *Axiom D is equivalent to the following. If a function f on X is finely l.s.c. q.e. in X, then f is quasi l.s.c.*

The above definition and results concerning quasi l.s.c. functions have counterparts in which "l.s.c." is replaced throughout by "continuous".

From the results in [11] one easily deduces that Axiom D is further equivalent to the following very useful relation between the two topologies on X, called the *Brelot property*:

Given a countable family (f_ν) of finely continuous functions $f_\nu : U \to [-\infty, \infty]$, all defined on the same finely open set $U \subset X$, every point of U has a fine neighbourhood V (compact if we like) in U such that the restriction of each f_ν to V is continuous (in the induced initial topology). See [34].

4. FINE POTENTIAL THEORY.

In view of the local connectedness of the fine topology on a harmonic space X satisfying axiom D, one might hope to extend the theory of harmonic and superharmonic functions for such a space so as to allow analogous functions to be defined even on *finely* open sets. And such an extension was indeed made soon after [27],[29].

At that stage the role of Axiom D appeared so crucial at many points that it was hard to imagine that this kind of fine potential theory could hold in harmonic spaces without Axiom D, for example in parabolic potential theory (Example II). Nevertheless it turned out, beginning in the mid seventies, that Axiom D is by no means indispensable for the basic theory, in particular for the fine Dirichlet problem, although new methods had to be added. This emerged through the work of Boboc, Bucur & Cornea [6],[7] on standard H-cones, the work of Bliedtner & Hansen on simplicial cones [3] and balayage spaces [4]; and the work of Lukeš & Malý [51],[52] on finely hyperharmonic functions on harmonic spaces without Axiom D.

Following the book of Lukeš, Malý & Zajíček [52] we now state the basic definitions in fine potential theory on a harmonic space X, possibly without Axiom D . Given a finely open subset U of X, write $\mathcal{B}(U)$ for the class of all finely open sets V of compact closure $\overline{V}$ (in the initial topology) contained in U.

Definition 1. A function $s : U \to]-\infty, \infty]$ is said to be *finely hyperharmonic* if s is finely l.s.c. and if

$$s(x) \geq \int s \, d\varepsilon_x^{\complement V} > -\infty, \quad \forall x \in V \in \mathcal{B}(U).$$

(It is part of the requirement that the integral exists.)

Definition 2. A function $s : U \to \mathbf{R}$ is called *finely harmonic* if s and $-s$ are both finely hyperharmonic, or equivalently if f is finely continuous and

$$s(x) = \int s \, d\varepsilon_x^{\complement V}, \quad \forall x \in V \in \mathcal{B}(U).$$

Definition 3. A function $s : U \to]-\infty, \infty]$ is called *finely superharmonic* if s is finely hyperharmonic on U and finite on a finely dense subset of U.

Naturally, $s : U \to [-\infty, \infty[$ is called *finely subharmonic* if $-s$ is finely superharmonic.

On an open set U (in the initial topology) the harmonic functions are the same as those finely harmonic functions which are locally bounded; and the hyperharmonic functions on U are precisely those finely hyperharmonic functions which are locally lower bounded.

The cone of finely hyperharmonic, resp. finely superharmonic functions on a finely open set U has lattice properties—including the Riesz decomposition property—similar to known properties of hyperharmonic, resp. superharmonic functions.

It turns out that every finely hyperharmonic function is *finely continuous* (not just finely l.s.c.). There is accordingly no new "fine-fine" topology.

Obviously, a finely hyperharmonic function s on a finely open set U has a finely hyperharmonic restriction to every finely open subset of U (the presheaf property). Conversely, suppose that a function s on U is finely hyperharmonic in some fine neighbourhood of each point of U. Does it follow that s is finely hyperharmonic in all of U? We say that the finely hyperharmonic functions on X have the *fine sheaf property* if the answer to this question is yes.

THEOREM 2. (Boboc-Bucur-Cornea [6],[7]; Lukeš-Malý [51],[52].) *The finely hyperharmonic functions on X have the fine sheaf property if and only if Axiom D holds.*

Similarly of course for the finely superharmonic functions. And quite recently, Chacrone [15] has shown the same for finely harmonic functions. (The "if" part is contained in [29] in all three cases.)

In the presence of Axiom D the following *fine boundary minimum principle* was established in [29, Theorem 9.1]. (For the sake of simplicity we suppose here that the finely open set U is relatively compact in X, and we confine the attention to lower bounded functions.)

Let s be a lower bounded, finely hyperharmonic function on U such that

$$\underset{U \ni x \to z}{\text{fine} \liminf} s(x) \geq 0 \quad \text{q.e. for } z \in \partial_f U.$$

Then $s \geq 0$ (in U).

Lukeš and Malý noticed that this breaks down if Axiom D is not imposed. They succeeded nevertheless in establishing a version of this boundary minimum principle which holds also in the absence of Axiom D, and which reduces to the above in the presence of Axiom D. Their idea was to replace the above fine lim inf by a quasi-topological notion of lim inf, but this requires considerable care. To state their result we need the following kind of "capacity" of subsets of U.

Fixing a finely open set $U \subset X$ we denote by $\mathcal{H}_+^*(U)$ the convex cone of all finely hyperharmonic functions ≥ 0 on U. The "capacity" $c(A)$ of a set $A \subset U$ is now defined as the following element of $\mathcal{H}_+^*(U)$:

$$c(A) := \bigwedge \{u \in \mathcal{H}_+^*(U) \mid u \text{ l.s.c.}, \quad u \geq 1 \text{ on } A\}.$$

Here $\bigwedge$ denotes infimum in the lower complete lattice $\mathcal{H}_+^*(U)$ endowed with the natural, pointwise order. Explicitly, $c(A)$ is therefore the greatest finely l.s.c. minorant of the pointwise infimum of the stated subfamily of $\mathcal{H}_+^*(U)$. Note that this subfamily consists of lower semicontinuous functions in the topology on U induced by the initial topology on X.

The fine boundary minimum principle now extends as follows to a general harmonic space X:

THEOREM 3. (Lukeš & Malý in [52].) *Let s be a lower bounded, finely hyperharmonic function on U. Consider the family $\mathcal{N}$ of all open sets G in X such that*

$$\liminf_{U \setminus G \ni x \to z} s(x) \geq 0, \quad \text{all } z \in \overline{U \setminus G} \setminus U.$$

Suppose that $\mathcal{N}$ contains sets G with arbitrarily small $c(G \cap U)$, in the sense that

$$\bigwedge\{c(G \cap U) \mid G \in \mathcal{N}\} = 0.$$

Then $s \geq 0$ (in U).

As usual in potential theory, such a boundary minimum principle is the key to the study of an associated Dirichlet problem, see [52,§14 C] for the present situation.

To see that Theorem 3 in the presence of Axiom D implies the original fine boundary minimum principle, note that, when s satisfies the hypotheses in the fine boundary minimum principle, the extension f of s by 0 in $\complement U$ is obviously finely l.s.c. q.e. According to Theorem 1 in §3, f is therefore quasi l.s.c. Any open set $G \subset X$ such that f has a l.s.c. restriction to $\complement G$ clearly belongs to $\mathcal{N}$, and hence, by the definition of f being quasi l.s.c.,

$$\bigwedge\{B^G 1 \mid G \in \mathcal{N}\} = 0.$$

But $c(G \cap U) \leq (B^G 1)_{|U}$ because every $v \in \mathcal{S}$ with $v \geq 1$ on G is l.s.c. and has a restriction to U of class $\mathcal{H}_+^*(U)$. It follows easily that also

$$\bigwedge\{c(G \cap U) \mid G \in \mathcal{N}\} = 0,$$

and consequently, by Theorem 3, that indeed $s \geq 0$ in U.

Among the most recent results in fine potential theory are the integral representation of fine potentials and the associated Riesz representation theorem for finely superharmonic functions.

Definition 4. A *fine potential* on a finely open set U is a finely superharmonic function $p \geq 0$ on U such that 0 is the greatest finely subharmonic minorant of p.

Consider first Example I, where X is a domain in $\mathbf{R}^n$ with a Green kernel G. For any finely open set $U \subset X$ the Green kernel G^U on $U \times U$ is defined [33] by

$$G^U(\cdot, y) = G(\cdot, y) - B^{\complement U} G(\cdot, y).$$

THEOREM 4. *A finely superharmonic function $p \geq 0$ on U is a fine potential if and only if p has the form*

$$p = \int G^U(\cdot, y)\, d\mu(y)$$

for some positive Borel measure μ on U, which is then uniquely determined.

The fine Riesz representation theorem involves the following generalization of non-negative finely harmonic functions. Assume, for simplicity, that the finely open set U is *regular*, that is, $\complement U$ is not thin at any of its points.

Definition 5. A finely superharmonic function $h \geq 0$ on U is *invariant* if and only if U is the union of an increasing sequence of finely open sets U_ν of fine closure contained in U such that

$$h(x) = \int h d\varepsilon_x^{\complement U_\nu}, \quad \text{all } x \in U_\nu,\ \nu \in \mathbf{N}.$$

Every invariant function h on U is finely harmonic in $[h < \infty]$ according to [29, Lemma 9.3]. A *finite valued* invariant function on U is therefore the same as a finely harmonic function ≥ 0 on U. The pointwise sum of any sequence of invariant functions on U is invariant, if finite q.e. It is an open problem whether every invariant function is the sum of a sequence of finely harmonic function ≥ 0; see Problem P2 in Collection of Problems, I, in these Proceedings. The mapping which to each finely open set $U \subset X$ assigns the convex cone of all invariant functions on U is a presheaf, but not a sheaf (follows easily from Lemma 2.5 and Remark 2, p.212 in [40]).

THEOREM 5. *Every finely superharmonic function $s \geq 0$ on a finely open set U in X decomposes uniquely as the sum*

$$s = p + h = G^U \mu + h$$

of a fine potential $p = G^U \mu$ and an invariant function h.

Theorems 4 and 5 were obtained in [40],[41] for a harmonic space satisfying not only Axiom D, but also other restrictive conditions (fullfilled in Example I). Shortly after, Boboc & Bucur [5] established a complete extension of these two theorems to the very general setting of a standard H-cone, with or without Axiom D.

5. DIFFERENTIABILITY PROPERTIES

We return to Example I. Combining a local extension property of finely superharmonic functions [29, Theorem 9.9] with the work of Mizuta [57] on fine differentiability of Riesz potentials, one obtains the following result, as noted in [38], see also Davie & Øksendal [17] for $n = 2$ with f finely harmonic.

THEOREM 6. *Every finely superharmonic function f on a finely open set $U \subset \mathbf{R}^n, n \geq 3$, has a fine differential $df(x) = \nabla f(x) \cdot dx$ at every point $x \in U$ not belonging to some set E of zero capacity with respect to the Riesz kernel $(x,y) \mapsto |x-y|^{1-n}$ (on $\mathbf{R}^n$). Explicitly, for every $x \in U \setminus E$ we have $f(x)$ finite, and there is a vector $\nabla f(x) \in \mathbf{R}^n$ such that*

$$\operatorname*{fine\,lim}_{y \to x,\, y \in U} \frac{f(y) - f(x) - \nabla f(x) \cdot (y - x)}{|y - x|} = 0.$$

In the case of the plane ($n = 2$) the exceptional set E has instead zero capacity with respect to the kernel $(x, y) \mapsto |x-y|^{-1} \log(|x-y|^{-1})$, x, y both in $\mathbf{R}^2$. This result is sharp (even for finely *harmonic* f) as shown by Davie and Øksendal [17].

In particular, finely (super)harmonic functions are finely differentiable in the above sense Lebesgue almost everywhere. A stochastic type of differentiability a.e. was obtained earlier (in the finely harmonic case) by Debiard & Gaveau [18]. There is no fine 2nd derivative in general.

As shown by Lyons [55], a finely harmonic function f in a fine domain U can be $\equiv 0$ in some fine neighbourhood of a point of U without being $\equiv 0$ in the whole of U.

6. Asymptotic paths for subharmonic funtions

A classical theorem of F. Iversen [50] from 1914 asserts that every non-constant entire function f admits a path Γ in $\mathbf{C}$ along which $f(z) \to \infty$ as $z \to \infty$. There is an extension of this in which $\log|f|$ is replaced by any *continuous* subharmonic function u on $\mathbf{R}^n$ $(n \geq 2)$. The proof involves the application of a well-known result of Phragmén-Lindelöf type to the components of the open sets $[u > \lambda]$, $\lambda \in \mathbf{R}$. Attempts were made during the sixties by W. Hayman and M.N.M. Talpur to remove the hypothesis of continuity, and they succeeded in the planar case $n = 2$; later also in the case $n > 2$, see [48]. The obstacle was that, for discontinuous u, the sets $[u > \lambda]$ are no longer open. However, all difficulties disappeared after the development of fine potential theory, because the sets $[u > \lambda]$ are always finely open—by definition of the fine topology. A sharpening of the fine boundary minimum principle, obtained in [31], led to the conclusion that u must be unbounded in each fine component of $[u > \lambda]$. The proof was completed in [32] by using the arcwise connectedness of fine domains, cf. §2 above. Subsequently, Carleson [13] showed by a direct, but complicated argument that Γ may even be taken as a sectionally polygonal line. This, however, is actually covered by the proof in [32] in view of the polygonal connectedness of fine domains as established afterwards by Lyons [53], cf. also [35].

7. Uniform approximation by harmonic or superharmonic functions

For any compact set $K \subset \mathbf{R}^n$ write

$$H_0(K) := \{u_{|K} \mid u \text{ harmonic in some open set } \supset K\},$$
$$S_0(K) := \{u_{|K} \mid u \text{ finite continuous superharmonic in some open set } \supset K\}.$$

Denoting by $C(K)$ the space of continuous functions $K \to \mathbf{R}$, and by K' the *fine interior* of K, we then have

THEOREM 7. *The uniform closure* $H(K) = \overline{H_0(K)}$, *resp.* $S(K) = \overline{S_0(K)}$, *consists of all* $u \in C(K)$ *such that* $u_{|K'}$ *is finely harmonic, resp. finely superharmonic.*

In the harmonic case this is due to Debiard and Gaveau [18], although it is implicitly contained in Deny [19] because the fine harmonicity of $u_{|K'}$ is equivalent to

$$u(x) = \int u \, d\varepsilon_x^{\complement K'} \quad \text{for all } x \in K',$$

by Lemma 9.3 and Theorem 9.4 in [29]. In the superharmonic case the theorem is due to Bliedtner & Hansen [3]; for an alternative proof see [39].

It is well known that $H(K)$ and $S(K)$ are *simplicial* (as a vector space, resp. a convex cone). The Choquet boundary is, in either case, the fine boundary $\partial_f K = K \setminus K'$; and the unique maximal measure majorizing a given measure μ on K in the Choquet ordering is the swept-out measure $\mu^{\complement K} = \mu^{\partial_f K}$. See e.g. Bliedtner & Hansen [3]. The above results have been extended to harmonic spaces and even to balayage spaces, see [3],[4].

Applying the local extension property [29, Theorem 9.9] and the Brelot property mentioned in §3 above, we derive from the above theorem the following corollary [30], valid also for a harmonic space satisfying Axiom D.

COROLLARY. *A function* u *defined in a finely open set* $U \subset \mathbf{R}^n, n \geq 2$, *is finely harmonic if and only if every point of* U *has a compact fine neighbourhood* $K \subset U$ *such that* $u_{|K} \in H(K) \quad (= \overline{H_0(K)})$.

There is a similar characterization of finite valued, finely superharmonic functions, obtained by replacing $H(K)$ by $S(K) \quad (= \overline{S_0(K)})$.

8. DUALITY AND APPROXIMATION IN DIRICHLET SPACE

In Example I, with $X =$ a Greenian domain in $\mathbf{R}^n$, we consider the classical Dirichlet space $L_1^2 = L_1^2(X)$, that is, the completion of $C_0^\infty(X)$ in the Dirichlet norm and associated inner product

$$\|f\|_1 = \|\nabla f\|_{L^2}, \quad (f \mid g)_1 = \int_X \nabla f \cdot \nabla g \, dx.$$

Thus L_1^2 is a real Hilbert space of distributions on X. Actually, each such distribution f is a locally Lebesgue integrable function which may even be chosen finely continuous q.e., and f is then uniquely determined q.e. See Deny and Lions [20]. For any set $E \subset X$ write

$$H_0(E) := \{f \in L_1^2 \mid f \text{ is harmonic in some open set} \supset E\},$$
$$S_0(E) := \{f \in L_1^2 \mid f \text{ is superharmonic in some open set} \supset E\}.$$

With closures in Dirichlet norm in L_1^2 indicated by a bar we then have [38]

$$\overline{H_0(E)} := H(E) = \{f \in L_1^2 \mid f \text{ is finely harmonic q.e. in } E'\},$$
$$\overline{S_0(E)} := S(E) = \{f \in L_1^2 \mid f \text{ is finely superharmonic q.e. in } E'\},$$

whereby E' denotes the fine interior of E.—Moreover: for $f \in L_1^2$ we have:

$$f \in H(E) \quad \Leftrightarrow \quad (f|\varphi)_1 = 0, \quad \forall \varphi \in L_1^2 \text{ with } \varphi = 0 \text{ q.e. in } X \setminus E,$$
$$f \in S(E) \quad \Leftrightarrow \quad (f|\varphi)_1 \geq 0, \quad \forall \varphi \in (L_1^2)_+ \text{ with } \varphi = 0 \text{ q.e. in } X \setminus E.$$

These latter results were initiated by Feyel & de La Pradelle [25],[26]. In the case $n = 2$ they lead to additional information regarding Khavin's theorem [47] on L^2-approximation by analytic functions, see [42].

9. FINELY HOLOMORPHIC FUNCTIONS

Let me close by describing the application of fine potential theory in the plane to an extension of classical complex analysis—the finely holomorphic functions, defined in finely open subsets of **C**. This was initiated in 1974, cf.[34]. It was furthered by Debiard & Gaveau [18] in the same year, and a break-through was achieved by Lyons [53], [54] in 1978. These three authors used probability. Alternative presentations—using only analytical tools, notably the Cauchy-Pompeiu transform, and containing further results—appeared shortly after [36], [37]; see also the quite recent survey [45].

Among several equivalent definitions of fine holomorphy the following [37] is perhaps the most natural one.

Definition 6. A function $f : U \to \mathbf{C}$ (U finely open in **C**) is said to be *finely holomorphic* if f is a fine C^1-function in the complex variable sense. Explicitly

$$f'(z) = \operatorname*{fine\,lim}_{w \to z} \frac{f(w) - f(z)}{w - z}$$

shall exist for every $z \in U$ and determine a finely continuous function $f' : U \to \mathbf{C}$, called the *fine derivative* of f.

It turns out that, in the above definition, one may equally well use the fine topology, not only for the independent variable $z \in U$, but also on the target **C**. The definition thus copies the classical definition by Cauchy, replacing now the standard topology on **C** throughout by the fine topology.

THEOREM 8. *If $f : U \to \mathbf{C}$ is finely holomorphic then so is the fine derivative f'.*

A finely holomorphic function f is thus infinitely finely differentiable, and all its fine derivatives $f^{(n)}$ are finely holomorphic.

This was first proved by Lyons [53], [54], who used 1. below as definition of fine holomorphy. With Definition 6 above, the theorem is found in [37].

The set of all finely holomorphic functions on a fine domain U is an *algebra*. If f is finely holomorphic in $U \subset \mathbf{C}$, and if v is finely harmonic in a finely open set $V \subset \mathbf{C}$, then $v \circ f$ is finely harmonic in $f^{-1}(V)$. From this it was deduced in [34] that the range of a non-constant finely holomorphic function is finely open; this was proved first by Debiard & Gaveau [18].

The *composite* of two finely holomorphic functions is clearly finely holomorphic, and the chain rule of (fine) differentiation applies. The *inverse* f^{-1} of an injective finely holomorphic function f is finely holomorphic with fine derivative $1/f' \circ f^{-1}$. Every finely holomorphic function f is *injective* in some fine neighbourhood of any point z at which $f'(z) \neq 0$ [45].

Each of the following alternative definitions of fine holomorphy of a function $f : U \to \mathbf{C}$ is equivalent to Definition 6 above. (A complex valued function f is said to be finely harmonic if Re f and Im f are both finely harmonic.)

1. f is finely harmonic and $\bar{\partial} f = 0$ a.e. (Debiard & Gaveau [18]; see also [36]).

2. f and $z \mapsto z f(z)$ are finely harmonic (Lyons [53]).

3. Every point of U has a compact fine neighbourhood $V \subset U$ such that $f_{|V} \in R(V)$ (the uniform closure of the space of restrictions to V of rational functions with poles off V).

4. Every point of U has a fine neighbourhood $V \subset U$ such that $f_{|V}$ can be extended to all of $\mathbf{C}$ as the Cauchy-Pompeiu transform

$$\hat{\varphi}(z) = \int \frac{1}{z - \varsigma} \varphi(\varsigma)\, dm(\varsigma)$$

of some $\varphi \in L^2(\mathbf{C})$ with $\varphi = 0$ a.e. in V (Fuglede [36]). (Here m denotes Lebesgue measure on $\mathbf{C}$). The space $L^2(\mathbf{C})$ can be replaced by $L^p(\mathbf{C})$ for any $p \in]1, \infty]$.

Of these definitions, 3. is the original definition from [34]. It easily implied the definitions 1. and 2. Until the work of Lyons it was not known that 1., 2., and 3. are in fact equivalent.

A very precise description of the local structure of a finely holomorphic function is given in the following theorem [36], [37].

THEOREM 9. *If $f : U \to \mathbf{C}$ is finely holomorphic then every point of U has a compact fine neighbourhood V in U satisfying a), b), c) below.*

a) (Approximation by rational functions). *There exists a sequence of rational functions f_j with poles off V such that, for each integer $n \geq 0$, the n'th derivative $f_j^{(n)}$ converges uniformly on V to the n'th fine derivative $f^{(n)}$ as $j \to \infty$.*

b) (Asymptotic Taylor expansion of f and its fine derivatives). *For any $m, n \geq 0$ there is a constant $A_{m,n}$ such that the inequality*

$$\left| f^{(n)}(w) - \sum_{k=0}^{m-1} \frac{1}{k!}(w-z)^k f^{(n+k)}(z) \right| \leq A_{m,n}|w-z|^m$$

holds for $(z, w) \in V \times V$.

c) (C^∞ extension). *There exists a usual C^∞ function $\tilde{f} : \mathbf{C} \to \mathbf{C}$ such that $\tilde{f} = f$ on V, and*

$$\partial^n \tilde{f} = f^{(n)}, \quad \partial^n \overline{\partial}^{n'} \tilde{f} = 0 \quad \text{in } V$$

for every $n \geq 0$ and $n' \geq 1$.

In contrast to finely harmonic functions (cf. end of §5), finely holomorphic functions have the following property of *quasi-analyticity* [36]:

THEOREM 10. *A finely holomorphic function f, defined on a fine domain U, is uniquely determined by the sequence of its fine derivatives $f^{(n)}(z_0), n \geq 0$, at any given point $z_0 \in U$.*

A finely holomorphic function f on a fine domain has at most countably many *zeros* (unless $f \equiv 0$). [36]

Certain classical results on *value distribution* of holomorphic functions, notably Radó's theorem, have been extended in a sharper form involving fine cluster sets. [43]

An extension of the above theory of finely holomorphic functions to functions of *several complex variables* has been made [44], but is not quite satisfactory because it uses the product fine topology, which is not biholomorphically invariant.

10. RELATION TO BOREL'S MONOGENIC FUNCTIONS

The first attempt at introducing generalized holomorphic functions, defined in sets that are not open, was made by Borel, beginning in 1892. His so-called monogenic functions were defined on sets D that are countable increasing unions of certain Swiss cheeses D_h. A complex valued function f, defined on such a *Borel domain* $D = \bigcup D_h$, is called *monogenic* if the restriction of f to each Swiss cheese D_h is a complex C^1 function, so that $\overline{\partial} f = 0$ on each D_h.

Borel established the infinite differentiability of a monogenic function relative to each D_h. His main result was that every monogenic function f on a Borel domain $D = \bigcup D_h$ is uniquely determined by the sequence of its derivatives $f^{(n)}(z_0)$, $n \geq 0$, at a single point z_0. Not every point of D can serve as such a point z_0 of uniqueness, however, but Borel constructed a certain dense subset of D—a so-called reduced domain—such that every point of that is a point of uniqueness for every monogenic function defined on D. It was this result which above all justified his theory.

Borel declares in his book [8] from 1917 that there undoubtedly exist domains more general than those considered by him on which a still further generalization of monogenic functions is possible, and that one will probably never succeed in fixing the exact limits beyond which such an extension is no longer possible.

The finely holomorphic functions, defined in domains for the Cartan fine topology on $\mathbf{C}$, do constitute such an extension—and a very natural one—of Borel's monogenic functions. It is in fact easily verified by use of the Wiener criterion (cf. Mastrangelo-Dehen & Dehen [56]) that

—Every Borel domain is a fine domain, indeed a very special kind of fine domain: a usual domain less a polar set.

—Every monogenic function on a Borel domain $D = \bigcup D_h$ is finely holomorphic in D less some polar set independent of the function (e.g. finely holomorphic in $\bigcup D'_h$, the union of the fine interiors of the Swiss cheeses D_h used in the construction of D).

From the point of view of the theory of finely holomorphic functions, the need for removing such a polar set explains why not every point of a Borel domain is a point of uniqueness for monogenic functions, although every point of a fine domain is a point of uniqueness for finely holomorphic functions.

REFERENCES

1. H. Bauer, "Harmonische Räume und ihre Potentialtheorie," Lecture Notes in Mathematics No. 22, Springer, Berlin/Heidelberg/New York, 1966.
2. H. Bauer, *Harmonic spaces - a survey*, "Conferenze del Seminario de Matematica dell' Università di Bari, No. 197," 1984.
3. J. Bliedtner & W. Hansen, *Simplicial cones in potential theory, I*, Inventiones Math. **29** (1975), 83-110; *II*, **46** (1978), 255-275.
4. J. Bliedtner & W. Hansen, "Potential Theory. An Analytic and Probabilistic Approach to Balayage," Springer, Berlin..., 1986.
5. N. Boboc & Gh. Bucur, *Green potentials on standard H-cones*, Revue Roumaine Math. Pures Appl. **32** (1987), 293-320.
6. N. Boboc, Gh. Bucur & A. Cornea, *H-cones and potential theory*, Ann. Inst. Fourier Grenoble **25: 3-4** (1975), 71-108.
7. N. Boboc, Gh. Bucur & A. Cornea, "Order and Convexity in Potential Theory: H-cones," Lecture Notes in Mathematics No. 853, Springer, Berlin/Heidelberg/New York, 1980.
8. É. Borel, "Leçons sur les fonctions monogènes uniformes d'une variable complexe," Gauthier-Villars, Paris, 1917.
9. M. Brelot, *Points irréguliers et transformations continues en théorie du potentiel*, J. de Math. **19** (1940), 319-337.
10. M. Brelot, *La convergence des fonctions surharmoniques et des potentiels généralisés*, C.R. Acad. Sci. Paris **246** (1958), 2709-2712.

11. M. Brelot, *Recherches axiomatiques sur un théorème de Choquet concernant l'effilement*, Nagoya J. Math. **30** (1967), 39-46.
12. M. Brelot, "On Topologies and Boundaries in Potential Theory," Lecture Notes in Mathematics No. 175, Springer, Berlin/Heidelberg/New York, 1971.
13. L. Carleson, *Asymptotic paths for subharmonic functions in R^n*, Ann. Acad. Sci. Fennicæ (A.I) **2** (1976), 35-39.
14. H. Cartan, *Théorie du potentiel Newtonien: énergie, capacité, suites de potentiels*, Bull. Soc. Math. France **73** (1945), 74-106.
15. S. Chacrone, (Personal communication).
16. C. Constantinescu & A. Cornea, "Potential Theory on Harmonic Spaces," Springer, Berlin/Heidelberg/New York, 1972.
17. A.M. Davie & B. Øksendal, *Analytic capacity and differentiability properties of finely harmonic functions*, Acta Math. **149** (1982), 127-152.
18. A. Debiard & B. Gaveau, *Potential fin et algèbres de fonctions analytiques, I*, J. Functional Anal. **16** (1974), 289-304; *II*, **17** (1974), 296-310.
19. J. Deny, *Systèmes totaux de fonctions harmoniques*, Ann. Inst. Fourier Grenoble **1** (1949), 103-113.
20. J. Deny & J.L. Lions, *Les espaces du type de Beppo Levi*, Ann. Inst. Fourier Grenoble **5** (1953-54), 305-370.
21. J.L. Doob, *Semimartingales and subharmonic functions*, Trans. Amer. Math. Soc. **77** (1954), 86-121.
22. J.L. Doob, *Probability methods applied to the first boundary value problem*, "Proc. 3rd Berkeley Symp. Math. Stat. Probab., 1954-55," Berkeley, 1956, pp. 49-80.
23. J.L. Doob, *Applications to analysis of a topological definition of smallness of a set*, Bull. Amer. Math. Soc. **72** (1966), 579-600.
24. J.L. Doob, "Classical Potential Theory and Its Probabilistic Counterpart," Springer, New York/Berlin/Heidelberg/Tokyo, 1984.
25. D. Feyel & A. de La Pradelle, *Espaces de Sobolev sur les ouverts fins*, C.R. Acad. Sci. Paris, Sér. A **280** (1975), 1125-1127.
26. D. Feyel & A. de La Pradelle, *Le rôle des espaces de Sobolev en topologie fine*, "Séminaire de Théorie du Potentiel, Paris, No. 2," Lecture Notes in Mathematics No. 563, Springer, Berlin/Heidelberg/New York, 1976, pp. 43-60.
27. B. Fuglede, *Fine connectivity and finely harmonic functions*, "Actes Congr. Internat. Math., Nice 1970, Tome 2," Gauthier-Villars, Paris, 1971, pp. 513-519.
28. B. Fuglede, *Connexion en topologie fine et balayage des mesures*, Ann. Inst. Fourier Grenoble **21.3** (1971), 227-244.
29. B. Fuglede, "Finely Harmonic Functions," Lecture Notes in Mathematics No. 289, Springer, Berlin/Heidelberg/New York, 1972.
30. B. Fuglede, *Fonctions harmoniques et fonctions finement harmoniques*, Ann. Inst. Fourier Grenoble **24.4** (1974), 77-91.
31. B. Fuglede, *Boundary minimum principles in potential theory*, Math. Ann. **210** (1974), 213-226.
32. B. Fuglede, *Asymptotic paths for subharmonic functions*, Math. Ann. **213** (1975), 261-274.
33. B. Fuglede, *Sur la fonction de Green pour un domaine fin*, Ann. Inst. Fourier Grenoble **25.3-4** (1975), 201-206.
34. B. Fuglede, *Finely harmonic mappings and finely holomorphic functions*, Ann. Acad. Sci. Fennicæ (A.I.) **2** (1976), 113-127.
35. B. Fuglede, *Asymptotic paths for subharmonic functions and polygonal connectedness of fine domains*, "Séminaire de Théorie du Potentiel, Paris, No. 5," Lecture Notes in Mathematics No. 814, Springer, Berlin/Heidelberg/New York, 1980, pp. 97-116.

36. B. Fuglede, *Sur les fonctions finement holomorphes*, Ann. Inst. Fourier Grenoble **31.4** (1981), 57-88.
37. B. Fuglede, *Fine topology and finely holomorphic functions*, "Proc. 18th Scand. Congr. Math., Aarhus 1980," Birkhäuser, 1981, pp. 22-38.
38. B. Fuglede, *Fonctions BLD et fonctions finement surharmoniques*, "Séminaire de Théorie du Potentiel, Paris, No. 6," Lecture Notes in Mathematics No. 906, Springer, Berlin/Heidelberg/New York, 1982, pp. 126-157.
39. B. Fuglede, *Localization in fine potential theory and uniform approximation by subharmonic functions*, J. Functional Anal. **49** (1982), 57-72.
40. B. Fuglede, *Integral representation of fine potentials*, Math. Ann. **262** (1983), 191-214.
41. B. Fuglede, *Représentation intégrale des potentiels fins.*, C.R. Acad. Sci. Paris **300** (1985), 129-132.
42. B. Fuglede, *Complements to Havin's theorem on L^2-approximation by analytic functions*, Ann. Acad. Sci. Fennicæ (A.I.) **10** (1985), 187-201.
43. B. Fuglede, *Value distribution of harmonic and finely harmonic morphisms and applications in complex analysis*, Ann. Acad. Sci. Fennicæ (A.I.) **11** (1986), 111-136.
44. B. Fuglede, *Fonctions finement holomorphes de plusieurs variables - un essai*, "Séminaire d'Analyse P. Lelong-P. Dolbeault-H. Skoda 1983/85," Lecture Notes in Mathematics No. 1198, Springer, Berlin/Heidelberg/New York/Tokyo, 1986, pp. 133-145.
45. B. Fuglede, *Finely holomorphic functions - a survey*, Revue Roumaine Math. Pures Appl. (to appear).
46. T.W. Gamelin & T.J. Lyons, *Jensen measures for R(K)*, J. London Math. Soc. **27** (1983), 317-330.
47. V.P. Havin, *Approximation in the mean by analytic functions*, Soviet Math. Dokl. **9** (1968), 245-248; (Russian), Dokl. Akad. Nauk SSSR **178-5** (1968).
48. W.K. Hayman & P.B. Kennedy, "Subharmonic Functions I," Academic Press, London/New York/San Francisco, 1976.
49. R.-M. Hervé, *Recherches axiomatiques sur la théorie des fonctions surharmoniques et du potentiel*, Ann. Inst. Fourier Grenoble **12** (1962), 415-571.
50. F. Iversen, "Recherches sur les fonctions inverses des fonctions méromorphes," Thèse, Helsingfors, 1914.
51. J. Lukeš & J. Malý, *Fine hyperharmonicity without axiom D*, Math. Ann. **261** (1982), 299-306.
52. J. Lukeš, J. Malý & L. Zajíček, "Fine Topology Methods in Real Analysis and Potential Theory," Lecture Notes in Mathematics No. 1189, Springer, Berlin..., 1986.
53. T.J. Lyons, *Finely holomorphic functions*, J. Functional Anal. **37** (1980), 1-18.
54. T.J. Lyons, *A theorem in fine potential theory and applications to finely holomorphic functions*, J. Functional Anal. **37** (1980), 19-26.
55. T. Lyons, *Finely harmonic functions need not be quasi-analytic*, Bull. London Math. Soc. **16** (1984), 413-415.
56. M. Mastrangelo-Dehen & D. Dehen, *Étude des fonctions finement harmoniques sur des ouverts fins*, C.R. Acad. Sci., Paris, Sér. A **275** (1972), 361-364.
57. Y. Mizuta, *Fine differentiability of Riesz potentials*, Hiroshima Math. J. **8** (1978), 505-514.
58. Nguyen-Xuan-Loc & T. Watanabe, *Characterization of fine domains for a certain class of Markov processes with applications to Brelot harmonic spaces*, Z. Wahrscheinlichkeitstheorie verw. Geb. **21** (1972), 167-178.
59. N. Wiener, *The Dirichlet problem*, J. Math. Phys. Mass. Inst. Tech. **3** (1924), 127-146.

BALAYAGE SPACES - A NATURAL SETTING FOR POTENTIAL THEORY

Wolfhard Hansen
Fakultät für Mathematik, Universität Bielefeld
Universitätsstraße, D-4800 Bielefeld 1

The introduction of harmonic spaces by M. BRELOT [21] - [24] and H. BAUER [2] - [4] was based on fundamental properties which solutions of certain linear differential equations of second order have in common. Papers of P.A. MEYER [42] and N. BOBOC - C. CONSTANTINESCU - A. CORNEA [17] showed that, by analogy with the relation between classical potential theory and Brownian motion, every harmonic space admits a corresponding Markov process with continuous paths. However, whether the paths are continuous or not plays no important role in the potential theory of Markov processes as presented in the book of R.M. BLUMENTHAL - R.K. GETOOR [15]. Moreover, even within the theory of harmonic spaces it turned out that in some respect hyperharmonic functions and potentials are more important than harmonic functions. In fact, the monograph of C. CONSTANTINESCU - A. CORNEA [29] on potential theory on harmonic spaces contains a chapter where basic properties of hyperharmonic functions are used in an axiomatic way to study important properties of balayage. The authors used an axiom of upper directed sets, an axiom of lower semicontinuous regularization, and an axiom of natural decomposition. On the other hand, G. MOKOBODZKI and D. SIBONY [43,45,47] studied cones of continuous potentials in an abstract setting where natural decomposition and additional continuity properties play a central role.

The concept of a balayage space is obtained by combining the stated three fundamental properties of positive hyperharmonic functions with the existence of a cone of continuous potentials such that every positive hyperharmonic function is the supremum of a sequence of continuous potentials. This notion was introduced in [12] where it helped to characterize the class of Markov processes associated with harmonic spaces. It was shown that balayage spaces are closely related to sub-Markov semigroups where every excessive function is the upper envelope of its continuous excessive minorants. However, there are additional important reasons for the consideration of balayage spaces. First of all, harmonic spaces, Riesz potentials and Markov chains are covered, and the class of balayage spaces has remarkable permanence properties with respect to restriction on subspaces, subordination by convolution semigroups, products, and images. Moreover, balayage spaces allow a clear and direct presentation of results on balayage of functions and balayage of measures. This can be carried out without losing known results for harmonic spaces and without more complicated proofs. In particular, different types of Dirichlet problems can be treated in an elegant manner.

The starting point of this article will be a discussion of balayage spaces and their relationship to resolvents, semigroups and Markov processes. It will then be pointed out how families of harmonic kernels associated with balayage spaces may be characterized and which additional properties yield harmonic spaces. Permanence properties leading to further examples will be looked at in detail. A short treatment of Dirichlet problems will then finish the paper. Almost all of the material is taken from the book of J. BLIEDTNER - W. HANSEN [14] where proofs and further details can be found.

1. Balayage Spaces

In the following let X be a locally compact space with countable base. Let $\mathcal{B}(X)$ ($\mathcal{C}(X)$ resp.) denote the set of all Borel measurable numerical (continuous real resp.) functions on X. We shall write $A \in \mathcal{B}(X)$ if $A \subset X$ and $1_A \in \mathcal{B}(X)$. Let $\mathcal{C}_o(X) = \{f \in \mathcal{C}(X) : f$ vanishes at infinity$\}$ and $\mathcal{K}(X) = \{f \in \mathcal{C}(X) : f$ has compact support supp $(f)\}$. Finally, let $\mathcal{M}_+(X)$ be the set of all (positive) Radon measures on X.

Let $\mathcal{W}$ be a convex cone of positive l.s.c. numerical functions on X. The coarsest topology on X which is finer than the initial topology and for which all functions of $\mathcal{W}$ are continuous will be called the ($\mathcal{W}$-) *fine topology*. Topological notions with respect to the fine topology are distinguished by the term "fine(ly)" or affix "f" from those pertaining to the initial topology on X. It can easily be seen that every $x \in X$ has a fundamental system of fine neighborhoods which are compact in the initial topology. In particular, X endowed with the fine topology is a Baire space.

$(X,\mathcal{W})$ is called a *balayage space* if the following axioms are satisfied:

(B_1) $\sup v_n \in \mathcal{W}$ for every increasing sequence (v_n) in $\mathcal{W}$.

(B_2) $\widehat{\inf \mathcal{V}}^f \in \mathcal{W}$ for every subset $\mathcal{V}$ of $\mathcal{W}$.

(B_3) If $u, v', v'' \in \mathcal{W}$ such that $u \leq v' + v''$, there exist $u', u'' \in \mathcal{W}$ such that $u = u' + u''$, $u' \leq v'$, $u'' \leq v''$.

(B_4) For every $v \in \mathcal{W}$, $v = \sup\{u \in \mathcal{W} \cap C(X) : u \leq v\}$. $\mathcal{W}$ is linearly separating, and there exist strictly positive $u_o, v_o \in \mathcal{W} \cap C(X)$ such that $\frac{u_o}{v_o} \in C_o(X)$.

The functions in $\mathcal{W}$ will be called positive *hyperharmonic functions* on X, the functions in

$$\mathcal{P} := \{p \in \mathcal{W} \cap C(X) : \tfrac{p}{v} \in C_o(X) \text{ for some } v \in \mathcal{W} \cap C(X),\ v > 0\}$$

will be called continuous real *potentials* on X.

A convex cone $F \subset C^+(X)$ is called a *function cone* if F is linearly separating $(x,y \in X,\ x \neq y,\ \lambda \in \mathbb{R}_+ \Rightarrow \exists f \in F : f(x) \neq \lambda f(y))$ and if for each $f \in F$ there exists $g \in F$ such that $g > 0$ and $\frac{f}{g} \in C_o(X)$. If $(X,\mathcal{W})$ is a balayage space, then $\mathcal{P}$ is a min-stable function cone,

$$\mathcal{W} = S(\mathcal{P}) := \{\sup p_n : (p_n) \subset \mathcal{P} \text{ increasing}\},$$

and $\mathcal{P}$ is obviously the greatest function cone in $\mathcal{W}$. Moreover, assuming (B_1), (B_2), (B_3) we conclude that $(X,\mathcal{W})$ satisfies (B_4) if and only if there exists a function cone F such that $\mathcal{W} = S(F)$. Let us note that given a min-stable function cone F and an additive, increasing functional $T : F \to \mathbb{R}_+$, there exists a unique measure $\mu \in M_+(X)$ such that $T(f) = \int f\, d\mu$ for every $f \in F$.

1.1. REMARKS. 1. (B_1) implies that $\sup \mathcal{V} \in \mathcal{W}$ for every increasingly filtered subset $\mathcal{V}$ of $\mathcal{W}$.

2. Suppose (B_2) and let $\mathcal{V} \subset \mathcal{W}$. Then $\widehat{\inf \mathcal{V}} = \widehat{\inf \mathcal{V}}^f$ and the set $\{\widehat{\inf \mathcal{V}} < \inf \mathcal{V}\}$ is finely meager.

3. Suppose (B_2) and define

$$R_f := \inf\{v \in \mathcal{W} : v \geq f\}$$

for every $f : X \to \overline{\mathbb{R}}$. Clearly, $R_f = \hat{R}_f \in \mathcal{W}$ if f is finely l.s.c. Moreover, (B_3) holds if and only if for any two functions $u,v \in \mathcal{W}$ there exists $w \in \mathcal{W}$ such that $u = R_f + w$ where $f = (u-v)^+$ on $\{v < \infty\}$, $f = 0$ on $\{v = \infty\}$.

The following results will show that many sub-Markov resolvents and sub-Markov semigroups are associated with balayage spaces. Let us first recall that a *sub-Markov resolvent* on X is a family $\mathbb{V} = (V_\lambda)_{\lambda > o}$ of kernels V_λ on X such that $\lambda V_\lambda 1 \leq 1$ and $V_\lambda - V_\mu = (\mu - \lambda) V_\lambda V_\mu$ for all $\lambda, \mu > 0$. Its potential kernel V_o is given by $V_o = \sup_{\lambda > o} V_\lambda$,

$$E_{\mathbb{V}} := \{u \in \mathcal{B}^+(X) : \sup_{\lambda > o} \lambda V_\lambda u = u\}$$

is the set of $\mathbb{V}$-excessive functions, and

$$S_{\mathbb{V}} := \{u \in \mathcal{B}^+(X) : \sup_{\lambda > o} \lambda V_\lambda u \leq u\}$$

is the set of $\mathbb{V}$-supermedian functions. A *sub-Markov semigroup* $\mathbb{P}$ on X

is a family $\mathbb{P} = (P_t)_{t>0}$ of kernels P_t on X such that $P_t 1 \leq 1$ and $P_s P_t = P_{s+t}$ for all $s,t > 0$. The corresponding set $E_{\mathbb{P}}$ of $\mathbb{P}$-excessive functions is defined by

$$E_{\mathbb{P}} = \{u \in \mathcal{B}^+(X) : \sup_{t>0} P_t u = u\} .$$

If $\mathbb{V}$ is the resolvent of $\mathbb{P}$, i.e., if for every $\lambda > 0$, $f \in \mathcal{B}^+(X)$ and $x \in X$,

$$V_\lambda f(x) = \int_0^\infty e^{-\lambda t} P_t f(x)\, dt ,$$

then $E_{\mathbb{P}} = E_{\mathbb{V}}$. Let us finally note that a *strong Feller kernel* on X is a bounded kernel V on X satisfying $V(\mathcal{B}_b(X)) \subset C_b(X)$.

If $(X, E_{\mathbb{V}})$ is a balayage space then $E_{\mathbb{V}}$ is min-stable by (B_2) and hence $\lim_{\lambda\to\infty} \lambda V_\lambda f = f$ for every $f \in K(X)$ by (B_4). Now suppose conversely that $\lim_{\lambda\to\infty} \lambda V_\lambda f = f$ for every $f \in K(X)$ and that $E_{\mathbb{V}}$ satisfies (B_4). Then $E_{\mathbb{V}}$ is a convex cone of l.s.c. functions which clearly satisfies (B_1). Proceeding as in [29] for harmonic measures it is easily shown that $\lim_{\lambda\to\infty} \lambda V_\lambda(x,U) = 1$ for every $E_{\mathbb{V}}$-open set $U \in \mathcal{B}(X)$ and every $x \in U$. Thus $E_{\mathbb{V}}$ is the set of all l.s.c. functions in $S_{\mathbb{V}}$. Fundamental properties of supermedian functions discovered by G. MOKOBODZKI and the topological lemma of Choquet finally yield that $E_{\mathbb{V}}$ satisfies (B_2) and (B_3). Whence the following result (J. BLIEDTNER - W. HANSEN [12,14]).

1.2. THEOREM. Let $\mathbb{V} = (V_\lambda)_{\lambda>0}$ be a sub-Markov resolvent on X. Then $(X, E_{\mathbb{V}})$ is a balayage space if and only if $\lim_{\lambda\to\infty} \lambda V_\lambda f = f$ for every $f \in K(X)$ and $E_{\mathbb{V}}$ satisfies (B_4).

1.3. COROLLARY. Let $\mathbb{P} = (P_t)_{t>0}$ be a sub-Markov semigroup on X. Then $(X, E_{\mathbb{P}})$ is a balayage space if and only if $\lim_{t\to 0} P_t f = f$ for every $f \in K(X)$ and $E_{\mathbb{P}}$ satisfies (B_4).

If $\mathbb{V} = (V_\lambda)_{\lambda>0}$ is a strong Feller resolvent then it is much easier to verify (B_4): It suffices to find $u,v \in E_{\mathbb{V}} \cap C(X)$ such that $u,v > 0$, $\frac{u}{v} \in C_0(X)$ and to show that $E_{\mathbb{V}}$ is linearly separating or the potential kernel V_0 of $\mathbb{V}$ is proper (i.e., $V_0 1_K$ is bounded for every compact K).

1.4. EXAMPLES. In particular, the following semigroups yield balayage spaces.

1. $X = \mathbb{R}^m$, $m \geq 3$, $\mathbb{P}$ Brownian semigroup, i.e. $P_t f(x) =$

$$P_t f(x) = \left(\frac{1}{2\pi t}\right)^{m/2} \int e^{-\frac{\|x-y\|^2}{2t}} f(y)\, dy) .$$

2. $X = \mathbb{R}$, $\mathbb{T} = (T_t)_{t>0}$ semigroup of uniform translation (to the left), i.e., $T_t f(x) = f(x-t)$.

3. X discrete, at most countable, P kernel on X such that $P1 \leq 1$ and $S_P := \{u \in B^+(X) : Pu \leq u\}$ separates the points of X, $\mathbb{P} = (P_t)_{t>0}$ with $P_t = e^{-t} \sum_{k=0}^{\infty} \frac{t^k}{k!} P^k$ (pseudo-Poisson semigroup). Note that $E_{\mathbb{P}} = S_P$!

Suppose now that $(X, \mathcal{W})$ is an arbitrary balayage space. Given $A \subset X$ and $u \in \mathcal{W}$ we define

$$R_u^A = R_{u1_A} = \inf\{v \in \mathcal{W} : v \geq u \text{ on } A\}, \quad \hat{R}_u^A = \widehat{R_u^A}.$$

Let us mention some basic facts on balayage of functions and measures.

<u>1.5. PROPOSITION.</u> For every subset A of X and all functions $u,v \in \mathcal{W}$, $R_{u+v}^A = R_u^A + R_v^A$, $\hat{R}_{u+v}^A = \hat{R}_u^A + \hat{R}_v^A$ (C. CONSTANTINESCU - A. CORNEA [28], N. BOBOC - C. CONSTANTINESCU - A. CORNEA [16]).

<u>1.6. PROPOSITION.</u> Let (A_n) be an increasing sequence of subsets of X and let (u_n) be an increasing sequence in $\mathcal{W}$, let $A = \bigcup_{n=1}^{\infty} A_n$ and $u = \sup u_n$. Then $\sup R_{u_n}^{A_n} = R_u^A$, $\sup \hat{R}_{u_n}^{A_n} = \hat{R}_u^A$ (C. CONSTANTINESCU - A. CORNEA [29]).

The properties of the mapping $u \mapsto \hat{R}_u^A$ stated in the preceding two propositions lead to balayage of measures. Let $\mathcal{M}(\mathcal{P})$ denote the set of all $\mu \in \mathcal{M}_+(X)$ such that $\int p \, d\mu < \infty$ for some strictly positive $p \in \mathcal{P}$. Given $\mu \in \mathcal{M}(\mathcal{P})$ and $A \subset X$, there exists a unique measure $\mu^A \in \mathcal{M}(\mathcal{P})$ such that $\int u \, d\mu^A = \int \hat{R}_u^A \, d\mu$ for every $u \in \mathcal{W}$. μ^A is called *balayage* of μ on A. The *base* of a subset A of X is defined by

$$b(A) = \{x \in X : \varepsilon_x^A = \varepsilon_x\}.$$

If $p \in \mathcal{P}$ is strict, then $b(A) = \{\hat{R}_p^A = p\}$, hence $b(A)$ is a G_δ-set.

Since the functions $\hat{R}_u^A$, $u \in \mathcal{W}$, depend only on the values of u on A it is easy to see that the measure μ^A is supported by the closure of A. In fact, a careful analysis shows that μ^A is even supported by the fine closure of A, a result which is an important tool in many proofs.

<u>1.7. PROPOSITION.</u> Let $A \subset X$. Then $\bar{A}^f = A \cup b(A)$ and $(\mu^A)_*(\complement \bar{A}^f) = 0$ for every $\mu \in \mathcal{M}(\mathcal{P})$ (M. BRELOT [20], R.M. HERVÉ [39], C. CONSTANTINESCU [26]).

<u>1.8. PROPOSITION.</u> Let $v \geq 0$ be a l.s.c. numerical function on X. Then $v \in \mathcal{W}$ if and only if for every $x \in X$ and every neighborhood U

of x there exists a subset V of U such that $x \in V$, $\varepsilon_x^{\complement V} \neq \varepsilon$ and $\int v \, d\varepsilon_x^{\complement V} \leq v(x)$.

As already for harmonic spaces we need several classes of exceptional sets. A subset A of X is called *polar* if $\varepsilon_x^A = 0$ for every $x \in X$ (M. BRELOT [19]). It is *totally thin* if $b(A) = \emptyset$, and it is *semipolar* if it is a countable union of totally thin sets (M. BRELOT [25]). In classical potential theory polar and semipolar sets coincide. However, already in example 1.4.2 there are many semipolar sets which are not polar (the empty set is the only polar subset, every finite set is totally thin, every countable set is semipolar). Semipolar sets occur in a natural way passing from R_u^A to $\hat{R}_u^A$: A set S is semipolar if and only if there exists a subset $\mathcal{V}$ of $\mathcal{W}$ such that $S = \{\widehat{\inf \mathcal{V}} < \inf \mathcal{V}\}$. In particular, for every $A \subset X$ the set $A \setminus b(A)$ is semipolar.

In order to treat the weak Dirichlet problem (see 6.5) the essential base $\beta(A)$ of subsets A of X is needed (J. BLIEDTNER - W. HANSEN [10], W. HANSEN [38]). By definition $\beta(A)$ is the set of all points x in X such that for every fine neighborhood V of x the set $A \cap V$ is not semipolar. Using the quasi-Lindelöf property of the fine topology it is easily seen that $\beta(A)$ is the smallest finely closed subset F of X such that $A \setminus F$ is semipolar. Moreover, $\beta(\beta(A)) = b(\beta(A)) = \beta(A)$. If A is finely closed, then $\beta(A)$ is the greatest subset B of A such that $B \subset b(B)$.

2. Resolvents and Semigroups on Balayage Spaces

In the preceding section we stated that many resolvents and semigroups are associated with balayage spaces. The converse holds as well, and many mathematicians have contributed to the main theorem: P.A. MEYER [42] for BRELOT's axiomatic theory, N. BOBOC - C. CONSTANTINESCU - A. CORNEA [17], W. HANSEN [34], [35], J.C. TAYLOR [48], [49] for general harmonic spaces, G. MOKOBODZKI - D. SIBONY [45] for adapted potential cones. A corresponding result for harmonic groups is due to J. BLIEDTNER [9].

Let $(X, \mathcal{W})$ be a balayage space. Given $p \in \mathcal{P}$, there exists a unique kernel V such that $V1 = p$ and $Vf \in \mathcal{P}$, $R_{Vf}^{\mathrm{supp}(Vf)} = Vf$ for every $f \in \mathcal{B}^+(X)$ (R.-M. HERVÉ [40], G. MOKOBODZKI - D. SIBONY [44]). V is called the *potential kernel* associated with p. Let us note that $V(\mathcal{B}_b(X)) \subset \mathcal{C}(X)$, hence V is a strong Feller kernel if p is bounded. The potential p is called a *strict* potential if any two measures $\mu,\nu \in M_+(X)$ satisfying $\mu(q) \leq \nu(q)$ for every $q \in P$ and $\mu(p) = \nu(p) < \infty$ coincide. Strict potentials are easily constructed forming sums $p = \sum_{n=1}^{\infty} \alpha_n p_n$ where (p_n) is a sequence in P separating $M_+(X)$. So the following results yield the existence of many resolvents and semigroups associated with (X,W) .

2.1. THEOREM. Let $1 \in W$ and $p \in P_b$. Then there exists a unique sub-Markov resolvent $\mathbb{V}$ such that $E_{\mathbb{V}} \subset W \subset S_{\mathbb{V}}$ and $V_o 1 = p$. V_o is the potential kernel of p . Moreover, $E_{\mathbb{V}} = W$ if and only if p is strict.

Corresponding semigroups can be obtained following R.K. GETOOR [33].

2.2. THEOREM. Let $\mathbb{V} = (V_\lambda)_{\lambda > o}$ be a sub-Markov resolvent on X such that $E_{\mathbb{V}} \subset S(F) \subset S_{\mathbb{V}}$ for some function cone F . Then there exists a unique sub-Markov semigroup $\mathbb{P} = (P_t)_{t > o}$ on X such that $\mathbb{V}$ is the resolvent of $\mathbb{P}$ and $t \mapsto P_t q$ is right continuous for every $q \in S(F)$.

2.3. COROLLARY. Let $1 \in W$ and let $p \in P_b$ be a strict potential. Then there exists a unique sub-Markov semigroup $\mathbb{P} = (P_t)_{t > o}$ on X such that $E_{\mathbb{P}} = W$ and $V_o 1 = p$. V_o is the potential kernel associated with p . Furthermore, $P_t(P) \subset P$ and $P_t(K(X)) \subset C_b(X)$ for every $t > 0$.

In many concrete examples it is fairly easy to construct associated semigroups $\mathbb{P}$ not even satisfying $V_\lambda(K(X)) \subset C(X)$. Nevertheless there will always be a corresponding Hunt process.

2.4. THEOREM. Let $\mathbb{P}$ be a sub-Markov semigroup on X such that $E_{\mathbb{P}} = W$. Then there exists a Hunt process X with state space X having $\mathbb{P}$ as transition function.

This result is based on the fact that the existence of many continuous excessive functions allows one to apply standard supermartingale arguments in order to obtain the necessary regularity properties of a corresponding process (J. BLIEDTNER - W. HANSEN [14]). Without going into any details let us mention some consequences for the probabilistic interpretation of potential-theoretic notions.

Let $X = (\Omega, \mathfrak{M}, \mathfrak{M}_t, X_t, \Theta_t, P^x)$ be a Hunt process with transition function $\mathbb{P}$ such that $E_{\mathbb{P}} = W$ and let $A \in \mathcal{B}(X)$. Then the *first hitting* time T_A defined by

$$T_A(\omega) := \inf\{t > 0 : X_t(\omega) \in A\} \quad (\omega \in \Omega)$$

is a stopping time and, for every $x \in X$ and $B \in \mathcal{B}(X)$,

$$P^x[X_{T_A} \in B] = \varepsilon_x^A(B) .$$

The set A is polar if and only if $T_A = \infty$ a.s. Moreover,

$$b(A) = \{x \in X : T_A = 0 \ \ P^x\text{-a.s.}\} .$$

In particular, $b(A)$ is totally thin if and only if $T_A > 0$ a.s. The *penetration time* τ_A is defined by

$$\tau_A(\omega) := \inf\{t > 0 : \{s \in [0,t] : X_s(\omega) \in A\} \text{ is uncountable}\} .$$

It can be shown that

$$\tau_A = T_{\beta(A)} \quad \text{a.s.}$$

In particular,

$$\beta(A) = \{x \in X : \tau_A = 0 \ \ P^x\text{-a.s.}\} .$$

Moreover, A is semipolar if and only if $\tau_A > 0$ a.s. or, equivalently, if and only if $\tau_A = \infty$ a.s.

3. Balayage Spaces and Harmonic Kernels

Since the axioms of a harmonic space may be expressed in terms of harmonic measures, it will be interesting to note that a similar decription can be given for balayage spaces. The main difference will be the fact that the harmonic measures are not necessarily supported by the boundary of the open set, their support may be the entire complement.

Let $\mathcal{U}$ be a base of relatively compact open sets in X. For every $x \in X$ and open subset V of X, let $\mathcal{U}_x = \{U \in \mathcal{U} : x \in U\}$, $\mathcal{U}(V) = \{U \in \mathcal{U} : \overline{U} \subset V\}$.

Let $(H_U)_{U \in \mathcal{U}}$ be a family of *sweeping kernels* on X, i.e., for each $U \in \mathcal{U}$ we have a kernel H_U on X satisfying $H_U(x,U) = 0$ for every $x \in U$ and $H_U(x,\cdot) = \varepsilon_x$ for every $x \in \complement U$. For every open subset V of X we define

$$*\mathcal{H}^+(V) := \{v \in \mathcal{B}^+(X) : v|_V \text{ l.s.c.}, \ H_U v \le v \ \forall U \in \mathcal{U}(V)\} ,$$

$$\mathcal{S}^+(V) := \{s \in *\mathcal{H}^+(V) : H_U s|_U \in \mathcal{C}(U) \ \forall U \in \mathcal{U}(V)\} ,$$

$$\mathcal{H}^+(V) := \{h \in \mathcal{S}^+(V) : H_U h = h \ \forall U \in \mathcal{U}(V)\} .$$

$*\mathcal{H}^+(V)$ (resp. $\mathcal{S}^+(V)$, $\mathcal{H}^+(V)$) is the set of all positive Borel functions which are hyperharmonic (resp. superharmonic, harmonic) on V. Given $U \in \mathcal{U}$, we shall say that a sequence (x_n) in U converging to a point $z \in U^*$ is *purely irregular* in U if there is no subsequence (y_n) of (x_n) satisfying $\lim_{n\to\infty} H_U(y_n, \cdot) = \varepsilon_z$.

$(H_U)_{U\in\mathcal{U}}$ is called a *family of harmonic kernels* (W. HANSEN [36]) if the following axioms are satisfied.

(H_1) For every $x \in X$, $\lim_{U,\mathcal{U}_x} H_U(x,\cdot) = \varepsilon_x$ or $R_1^{\{x\}}$ is l.s.c. at x.

(H_2) $H_V H_U = H_U$ for all $V, U \in \mathcal{U}$ with $\overline{V} \subset U$.

(H_3) $H_U f|_U \in \mathcal{C}_b(U)$ for every $U \in \mathcal{U}$ and $f \in \mathcal{B}_b(X)$ with compact support.

(H_4) For every $U \in \mathcal{U}$ and $x \in U$ there exists $w \in *\mathcal{H}^+(U)$ such that $w(x) < \infty$ and $\lim_{n\to\infty} w(x_n) = \infty$ for every purely irregular sequence (x_n) in U.

(H_5) $*\mathcal{H}^+(X)$ is linearly separating, there exists $s \in \mathcal{S}^+(X) \cap \mathcal{C}(X)$, $s>0$.

3.1. EXAMPLES. It is easy to recover our first standard examples.

1. Classical theory. $\mathcal{U}$ family of all open balls in $\mathbb{R}^m$, H_U given by the Poisson integral.

2. Uniform translation on $\mathbb{R}$. $\mathcal{U} = \{]\alpha,\beta[\ : -\infty < \alpha < \beta < \infty\}$, $H_{]\alpha,\beta[}(x,\cdot) = \varepsilon_\alpha$ for $\alpha < x < \beta$ (take $w(x) = \frac{1}{\beta - x}$).

3. Discrete theory. If P is a sub-Markov kernel on a countable, discrete space X we take $\mathcal{U} = \{\{x\} : x \in X\}$ and define

$$H_{\{x\}}(x,A) = \begin{cases} \dfrac{P(x,A\setminus\{x\})}{1-P(x,\{x\})}, & P(x,\{x\}) < 1, \\ 0, & P(x,\{x\}) = 1. \end{cases}$$

Then $*\mathcal{H}^+(X) = S_P$ and (H_1) - (H_4) are trivially satisfied. (H_5) holds if S_P separates the points of X.

3.2. THEOREM. Let $(X,\mathcal{W})$ be a balayage space, let $\mathcal{U}$ be the family of all relatively compact open subsets of X and define sweeping kernels H_U by $H_U(x,\cdot) = \varepsilon_x^{\complement U}$ if $x \in U$ (and $H_U(x,\cdot) = \varepsilon_x$ if $x \in \complement U$). Then $(H_U)_{U\in\mathcal{U}}$ is a family of harmonic kernels on X, $\mathcal{W} = *\mathcal{H}^+(X)$, $\mathcal{W} \cap \mathcal{C}(X) = \mathcal{S}^+(X) \cap \mathcal{C}(X)$, and

$$\mathcal{P} = \{p \in \mathcal{W} \cap \mathcal{C}(X) : \inf\{R_p^{\complement K} : K \text{ cp.} \subset X\} = 0\}.$$

$$= \{p \in \mathcal{W} \cap \mathcal{C}(X) : h \in \mathcal{H}^+(X),\ 0 \le h \le p \Rightarrow h = 0\}.$$

3.3. THEOREM. Let $(H_U)_{U\in\mathcal{U}}$ be a family of harmonic kernels on X. Then $(X, *\mathcal{H}^+(X))$ is a balayage space and $H_U(x,\cdot) = \varepsilon_x^{\complement U}$ for every $U \in \mathcal{U}$ and $x \in U$.

Let us finally note that we have the following sheaf properties.

3.4. PROPOSITION. Let $(H_U)_{U \in \mathcal{U}}$ be a family of harmonic kernels on X and let $(V_i)_{i \in I}$ be a family of open subsets of X . Then

$$*\mathcal{H}^+(\bigcup_{i\in I} V_i) = \bigcap_{i\in I} *\mathcal{H}^+(V_i) , \quad \mathcal{H}^+(\bigcup_{i\in I} V_i) = \bigcap_{i\in I} \mathcal{H}^+(V_i) .$$

4. Harmonic Spaces

If $(X,\mathcal{H})$ is a $\mathcal{P}$-harmonic space then $(X,*\mathcal{H}^+(X))$ is a balayage space and the sheaf $\mathcal{H}$ is uniquely determined by $*\mathcal{H}^+(X)$. Hence we may identify $\mathcal{P}$-harmonic spaces with a subclass of balayage spaces. Having several equivalent ways to describe balayage spaces we have different possibilities to characterize the subclass of $\mathcal{P}$-harmonic spaces.

Let $(X,\mathcal{W})$ be a balayage space and let $(H_U)_{U \in \mathcal{U}}$ be a family of harmonic kernels such that $*\mathcal{H}^+(X) = \mathcal{W}$. We shall say that $\mathcal{W}$ has the *local truncation property* if for every open U in X and all $u,v \in \mathcal{W}$ satisfying $u \geq v$ on U^* the function w defined by $w = \inf(u,v)$ on U , $w = v$ on $\complement U$ is contained in $\mathcal{W}$.

4.1. THEOREM. The following statements are equivalent:

(1) $(X,\mathcal{W})$ is a harmonic space.

(2) $\mathcal{W}$ has the local truncation property and there are no finely isolated points in X .

(3) For every $U \in \mathcal{U}$ and $x \in \mathcal{U}$, $H_U(x,\complement U^*) = 0$. For every $x \in X$ there exists $U \in \mathcal{U}_x$ such that $H_U(x,U^*) > 0$.

(4) There exists an associated sub-Markov semigroup $\mathbb{P}$ on X having no absorbing points (points x with $P_t(x,\complement\{x\}) = 0$) and such that, for every $f \in \mathcal{K}(X)$, the function $\frac{1}{t}P_t f$ tends to zero locally uniformly on $\complement\,\mathrm{supp}(f)$ as t tends to zero.

(5) There exists an associated Hunt process with continuous paths which has no absorbing points.

5. Permanence Properties, Further Examples

Balayage spaces have nice permanence properties which may serve to construct many new examples. In particular, the list of standard examples will be completed by balayage spaces corresponding to Riesz potentials and the heat equation.

a) Restriction on subsets

Two entirely different types of restriction are possible: restriction on an open subset by restriction of the harmonic kernels, restriction on a closed basic subset by simply restricting the global positive hyperharmonic functions. Given a P-harmonic space, the first kind of restriction yields of course again a P-harmonic space whereas the second one usually produces a balayage space which is not a harmonic space. Note, however, that restriction on an absorbing subset belongs to both categories.

Let $(X,\mathcal{W})$ be a balayage space and let $(H_U)_{U\in\mathcal{U}}$ be a corresponding family of harmonic kernels.

5.1. PROPOSITION. For every open subset U of X, $(U,{}^*\mathcal{H}^+(U)_{|U})$ is a balayage space, $(H_V|_U)_{V\in\mathcal{U}(U)}$ is a corresponding family of harmonic kernels.

5.2. PROPOSITION. Let A be an absorbing subset of X (i.e., $A=\{u=0\}$ for some $u\in\mathcal{W}$). Then $(A,\mathcal{W}_{|A})$ is a balayage space, a corresponding family of harmonic kernels is given by ${}^AH_U(x,B)=H_U(x,B)$ $=\varepsilon_x^{\complement U}(B)$, $U\in\mathcal{U}$ with $A\cap U\neq\emptyset$, $x\in A\cap U$, $B\in\mathcal{B}(A)$.

5.3. PROPOSITION. Let A be closed subset of X such that $b(A)=A$. Then $(A,\mathcal{W}_{|A})$ is a balayage space, a corresponding family of harmonic kernels is given by ${}^AH_U(x,B)=\varepsilon_x^{A\setminus U}(B)$, $U\in\mathcal{U}$ with $A\cap U\neq\emptyset$, $x\in A\cap U$, $B\in\mathcal{B}(A)$.

If we apply (5.3) to the harmonic space $(\mathbb{R}^m,{}^*\mathcal{H}^+(\mathbb{R}^m))$, $m\geq 3$, of classical potential theory restricting to the hyperplane $\mathbb{R}^{m-1}\times\{0\}$ we obtain the balayage space of Riesz potentials of index 1 on $\mathbb{R}^{m-1}$ (see section b). This is a consequence of the trivial observation $\|x-y\|^{2-m}=\|x-y\|^{1-(m-1)}$.

b) Subordination by convolution semigroups, Riesz potentials

A family $(\mu_t)_{t>0}$ of measures on $\mathbb{R}_+^*$ is called a (vaguely continuous) *convolution semigroup* on $\mathbb{R}_+^*$ if $\mu_t(1)\leq 1$, $\mu_s*\mu_t=\mu_{s+t}$ for all $s,t\in\mathbb{R}_+^*$, and $\lim_{t\to 0}\mu_t=\varepsilon_0$ (vaguely). Given a convolution semigroup $(\mu_t)_{t>0}$ on $\mathbb{R}_+^*$ and a measurable sub-Markov semigroup $\mathbb{P}=(P_t)$ on X, we obtain a sub-Markov semigroup $\mathbb{P}^\mu=(P_t^\mu)$ on X defining

$$P_t^\mu f(x)=\int P_s f(x)\,\mu_t(ds)\ .$$

$\mathbb{P}^\mu$ is called the sub-Markov *semigroup subordinated to* $\mathbb{P}$ *by means of* $(\mu_t)_{t>0}$. Interesting examples are furnished by the *one-sided* *stable semigroups* $(\eta_t^\alpha)_{t>0}$, $0 < \alpha < 2$, having Laplace transforms $L\eta_t^\alpha(s) = \exp(-ts^{\alpha})$. If $\mathbb{P}$ is the Brownian semigroup on $\mathbb{R}^m$ and $0 < \alpha < 2$, the sub-Markov semigroup subordinated to $\mathbb{P}$ by means of $(\eta_t^\alpha)_{t>0}$ is the *symmetric stable semigroup* $\mathbb{P}^{\eta^\alpha} = (P_t^{\eta^\alpha})$ *of index* α on $\mathbb{R}^m$. Suppose that $\alpha < 1$ if $m = 1$. It is easily verified that the potential kernel V^α of $\mathbb{P}^{\eta^\alpha}$ is given by $V^\alpha f = c_m^\alpha k_\alpha * f$ where c_m^α is a constant and $k_\alpha(x) = \|x\|^{\alpha-m}$. So the excessive functions of $\mathbb{P}^{\eta^\alpha}$ are the Riesz potentials of order α.

General references are C. BERG - G. FORST [8] and F. HIRSCH [41]. The following result is an immediate consequence of (1.3).

5.4. PROPOSITION. Let $\mathbb{P}$ be a strong Feller semigroup on X such that $(X, E_{\mathbb{P}})$ is a balayage space and let $(\mu_t)_{t>0}$ be a convolution semigroup on $\mathbb{R}_+^*$. Then $\mathbb{P}^\mu$ is a strong Feller semigroup, $(X, E_{\mathbb{P}^\mu})$ is a balayage space, and $E_{\mathbb{P}} \subset E_{\mathbb{P}^\mu}$.

In particular, we obtain that for each $\alpha \in]0,2[$ and natural $m \geq \alpha$ the Riesz potentials of order α on $\mathbb{R}^m$ form a balayage space $(\mathbb{R}^m, E_{\mathbb{P}^{\eta^\alpha}})$ (if $m \leq 2$, we have to go back to (1.5)).

The preceding proposition states that any convolution semigroup will do if we have a balayage space $(X, E_{\mathbb{P}})$ given by a strong Feller semigroup. Let us now ask which convolution semigroups will permit a subordination on any balayage space where the positive constants are hyperharmonic. To that end we recall from (2.3) that for every balayage space $(X, \mathcal{W})$ with $1 \in \mathcal{W}$ there exist many sub-Markov semigroups $\mathbb{P}$ such that $E_{\mathbb{P}} = \mathcal{W}$ and $\mathbb{P}$ has a strong Feller resolvent. So the following result is an answer to our question (private communication by F. HIRSCH).

5.5. THEOREM. Let $(\mu_t)_{t>0}$ be a convolution semigroup on $\mathbb{R}_+^*$ and $\kappa = \int \mu_t \, dt$. Then the following statements are equivalent:

(1) If $(X, E_{\mathbb{P}})$ is a balayage space where $\mathbb{P}$ is a sub-Markov semigroup having a strong Feller resolvent, then $(X, E_{\mathbb{P}^\mu})$ is a balayage space.

(2) $(\mathbb{R}, E_{\mathbb{T}^\mu})$ is a balayage space.

(3) κ is absolutely continuous with respect to Lebesgue measure $\lambda_{\mathbb{R}_+^*}$ on $\mathbb{R}_+^*$.

Note that the semigroup $\mathbb{T}$ of uniform translation on $\mathbb{R}$ has a strong Feller resolvent. So the balayage space $(\mathbb{R}, E_{\mathbb{T}})$ may serve as a test for subordination. We shall see in (5.7) that $(\mathbb{R}, E_{\mathbb{T}})$ plays the same crucial role for products of balayage spaces.

Let us finally mention that, given any $\alpha \in]0,2[$, the measure $\int_0^\infty \eta_t^\alpha dt$ has the density $t \mapsto (\Gamma(\frac{\alpha}{2}))^{-1} t^{\frac{\alpha}{2}-1}$ with respect to $\lambda_{\mathbb{R}_+^*}$, hence $(\mathbb{R}, E_{\mathbb{T}\eta^\alpha})$ is a balayage space with $T_t^{\eta^\alpha} f = f * \eta_t^\alpha$.

c) Products, heat semigroup

Let $\mathbb{P} = (P_t)_{t>0}$ be a sub-Markov semigroup on X and let $\tilde{\mathbb{P}} = (\tilde{P}_t)_{t>0}$ be a sub-Markov semigroup on a space $\tilde{X}$. Defining

$$(P_t \otimes \tilde{P}_t) f(x,\tilde{x}) = \iint f(y,\tilde{y}) P_t(x,dy)\, \tilde{P}_t(\tilde{x},d\tilde{y})$$

we obtain a sub-Markov semigroup $\mathbb{P} \otimes \tilde{\mathbb{P}} = (P_t \otimes \tilde{P}_t)_{t>0}$ on $X \times \tilde{X}$. Supposed that $(X, E_{\mathbb{P}})$ and $(\tilde{X}, E_{\tilde{\mathbb{P}}})$ are balayage spaces will $(X \times \tilde{X}, E_{\mathbb{P}\otimes\tilde{\mathbb{P}}})$ be a balayage space? It suffices to consider $X = \tilde{X} = \mathbb{R}$, $\mathbb{P} = \tilde{\mathbb{P}} = \mathbb{T}$ in order to see that further conditions are necessary. A first result based on (1.5) is the following.

5.6. PROPOSITION. Let $(X, E_{\mathbb{P}})$ and $(\tilde{X}, E_{\tilde{\mathbb{P}}})$ be balayage spaces. Then $(X \times \tilde{X}, E_{\mathbb{P}\otimes\tilde{\mathbb{P}}})$ is a balayage space if and only if $E_{\mathbb{P}\otimes\tilde{\mathbb{P}}} = S(E_{\mathbb{P}\otimes\tilde{\mathbb{P}}} \cap C(X \times \tilde{X}))$. In particular, $(X \times \tilde{X}, E_{\mathbb{P}\otimes\tilde{\mathbb{P}}})$ is a balayage space if $\mathbb{P} \otimes \tilde{\mathbb{P}}$ has a strong Feller resolvent.

The next theorem is of the same type as (5.5) (U. SCHIRMEIER [46]).

5.7. THEOREM. Let $(X, E_{\mathbb{P}})$ be a balayage space. Then the following statements are equivalent:

(1) $(X \times \tilde{X}, E_{\mathbb{P}\otimes\tilde{\mathbb{P}}})$ is a balayage space for every balayage space $(\tilde{X}, E_{\tilde{\mathbb{P}}})$ such that $\tilde{\mathbb{P}}$ is a sub-Markov semigroup having a strong Feller resolvent.

(2) $(X \times \mathbb{R}, E_{\mathbb{P}\otimes\mathbb{T}})$ is a balayage space.

(3) $\mathbb{P}$ is a strong Feller semigroup.

Taking the Brownian semigroup $\mathbb{P}$ on $\mathbb{R}^m$, $m \geq 1$, we obtain the *heat semigroup* $\mathbb{P} \otimes \mathbb{T}$ on $\mathbb{R}^{m+1}$. The preceding result implies that $(\mathbb{R}^{m+1}, E_{\mathbb{P}\otimes\mathbb{T}})$ is a balayage space if $m \geq 3$. Going back to (1.5) and using the fact that $\mathbb{P} \otimes \mathbb{T}$ has a strong Feller resolvent we conclude that $(\mathbb{R}^{m+1}, E_{\mathbb{P}\otimes\mathbb{T}})$ is a balayage space for any $m \geq 1$. By (4.1), it is even a harmonic space.

d) Brownian semigroups on the infinite dimensional torus

Brownian semigroups on the infinite dimensional torus have been investigated by C. BERG [6,7] (see also BENDIKOV [5]). Using the connection between semigroups and harmonic spaces established in the previous sections his main result is now accessible in a very direct and elementary way.

The Brownian semigroup $\overline{\mathbb{P}}$ on the torus $T = {}^{\mathbb{R}}/_{2\pi\mathbb{Z}}$ is obtained by considering the Brownian semigroup $\mathbb{P}$ on $\mathbb{R}$ modulo 2π, i.e., by defining $\overline{P}_t f(\overline{x}) = P_t(f\circ j)(x)$ where $j : x \mapsto \overline{x}$ denotes the quotient map from $\mathbb{R}$ on ${}^{\mathbb{R}}/_{2\pi\mathbb{Z}}$. Using Fourier transforms it is easily verified that

$$\overline{P}_t f(\overline{x}) = \int_0^{2\pi} \overline{g}_t(x-y)(f\circ j)(y)\,dy$$

where the density $\overline{g}_t$ is given by

$$\overline{g}_t(x) = \frac{1}{2\pi} + \frac{1}{\pi}\sum_{n=1}^{\infty} e^{-\frac{t}{2}n^2} \cos nx \ .$$

Given an infinite sequence $A = (a_k)$ in $\mathbb{R}_+^*$ we define probability measures $P_t^A(x,\cdot)$, $t > 0$, $x = (x_k) \in T^\infty$, on T^∞ by

$$P_t^A(x,\cdot) = \bigotimes_{k=1}^{\infty} \overline{P}_{a_k t}(x_k,\cdot) \ .$$

Using the fact that $|\overline{g}_t - \frac{1}{2\pi}| \leq \frac{1}{\pi}\sum_{n=1}^{\infty} e^{-\frac{t}{2}n^2}$ and applying (1.3), (4.1), and (5.7) we obtain the following result.

5.8. THEOREM. Let $A = (a_k)$ be an infinite sequence in $\mathbb{R}_+^*$ such that $\sum_{k=1}^{\infty} e^{-a_k t} < \infty$ for every $t > 0$ (e.g. $\sum_{k=1}^{\infty} \frac{1}{a_k^m} < \infty$ for some $m \in \mathbb{N}$).
Then $\mathbb{P}^A = (P_t^A)_{t>0}$ is a strong Feller semigroup on T^∞ such that $\lim_{t\to\infty} \frac{1}{t} P_t^A f = 0$ locally uniformly on $\complement\operatorname{supp}(f)$ for every $f \in C(T^\infty)$.
In particular, $(T^\infty, E_{(\mathbb{P}^A)\alpha})$ is a harmonic space for every $\alpha > 0$.
Moreover, $(T^\infty \times \mathbb{R}, E_{\mathbb{P}^A \otimes \mathbb{T}})$ is a harmonic space.

e) Images

Images have previously been studied by C. CONSTANTINESCU - A. CORNEA [27], E.B. DYNKIN [30] and W. HANSEN [37]. Little is needed to ensure that the image of a balayage space is again a balayage space.

5.9. THEOREM. Let $(X,\mathcal{W})$ be a balayage space and let π be an open continuous mapping of X onto a locally compact space X_π with countable base such that, given a l.s.c. function $f : X_\pi \to [0,\infty]$ there exists a function $g : X_\pi \to [0,\infty]$ satisfying $R_{f\circ\pi} = g\circ\pi$. Define

$$\mathcal{W}_\pi = \{w \in \mathcal{B}^+(X_\pi) : w\circ\pi \in \mathcal{W}\} \ .$$

Then $(X_\pi,\mathcal{W}_\pi)$ is a balayage space if and only if $\mathcal{W}_\pi$ is linearly separating and there exists a strictly positive function in $\mathcal{W}_\pi \cap C(X_\pi)$.

Clearly $1 \in \mathcal{W}_\pi \cap C(X_\pi)$ if $1 \in \mathcal{W}$, and then $\mathcal{W}_\pi$ is linearly separating if it is separating. If $(X,\mathcal{W})$ is a $\mathcal{P}$-harmonic space then the additional assumption that none of the sets $\pi^{-1}(x)$, $x \in \mathcal{W}$, is an absorbing subset of X is necessary and sufficient to have a $\mathcal{P}$-harmonic space $(X_\pi, \mathcal{W}_\pi)$.

We shall say that a homeomorphism $\sigma : X \to X$ is an *automorphism* of the balayage space $(X,\mathcal{W})$ if $\{w \circ \sigma : w \in \mathcal{W}\} = \mathcal{W}$.

5.10. COROLLARY. Let G be a group of automorphisms of a balayage space $(X,\mathcal{W})$ and let $\pi : X \to {}^{X}/_{G} = X_\pi$ denote the quotient mapping. Suppose that ${}^{X}/_{G}$ is a Hausdorff space, that points lying on different orbits in X are linearly separated by the set $\mathcal{W}^G$ of all G-invariant functions in $\mathcal{W}$, and that there exists a strictly positive function in $\mathcal{W}^G \cap C(X)$. Then $w \to w \circ \pi$ is a one-to-one correspondence between $\mathcal{W}_\pi$ and $\mathcal{W}^G$, and $(X_\pi,\mathcal{W}_\pi)$ is a balayage space.

Evidently, the preceding result allows the construction of many new examples from the classical harmonic space, the space of Riesz potentials, and the hàrmonic space associated with the heat equation using reflections, translations, and rotations.

6. Dirichlet Problem

In the following let $(X,\mathcal{W})$ be a balayage space and let $\mathcal{U}$ be the set of all relatively compact open subsets of X. Given an open subset U of X, we define the set $^*\mathcal{H}(U)$ of all *hyperharmonic* functions on U by

$$^*\mathcal{H}(U) = \{u \in \mathcal{B}(X) : u|_U \text{ l.s.c.}, -\infty < H_V u(x) \leq u(x) \ \forall x \in V \in \mathcal{U}(U)\} ,$$

the set $\mathcal{H}(U)$ of all *harmonic* functions on U by

$$\mathcal{H}(U) = {}^*\mathcal{H}(U) \cap (-{}^*\mathcal{H}(U)) = \{h \in \mathcal{B}(X) : h|_U \in C(U), H_V h = h \ \forall V \in \mathcal{U}(U)\} ,$$

and the set $\mathcal{S}(U)$ of all *superharmonic* functions on U by

$$\mathcal{S}(U) = \{s \in {}^*\mathcal{H}(U) : H_V s|_V \in C(V) \ \forall V \in \mathcal{U}(U)\}$$

$$= \{s \in {}^*\mathcal{H}(U) : H_V s \in \mathcal{H}(V) \ \forall V \in \mathcal{U}(U)\} .$$

Let us note that of course for all these functions the values on $\complement U$ are very important unless $H_V 1_{\complement U} = 0$ for all $V \in \mathcal{U}(U)$.

A numerical function f on X is called *P-bounded* (resp. *lower P-bounded, upper P-bounded*) if there exists a potential $p \in \mathcal{P}$ such that $|f| \leq p$ (resp. $f \geq -p$, $f \leq p$).

a) Generalized Dirichlet problem

In this section we study the Dirichlet problem by the Perron-Wiener-Brelot method. However, in our situation of a balayage space the boundary of the open set U has to be replaced by the complement of U (W. HANSEN [36]).

Let U be an open subset of X. For every $f : X \to \overline{\mathbb{R}}$ we define the set

$$\mathcal{U}_f^U := \{u \in {}^*\mathcal{H}(U) : u \text{ l.s.c. and lower } \mathcal{P}\text{-bounded on } X,\ u \geq f \text{ on } \complement U\}$$

of all *upper functions* of f; the set of all *lower functions* of f is defined by $\mathcal{L}_f^U := -\mathcal{U}_{-f}^U$. Then

$$\overline{H}_f^U = \inf \mathcal{U}_f^U \quad (\underline{H}_f^U = \sup \mathcal{L}_f^U \text{ resp.})$$

is called the *upper (lower* resp.) *solution of the generalized Dirichlet problem* of f on U. The function f is called *resolutive* if $\overline{H}_f^U = \underline{H}_f^U =: H_f^U$ and $H_f^U \in \mathcal{H}(U)$.

6.1. REMARKS. 1. $\overline{H}_f^U$, $\underline{H}_f^U$ depend only on $f|_{\complement U}$, $\overline{H}_f^U = \underline{H}_f^U = f$ on $\complement U$.

2. If $(X,\mathcal{W})$ is a $\mathcal{P}$-harmonic space, then $\mathcal{U}_f^U|_U$ is the set of all l.s.c., lower $\mathcal{P}$-bounded numerical functions u on U such that $H_V u(x) \leq u(x)$ for all $x \in V \in \mathcal{U}(U)$ and $\liminf\limits_{x\to z} u(x) \geq f(z)$ for every $z \in U^*$, hence $\overline{H}_f^U|_U$ (resp. $\underline{H}_f^U|_U$) is the upper (lower resp.) solution familiar from harmonic spaces for the boundary function $f|_{U^*}$.

6.2. THEOREM. For every $f : X \to \overline{\mathbb{R}}$ and every $x \in U$, $\overline{H}_f^U(x) = (\varepsilon_x^{\complement U})^*(f)$, $\underline{H}_f^U(x) = (\varepsilon_x^{\complement U})_*(f)$. In particular, a $\mathcal{P}$-bounded function $f : X \to \mathbb{R}$ is resolutive if and only if it is $\varepsilon_x^{\complement U}$-integrable for every $x \in U$.

So we may write $H_U f$ instead of H_f^U whenever $f \in \mathcal{B}(X)$ and $\varepsilon_x^{\complement U}(f)$ is defined for every $x \in U$.

A boundary point $z \in U^*$ is called *regular* (with respect to U) if $\lim\limits_{x\to z, x\in U} H_U f(x) = f(z)$ for every $f \in \mathcal{K}(X)$. Exactly as for harmonic spaces we obtain the following result.

6.3. PROPOSITION. For every $z \in U^*$, the following properties are equivalent:

(1) z is regular.

(2) $\complement U$ is not thin at z, i.e., $\varepsilon_z^{\complement U} = \varepsilon_z$.

(3) z admits a *barrier*, i.e., there exist an open neighborhood V of z and a function $w \in {}^*\mathcal{H}^+(U \cap V)$ such that $w > 0$ on $U \cap V$ and $\lim\limits_{x\to z, x\in U} w(x) = 0$.

6.4. COROLLARY. The set of irregular boundary points is semipolar.

b) Weak Dirichlet problem

As before let U be an open subset of a balayage space $(X,\mathcal{W})$.

Let $C_P(X)$ denote the set of all P-bounded continuous real functions on X and define

$$H(U) = \mathcal{H}(U) \cap C_P(X) \ , \ S(U) = \mathcal{S}(U) \cap C_P(X) \ .$$

Obviously, $H(U) + P \subset S(U)$ and $S(U) \cap (-S(U)) = H(U)$. For every $x \in X$ let

$$M_x(S(U)) = \{\mu \in M_+(X) : \mu(s) \leq s(x) \ \forall s \in S(U)\} \ .$$

The *Choquet boundary* of X with respect to $S(U)$ is defined by

$$\mathrm{Ch}_{S(U)}X = \{x \in X : M_x(S(U)) = \{\varepsilon_x\}\} \ .$$

Clearly, $\mathrm{Ch}_{S(U)}X \subset \complement U$ since $\varepsilon_x^{\complement V} \in M_x(S(U))$, $\varepsilon_x^{\complement V} \neq \varepsilon_x$ for every $x \in V \in \mathcal{U}(U)$.

By a general minimum principle, a function $s \in S(U)$ is positive if $s \geq 0$ on $\mathrm{Ch}_{S(U)}X$. In particular, each function $h \in H(U)$ is uniquely determined by its restriction to $\mathrm{Ch}_{S(U)}X$ (H. BAUER [1]).

The *weak Dirichlet problem* is now the following: Given a closed subset A_o of $\mathrm{Ch}_{S(U)}X$ and a continuous P-bounded function f on A_o, is there a continuous extension of f to a function $h \in H(U)$?

For harmonic spaces satisfying the axiom of domination a positive answer has been given by N. BOBOC - A. CORNEA [18]. Examples for the heat equation are discussed by E.G. EFFROS - J.L. KAZDAN [31,32]. The final breakthrough is due to J. BLIEDTNER - W. HANSEN [10].

In fact, the weak Dirichlet problem can be solved even in an additive and increasing way using P-dilations obtained by a "smearing" of balayage on suitable families of closed sets (J. BLIEDTNER - W. HANSEN [13]). By definition a *dilation* D is a kernel D on X such that $Dp \leq p$ for every $p \in P$, and a dilation D is called a *P-dilation* provided $D(P) \subset P$. In order to obtain suitable dilations we use the following strong result on the essential base.

6.5. THEOREM. Let $A \in \mathcal{B}(X)$ and let $A_o \subset A \cap \beta(A)$ be closed. Then there exists an increasing family $(A_t)_{0<t<1}$ of closed sets contained in $A \cap \beta(A)$ such that $A_o = \bigcap_{0<t<1} A_t$, $\beta(A) = \beta(\bigcup_{0<t<1} A_t)$, and $A_s \subset \beta(A_t)$ for all $0 < s < t < 1$.

Suppose now that $(A_t)_{a \leq t \leq b}$, $a,b \in \mathbb{R}$, $a < b$, is a family of closed subsets of X such that $A_s \subset \beta(A_t)$ for all $a \leq s < t \leq b$. Then it is easily verified that

$$D : (x,B) \longrightarrow \frac{1}{b-a} \int_a^b \varepsilon_x^{A_t}(B) \, dt$$

is a P-dilation such that $Dp = p$ on A_a and $Dp \in H(\complement A_b)$ for every $p \in P$. This leads to the following result.

6.6. THEOREM. $Ch_{S(U)}X = \beta(\complement U)$ and for every closed subset A_o of $\beta(\complement U)$ there exists a P-dilation D such that $Df \in H(U)$, $Df = f$ on A_o for every $f \in C_P(X)$ and all the measures $D(x,\cdot)$, $x \in X$, are supported by a closed subset A_1 of $\beta(\complement U)$.

Moreover, for every $x \in X$ the measure $D_U(x,\cdot) = \varepsilon_x^{\beta(\complement U)}$ is the only measure $D_U(x,\cdot) \in M_x(S(U))$ such that $D_U(x, X \setminus Ch_{S(U)}X) = 0$. In particular, $S(U)$ is a simplicial cone.

6.7. REMARK. A dilation K on X is called a *Keldych operator* if $K(C_P(X)) \subset H(U)$ and $Kh = h$ for every $h \in H(U)$. The kernels D_U and H_U are Keldych operators and $D_U p \leq Kp \leq H_U p$ for every Keldych operator K and every $p \in P$.

Let us finally note that the preceding considerations on the weak Dirichlet problem may be extended to functions in $C_P(X)$ which are finely harmonic (finely superharmonic resp.) on a finely open set G. Sets $H(G)$ ($S(G)$ resp.) where G is finely open, but in general not open, arise in a natural way if we are interested in functions in $C_P(X)$ which are harmonic (superharmonic resp.) in a neighborhood of a given closed set F: The set of these functions is dense in $H(G)$ ($S(G)$ resp.) where G is the fine interior of F (J. BLIEDTNER - W. HANSEN [11]).

References

1. Bauer, H.: Minimalstellen von Funktionen und Extremalpunkte II. Arch. Math. 11 (1960), 200-205.
2. Bauer, H.: Axiomatische Behandlung des Dirichletschen Problems für elliptische und parabolische Differentialgleichungen. Math. Ann. 146 (1962), 1-59.
3. Bauer, H.: Weiterführung einer axiomatischen Potentialtheorie ohne Kern (Existenz von Potentialen). Z. Wahrscheinlichkeitstheorie verw. Gebiete 1 (1963), 197-229.
4. Bauer, H.: Harmonische Räume und ihre Potentialtheorie. LN in Math. 22, Springer-Verlag (1966).
5. Bendikov, A.D.: Functions that are harmonic for a certain class of diffusion processes on a group. Teor. Verojatnost i Primenen 20 (1975), 773-784.
6. Berg, C.: Potential theory on the infinite dimensional torus. Inventiones math. 32 (1976), 49-100.
7. Berg, C.: On Brownian and Poisonnian convolution semigroups on the infinite dimensional torus. Inventiones math. 38 (1977), 227-235.

8. Berg, C., Forst, G.: Potential theory on locally compact abelian groups. Ergebn. d. Math. 87, Springer-Verlag (1975).

9. Bliedtner, J.: Harmonische Gruppen und Huntsche Faltungskerne. In: "Seminar über Potentialtheorie", LN in Math. 69, Springer-Verlag (1968), 69-102.

10. Bliedtner, J., Hansen, W.: Simplicial cones in potential theory. Inventiones math. 29 (1975), 83-110.

11. Bliedtner, J., Hansen, W.: Simplicial cones in potential theory II (Approximation theorems). Invent. math. 46 (1978), 255-275.

12. Bliedtner, J., Hansen, W.: Markov processes and harmonic spaces. Z. Wahrscheinlichkeitstheorie verw. Gebiete 42 (1978), 309-325.

13. Bliedtner, J., Hansen, W.: The weak Dirichlet problem. J. Reine Angew. Math. 348 (1984), 34-39.

14. Bliedtner, J., Hansen, W.: Potential theory - an analytic and probabilistic approach to balayage. Universitext, Springer-Verlag (1986).

15. Blumenthal, R.M., Getoor, R.K.: Markov processes and potential theory. Academic Press (1968).

16. Boboc, N., Constantinescu, C., Cornea, A.: Axiomatic theory of harmonic functions. Balayage. Ann. Inst. Fourier 15 (1965), 37-70.

17. Boboc, N., Constantinescu, C., Cornea, A.: Semigroups of transitions on harmonic spaces. Rev. Roumaine Math. Pures Appl. 12 (1967), 763-805.

18. Boboc, N., Cornea, A.: Convex cones of lower semicontinuous functions on compact spaces. Rev. Roumaine Math. Pures Appl. 12 (1967), 471-525.

19. Brelot, M.: Sur la théorie autonome des fonctions sousharmoniques. Bull. Sci. Math. France 65 (1941), 78-91.

20. Brelot, M.: Minorantes sousharmoniques, extrémales et capacités. Journ. Math. Pures Appl. 24 (1945), 1-32.

21. Brelot, M.: Extension axiomatique des fonctions sousharmoniques I. C.R. Acad. Sci. Paris 245 (1957), 1688-1690.

22. Brelot, M.: Extension axiomatique des fonctions sousharmoniques II. C.R. Acad. Sci. Paris 246 (1958), 2334-2337.

23. Brelot, M.: Une axiomatique générale du problème de Dirichlet dans les espaces localement compacts. In: Sém. Brelot-Choquet "Théorie du potentiel" 1 (1958), 6.01-6.16.

24. Brelot, M.: Axiomatique des fonctions harmoniques et surharmoniques dans un espace localement compact. In: Sém. Brelot-Choquet-Deny "Théorie du potentiel" 2 (1959), 1.01-1.40.

25. Brelot, M.: Quelques propriétés et applications nouvelles de l'effilement. in: Sém. Brelot-Choquet-Deny "Theorie du potentiel" 6 (1962), 1.27-1.40.

26. Constantinescu, C.: Some properties of the balayage of measures on a harmonic space. Ann. Inst. Fourier 17 (1967), 273-293.

27. Constantinescu, C., Cornea, A.: Ideale Ränder Riemannscher Flächen. Ergebn. d. Math. 32, Springer-Verlag (1963).

28. Constantinescu, C., Cornea, A.: On the axiomatic theory of harmonic functions I, II. Ann. Inst. Fourier 13 (1963), 373-388, 389-394.

29. Constantinescu, C., Cornea, A.: Potential theory on harmonic spaces. Grundl. d. math. Wiss. 158, Springer-Verlag (1972).

30. Dynkin, E.B.: Markov processes I, II. Grundl. d. math. Wiss. 121, 122, Springer-Verlag (1965).

31. Effros, E.G., Kazdan, J.L.: Applications of Choquet simplexes to elliptic and parabolic boundary value problems. J. Differential Equations 8 (1970), 95-134.

32. Effros, E.G., Kazdan, J.L.: On the Dirichlet problem for the heat equation. Indiana Univ. Math. J. 20 (1971), 683-693.

33. Getoor, R.K.: Markov processes: Ray processes and right processes. LN in Math. 440, Springer-Verlag (1975).

34. Hansen, W.: Konstruktion von Halbgruppen und Markoffschen Prozessen. Inventiones math. 3 (1967), 179-214.

35. Hansen, W.: Charakterisierung von Familien exzessiver Funktionen. Inventiones math. 5 (1968), 335-348.

36. Hansen, W.: Potentialtheorie harmonischer Kerne. In: "Seminar über Potentialtheorie", LN in Math. 69, Springer-Verlag (1968), 103-159.

37. Hansen, W.: Abbildungen harmonischer Räume mit Anwendung auf die Laplace- und Wärmeleitungsgleichung. Ann. Inst. Fourier 21 (1971), 203-216.

38. Hansen, W.: Semi-polar sets and quasi-balayage. Math. Ann. 257 (1981), 495-517.

39. Hervé, R.M.: Développements sur une théorie axiomatique des fonctions surharmoniques. C.R. Acad. Sci. Paris 248 (1959), 179-181.

40. Hervé, R.M.: Recherches axiomatiques sur la théorie des fonctions surharmoniques et du potentiel. Ann. Inst. Fourier 12 (1962), 415-571.

41. Hirsch, F.: Familles d'opérateurs potentiels. Ann. Inst. Fourier 25 (1975), 263-288.

42. Meyer, P.A.: Brelot's axiomatic theory of the Dirichlet problem and Hunt's theory. Ann. Inst. Fourier 13 (1963), 357-372.

43. Mokobodzki, G., Sibony, D.: Cônes adaptés de fonctions continues et théorie du potentiel. In: Sém. Choquet "Initiation à l'analyse" 6 (1968), 5.01-5.35.

44. Mokobodzki, G., Sibony, D.: Cônes de fonctions et théorie du potentiel I: Les noyaux associés à un cône de fonctions. In: Sém. Brelot-Choquet-Deny "Théorie du potentiel" 11 (1968), 8.01-8.35.

45. Mokobodzki, G., Sibony, D.: Cônes de fonctions et théorie du potentiel I, II. In: Sém. Brelot-Choquet-Deny "Théorie du potentiel" 11 (1968), 8.01-8.35, 9.01-9.29.

46. Schirmeier, U.: Produkte harmonischer Räume. Sitz.ber. math. nat. Kl. Bayer. Akad. Wiss. 1978 (1979), 5-22.

47. Sibony, D.: Cônes de fonctions et potentiels. Lecture Notes Mc Gill Univ. Montreal (1968).

48. Taylor, J.C.: Strict potentials and Hunt processes. Inventiones math. 16 (1972), 249-259.

49. Taylor, J.C.: Potential kernels of Hunt processes. Indiana Math. J. 22 (1973), 1091-1102.

AXIOMATIC NON–LINEAR POTENTIAL THEORIES

Ilpo Laine
Department of Mathematics, University of Joensuu
P.O.Box 111, SF–80101 Joensuu, Finland

1. Introduction

Axiomatic potential theory has been described as a crossing–point of several areas of mathematics. Such areas are at least function theory, partial differential equations, functional analysis and probability. Therefore, one should perhaps remark that the point of view of this article comes from complex analysis.

The familiar connection between the classical function theory and the classical potential theory reflects the important role of harmonic functions in many properties of analytic functions. A special position here is deserved by the basic result that $u \circ f$ remains harmonic whenever u is harmonic and f analytic. In the recent development of extending the classical function theory to several dimensions, notably via the theory of quasiregular mappings, this basic result becomes non–relevant. Moreover, the corresponding relevant invariance property, see Theorem 2.2 below, is non-linear by its intrinsic nature. On the other hand, several familiar phenomena from potential theory still hold in the frame of quasiregular mappings. Certainly, this is certainly one important motivation for non–linear considerations of potential theoretic nature which have been more or less frequent recently.

This article has been organized as follows. In Section 2 we describe more precisely the above background from complex analysis. To axiomatize such situations, two possible approaches seem to be natural. The first approach, to find an axiomatic system as light as possible, to describe concrete non–linear situations from Section 2, seems to result in Brelot–type axiomatic settings. Such an approach, due to P. Lehtola [10], will be shortly presented in Section 3. The second natural approach is to modify the classical linear axioms, say from [4], to obtain a fairly general non–linear axiomatic theory, covering at the same time as many concrete non–linear situations as possible. Such an approach will be described in Sections 4 and 5. The basic axioms of these sections have been published in [9]. Moreover, we propose here an additional axiom for handling MP–sets. This additional axiom enables us to consider the quasi-linear Dirichlet problem in the resolutive setting [4] rather than the regular setting which is more natural in Brelot–type theories, see [10]. Finally, we add some remarks concerning the global balayage in the axiomatic frame of Section 4.

2. Some background from complex analysis

We first recall the definitions of quasiregular mappings. Let $f:G \to \mathbb{R}^n$, $n \geq 2$, be a mapping in a domain $G \subset \mathbb{R}^n$ such that $f \in \mathcal{C}(G) \cap \mathrm{loc}\, W_n^1(G)$. Here $W_n^1(G)$ means the Sobolev space of all L^n–integrable functions in G whose distributional gradient ∇u is L^n–integrable in G. Now, f is quasiregular, if the supremum norm of the linear map f' satisfies

$$|f'(x)|^n \leq KJ(x,f)$$

a.e. in G for some $K \geq 1$, $J(x,f)$ being the Jacobian of f at x.

Next, we define variational kernels in a domain $G \subset \mathbb{R}^n$ following S. Granlund, P. Lindqvist and O. Martio, see e.g. [5]. This is a mapping $F:G{\times}\mathbb{R}^n \to \mathbb{R}$ satisfying certain regularity conditions, say for instance the following ones, see [13], p. 67:

(a) For any $\epsilon > 0$, there is a closed set $K_\epsilon \subset G$ such that the Lebesgue measure $\mu(G \setminus K_\epsilon) < \epsilon$ and the restriction $F|K_\epsilon{\times}\mathbb{R}^n$ is continuous.

(b) For a.e. $x \in G$, the function $h \mapsto F(x,h)$ is strictly convex and differentiable.

(c) For some constants α,β, $0 < \alpha \leq \beta < \infty$, the inequalities

$$\alpha|h|^n \leq F(x,h) \leq \beta|h|^n$$

hold for a.e. $x \in G$ and all $h \in \mathbb{R}^n$.

(d) For a.e. $x \in G$, the equality

$$F(x,\lambda h) = |\lambda|^n F(x,h)$$

holds for all $\lambda \in \mathbb{R}$ and all $h \in \mathbb{R}^n$.

For $u \in \mathrm{loc}\, W_n^1(G)$ and A measurable in G, we are now able to consider the variational integral

$$I_F(u,A) := \int_A F(x,\nabla u(x))d\mu(x).$$

A function $u \in \mathscr{C}(G) \cap \mathrm{loc}\, W_n^1(G)$ is called an <u>F–extremal</u> in G, if

$$I_F(u,D) = \inf\{I_F(w,D) \,|\, w \in \mathscr{C}(\mathrm{cl}D) \cap W_n^1(D),\ w = u \text{ in } \partial D\}$$

for all domains D relatively compact in G. F–extremals connect now to non–linear partial differential equations via the following

<u>Theorem 2.1.</u> A function $u \in \mathscr{C}(G) \cap W_n^1(G)$ is an F–extremal in G if and only if u is a weak solution, in the distributional sense, of

$$\nabla \cdot \nabla_h F(x,h(x)) = 0.$$

A lower semi–continuous function $u: G \to (-\infty,+\infty]$ is a <u>super–F–extremal</u>, if it satisfies the following comparison principle in each domain D relatively compact in G: For every F–extremal h in D such that $h \in \mathscr{C}(\mathrm{cl}D) \cap \mathrm{loc}\, W_n^1(D)$ and $h \le u$ in ∂D, $h \le u$ holds in D. It is now possible to prove that F–extremals and super–F-extremals share several familiar potential theoretic properties, see e.g. [5] and [6]. Unfortunately, linearity does not hold in general.

Let now $f: G \to \mathbb{R}^n$ be a non–constant quasiregular mapping and $F: f(G) \times \mathbb{R}^n \to \mathbb{R}$ be a variational kernel satisfying the conditions (a) – (d) in $f(G)$. Define a new kernel $f^*F: G \times \mathbb{R}^n \to \mathbb{R}$ as

$$f^*F(x,h) := \begin{cases} F(f(x), J(x,f)^{1/n}(f'(x)^{-1})^* h), & \text{if } J(x,f) \neq 0 \\ |h|^n, & \text{if } J(x,f) \text{ does not exist or if } J(x,f) = 0. \end{cases}$$

The following theorem now holds, see [13], Theorem 6.18:

<u>Theorem 2.2.</u> Let $f: G \to \mathbb{R}^n$ be a quasiregular mapping and let u be an F–extremal, resp. a super–F–extremal, in a domain $G' \supset f(G)$. Then $u \circ f$ is an f^*F–extremal, resp. a super–f^*F–extremal, in G.

Before proceeding, we still remark that many of the potential theoretic aspects of (super–)F–extremals may be proved to hold for some other non–linear partial differential equations too. For instance,

$$\nabla \cdot (|\nabla u|^{p-2}\nabla u) = 0$$

has been studied from this point of view by P. Lindqvist [11]. More generally,

$$\nabla \cdot A(x,\nabla u) = 0,$$

where $A:G\times\mathbb{R}^n \to \mathbb{R}^n$ is a strictly monotone elliptic differential operator in an open set $G \subset \mathbb{R}^n$ satisfying certain regularity conditions [7] and originally investigated by V. G. Maz'ya in [14], was studied from the above point of view by J. Heinonen and T. Kilpeläinen in [7] and [8].

3. Brelot–type non–linear axioms.

To describe the Brelot–type non–linear axioms due to Lehtola in [10], let X be a locally compact, locally connected, connected and non–compact topological space and let $\mathfrak{H}$ be a sheaf of continuous functions on X. The functions in $\mathfrak{H}(G)$ on an open set G are called harmonic. A relatively compact open set G is called Dirichlet regular, if for every $f \in \mathscr{C}(\partial G)$ there is a unique function $h = H_f^G \in \mathfrak{H}(G) \cap \mathscr{C}(\mathrm{cl}G)$ such that $f = h|\partial G$. A Dirichlet regular set G is called regular, if for every pair $f,g \in \mathscr{C}(\partial G)$ the condition $f < g$ implies $H_f^G < H_g^G$. Observe that the strict comparison principle in the definition of a regular set is more or less technical only, see [10], Theorem 2.4.

We are now ready to give the following

Definition 3.1. The pair $(X,\mathfrak{H})$ is a non–linear Brelot space, if the following three axioms hold:

Axiom of scalar multiplication: $\lambda h \in \mathfrak{H}(G)$ for every $h \in \mathfrak{H}(G)$ and every $\lambda \in \mathbb{R}$;

Axiom of regularity: For every open set $U \subset X$ and every compact set $K \subset U$ there is a regular set G such that $K \subset G$ and G is relatively compact in U;

Axiom of convergence: The harmonic sheaf $\mathfrak{H}$ possesses the Brelot convergence property.

Observe that the original axioms given in [10] are a bit more general. We prefer here the above modified form to make the comparison with the next section easier. Concerning details of non–linear Brelot spaces, we refer to [10]. We just mention here that the above axioms are sufficient to develop Poisson modifications, Perron families and the usual theory of the Dirichlet problem by the Perron method. However, to develop standard results about resolutive functions, one needs an additional axiom stating that $h + \lambda \in \mathfrak{H}(G)$ for every $h \in \mathfrak{H}(G)$ and every $\lambda \in \mathbb{R}$. Therefore, one is tempted to include this axiom in the above Definition 3.1.

4. Bauer–Constantinescu–Cornea–type quasi–linear axioms

A Bauer–Constantinescu–Cornea–type set of quasi–linear axioms was proposed in [9]. Essential differences to Brelot–type axioms described in the preceding section are similar to the linear theory, namely a more general convergence axiom and a different approach to the Dirichlet problem. In fact, we propose here a quasi-linear method which makes it possible to rely on resolutive rather than regular sets while considering the Poisson modification and the Perron method for the corresponding non–linear Dirichlet problem. This approach differs slightly from what was anticipated in [9], p. 339.

To recall the axioms for a quasi–linear harmonic space [9], p. 340–342, we consider a locally compact space X together with a hyperharmonic sheaf $\mathfrak{U}$. An open set $U \subset X$ is called an MP–set (relative to $\mathfrak{U}$), if $u \geq v$ holds for any lower bounded hyperharmonic function $u \in \mathfrak{U}(U)$ and any upper bounded hypoharmonic function $v \in -\mathfrak{U}(U)$ as soon as (1) $\liminf_{x \in U, x \to y} u(x) \geq \limsup_{x \in U, x \to y} v(x)$ for every $y \in \partial U$ and (2) there exists a compact set $K(u,v)$ in X such that $u \geq v$ holds on $U \setminus K(u,v)$. This definition enables us to speak about resolutive sets relative to $\mathfrak{U}$ in the similar way as in the usual linear theory. We now recall the quasi–linear axioms from [9]. They should be compared with the corresponding linear axioms, see [4].

Axiom of quasi–linearity: The hyperharmonic sheaf $\mathfrak{U}$ contains a non–empty harmonic subsheaf $\mathfrak{V} \subset \mathfrak{H}_{\mathfrak{U}} := \mathfrak{U} \cap (-\mathfrak{U})$ such that for every open set $U \subset X$,

(1) $\alpha v \in \mathfrak{V}(U)$ for every $v \in \mathfrak{V}(U)$ and every $\alpha \in \mathbb{R}$,

(2) $u + v \in \mathfrak{U}(U)$ for every $u \in \mathfrak{U}(U)$ and every $v \in \mathfrak{V}(U)$.

Axiom of resolutivity: The open sets resolutive to $\mathfrak{U}$ form a base for the topology of X.

Axiom of quasi–linear positivity: The quasi–linear subsheaf $\mathfrak{V}$ determined above is non–degenerated at every point $x \in X$.

Before proceeding, we recall here that an open set U is called sufficiently small, if $\mathrm{cl}U$ is contained in an open set V such that there exists a strictly positive function $v \in \mathfrak{V}(V)$.

Axiom of completeness: A lower semicontinuous function $u:U \to (-\infty,+\infty]$ on an open set U is hyperharmonic on U, if for every relatively compact, resolutive set V such that $\mathrm{cl}V \subset U$, the inequality $\underline{H}_u^V \le u$ holds.

Axiom of convergence: The harmonic sheaf $\mathfrak{H}_{\mathfrak{U}} = \mathfrak{U}\cap(-\mathfrak{U})$ satisfies the Bauer convergence property.

$(X,\mathfrak{U})$ satisfying the above axioms is called a quasi–linear harmonic space. Concerning elementary properties of these spaces, see [9]. We recall here the following lemma only, which is an essential device to overcome some difficulties arising from the fact that the functional $f \mapsto H_f^U(x)$ is not any more linear:

Lemma 4.1. ([9], Lemma 3.1 and [10], Theorem 3.7) Let V be a sufficiently small open set. If U is an MP–set relatively compact in V and $\{f_\alpha\}_{\alpha\in I}$ is an upper directed family of lower semicontinuous functions on ∂U, then

$$\underline{H}^U_{\sup_{\alpha\in I} f_\alpha} = \sup_{\alpha\in I} \underline{H}^u_{f_\alpha} .$$

To proceed now to the quasi–linear Dirichlet problem, we propose an additional axiom which we assume to hold without any further mention:

Axiom of MP–sets: Any open subset of an MP–set is also an MP–set.

This axiom enables us to prove (in our quasi–linear context) the standard truncation property:

Lemma 4.2. Let $U, U' \subset U$ be two open sets of X and let $u \in \mathfrak{U}(U)$, $u' \in \mathfrak{U}(U')$ be hyperharmonic functions. If the function $u^*:U \to (-\infty,+\infty]$ defined by

$$u^*(x) := \begin{cases} \inf(u,u') & \text{on } U' \\ u & \text{on } U \setminus U' \end{cases}$$

is lower semicontinuous, then it is hyperharmonic on U.

Proof. Let V be a resolutive set relativewly compact in U and take

$f \in \mathscr{C}(\partial V)$, $f \le u^*$. Since $u \in \overline{\mathfrak{U}}_f(V)$, we have $u^* = u \ge \overline{H}_f^V = H_f^V$ on $V \setminus U'$. Take now $v \in \underline{\mathfrak{U}}_f(V)$, hence $H_f^V \ge v$ on V and $u^* \ge v$ on $V \setminus U'$. By semicontinuity,

$$\liminf_{x \in U' \cap V, x \to y} u^*(x) \ge \liminf_{x \to y} u^*(x) \ge u^*(y) \ge v(y) \ge$$

$$\limsup_{x \to y} v(x) \ge \limsup_{x \in U' \cap V, x \to y} v(x)$$

for all $y \in \partial U' \cap V \subset V \setminus U'$. Since

$$\partial(U' \cap V) = (\partial U' \cap V) \cup (\partial V \cap \mathrm{cl}U'),$$

it remains to consider $y \in \partial V \cap \mathrm{cl}U'$. Then

$$\liminf_{x \in U' \cap V, x \to y} u^*(x) - \limsup_{x \in U' \cap V, x \to y} v(x) \ge$$

$$\liminf_{x \to y} u^*(x) - \limsup_{x \to y} v(x) \ge u^*(y) - f(y) \ge f(y) - f(y) = 0,$$

since u^* is lower semicontinuous and $v \in \underline{\mathfrak{U}}_f(V)$. Since $U' \cap V$ is an MP–set, $u^* \ge v$ on $U' \cap V$, hence on the whole V. This implies $u^* \ge H_f^V = \underline{H}_f^V$. Since

$$u^* = \sup\{f \in \mathscr{C}(\partial V) \,|\, f \le u^* \,|\, \partial V\}$$

on ∂V by lower semicontinuity, we obtain

$$u^* \ge \sup\{\underline{H}_f^V \,|\, f \in \mathscr{C}(\partial V), f \le u^* \,|\, \partial V\} = \underline{H}^V_{\sup\{f \in \mathscr{C}(\partial V) | f \le u^* | \partial V\}} = \underline{H}^V_{u^*}$$

by Lemma 4.1. By the axiom of completeness, $u^* \in \mathfrak{U}(U)$.

The prototype to consider the Poisson modification in a resolutive setting is, of course, Proposition 2.1.3 in [4]. As we see immediately, the (standard) proof in [4], p. 32 – 33, relies on linearity. Therefore, we have to find a different approach to obtain similar quasi–linear results. Let u be hyperharmonic on X and let U be a resolutive, relatively compact and sufficiently small set. We define the <u>abstract Poisson modification</u> of u on U by

$$P(u,U) := \begin{cases} u & \text{on } X \setminus \mathrm{cl}U \\ \inf(u, \underline{H}_u^U) & \text{on } \mathrm{cl}U \end{cases},$$

where $\underline{H}_u^U$ has been extended to ∂U by its lower limit. Clearly, P(u,U) is lower semicontinuous. Let $\{f_\alpha\}_{\alpha\in I}$ denote the upper directed family of continuous minorants of u on ∂U, hence $u = \sup_{\alpha\in I} f_\alpha$. By Lemma 4.1,

$$P(u,U) = \sup_{\alpha\in I} P(f_\alpha,u,U),$$

where

$$P(f_\alpha,u,U) := \begin{cases} u & \text{on } X \setminus \mathrm{cl}U \\ \inf(u,\underline{H}_{f_\alpha}^U) & \text{on } \mathrm{cl}U \end{cases}$$

produces an upper directed family of lower semicontinuous functions.

Theorem 4.3. The abstract Poisson modification P(u,U) is hyperharmonic on X.

Proof. Let V be a relatively compact, resolutive, sufficiently small set such that $U\cap\partial V \neq \phi$, v_0 being a strictly positive, quasi–linear harmonic function in an open neighbourhood of clV, see [9],p. 344.

Let now $\epsilon > 0$ be given and take

$$w \in \overline{\mathcal{U}}_{f_\beta}^U,\ v \in \underline{\mathcal{U}}_{P(f_\alpha,u,U)}^V, \tag{4.1}$$

considering the MP–set $U\cap V$ whose boundary divides into

$$\partial(U\cap V) = (U\cap\partial V)\cup(\partial U\cap \mathrm{cl}V).$$

Here f_α is first kept fixed as well v and $f_\beta = f_{\beta(\alpha)} \geq f_\alpha$ is to be specified below. For $y \in U\cap\partial V$, we obtain

$$\begin{aligned}\limsup_{x\in U\cap V, x\to y} v(x) &\leq P(f_\alpha,u,U)(y) = \inf(u(y),\underline{H}_{f_\alpha}^U(y)) \leq \\ \underline{H}_{f_\alpha}^U(y) &\leq \underline{H}_{f_\beta}^U(y) \leq w(y) \leq \liminf_{x\in U\cap V, x\to y} w(x).\end{aligned} \tag{4.2}$$

To consider $y \in \partial U\cap \mathrm{cl}V$, clearly $f_{\beta_y}(y) + \epsilon v_0(y) > u(y)$ for some $\beta_y \in I$, $f_{\beta_y} \geq f_\alpha$. Then

$$\begin{aligned}\limsup_{x\in U\cap V, x\to y} v(x) &\leq \limsup_{x\in V, x\to y} v(x) =: v(y) \leq \\ P(f_\alpha,u,U)(y) &\leq u(y) < f_{\beta_y} + \epsilon v_0(y)\end{aligned}$$

Now, $f_{\beta_y} + \epsilon v_o(y) - v$ is lower semicontinuous in $\partial U \cap \mathrm{cl} V$, hence $f_{\beta_y} + \epsilon v_o(y) - v > 0$ holds in an open neighbourhood of y in $\partial U \cap \mathrm{cl} V$. By compactness, we find $f_{\beta(\alpha)} \geq f_\alpha$ in (4.1) such that $f_{\beta(\alpha)} + \epsilon v_o - v \geq 0$ in $\partial U \cap \mathrm{cl} V$. Assume now that we have selected $f_\beta = f_{\beta(\alpha)}$ in (4.1). Hence

$$\limsup_{x \in U \cap V, x \to y} v(x) \leq v(y) \leq f_\beta(y) + \epsilon v_o(y) \leq \liminf_{x \in U \cap V, x \to y} w(x) + \epsilon v_o(y) \quad (4.3)$$

holds for all $y \in \partial U \cap \mathrm{cl} V$. Combining (4.2) and (4.3), we see that

$$\limsup_{x \in U \cap V, x \to y} v(x) \leq \liminf_{x \in U \cap V, x \to y} (w(x) + \epsilon v_o(x))$$

holds for all $y \in \partial(U \cap V)$. Since $U \cap V$ is an MP–set, $v \leq w + \epsilon v_o$. Letting now w vary we obtain, by applying Lemma 4.1,

$$v \leq \overline{H}^U_{f_\beta} + \epsilon v_o = \underline{H}^U_{f_\beta} + \epsilon v_o \leq \sup_\gamma \underline{H}^U_{f_\gamma} + \epsilon v_o = \underline{H}^U_u + \epsilon v_o$$

in $U \cap V$. Since $v \in \underline{\mathfrak{U}}^V_{P(f_\alpha, u, U)}$ was arbitrary, we get

$$\underline{H}^V_{P(f_\alpha, u, U)} \leq \underline{H}^U_u + \epsilon v_o$$

on $U \cap V$ and, by Lemma 4.1 again,

$$\underline{H}^V_{P(u,U)} = \underline{H}^V_{\sup_{\alpha \in I} P(f_\alpha, u, U)} = \sup_{\alpha \in I} \underline{H}^V_{P(f_\alpha, u, U)} \leq \underline{H}^U_u + \epsilon v_o$$

on $U \cap V$, hence also on $V \cap \partial U$. Since $P(u,U) \leq u$,

$$\underline{H}^V_{P(u,U)} \leq \underline{H}^V_u \leq u < u + \epsilon v_o$$

holds on V. Therefore

$$\underline{H}^V_{P(u,U)} \leq P(u,U) + \epsilon v_o$$

holds on V and, letting $\epsilon \to 0$,

$$\underline{H}^V_{P(u,U)} \leq P(u,U)$$

now follows.

By the above reasoning, $P(u,U)$ is hyperharmonic on all sufficiently small sets. By the sheaf property of hyperharmonic functions, $P(u,U)$ is hyperharmonic on

X.

The preceding theorem now enables us to define the quasi–linear Perron families:

Definition 4.4. A lower directed family $\mathfrak{W}$ of hyperharmonic functions on a quasi–linear harmonic space X satisfying the axiom of MP–sets is called a Perron family, if it admits a subharmonic minorant and if every point of X has a relatively compact, resolutive, sufficiently small neighbourhood V such that a) for every $v \in \mathfrak{W}$, also $P(v,V) \in \mathfrak{W}$, b) there exists $v \in \mathfrak{W}$ such that $\underline{H}_v^V$ is harmonic.

Remark. Definition 4.4 extends in a natural way to define the Perron family generated by a superharmonic function u on X possessing a subharmonic minorant and by a covering $\mathfrak{T}$ of X by relatively compact, resolutive, sufficiently small sets.

Proposition 4.5. The infimum of a Perron family $\mathfrak{W}$ is a harmonic function.

Proof. Let us write $v := \inf \mathfrak{W}$ and let V be a resolutive set described in Definition 4.4. Since $P(u,V) \le u$ for all $u \in \mathfrak{W}$, $v = \inf\{P(u,V) \mid u \in \mathfrak{W}\} = \inf\{\underline{H}_u^V \mid u \in \mathfrak{W}\}$ holds on V. Therefore we may assume that $\mathfrak{W}$ is a lower directed family of functions harmonic on V. Let s be a subharmonic minorant of $\mathfrak{W}$. Then $\overline{H}_s^V$ is a harmonic minorant on V of $\{\underline{H}_u^V \mid u \in \mathfrak{W}\}$. Hence v is harmonic on V by the axiom of convergence. By the sheaf property of harmonic functions, v is harmonic on X.

5. The quasi–linear Dirichlet problem in a quasi–linear harmonic space

The preceding developments enable us to consider the Dirichlet problem, at least in a sufficiently small, resolutive set. In this section we now assume throughout that U is a relatively compact, resolutive, sufficiently small set. Also, [9], Lemma 2.1 and the above Lemma 4.1 will be applied below.

Proposition 5.1. Let $f: \partial U \to \mathbb{R}$ be bounded. Then $\overline{H}_f^U$ and $\underline{H}_f^U$ are harmonic functions.

Proof. Since U is sufficiently small, there exists $v \in \mathfrak{V}(V)$, $V \supset \mathrm{cl}U$, such that $v > |f|$ on ∂U. Clearly, $\mathfrak{U}_f(U)$ is now a Perron family, hence $\overline{H}_f^U$ is harmonic. From $\underline{H}_f^U = -\overline{H}_f^U$ it follows that $\underline{H}_f^U$ is harmonic, too.

Proposition 5.2. Let $f_n: \partial U \to \mathbb{R}$, $n \in N$, be resolutive functions with a bounded uniform limit $f := \lim_{n\to\infty} f_n$ on ∂U. Then f is resolutive and $H_f^U = \lim_{n\to\infty} H_{f_n}^U$ uniformly on U.

Proof. Assume $v_0 \in \mathfrak{V}(V)$, $V \supset \mathrm{cl}U$, is such that $\inf v_0(U) \geq 1$. Clearly $v_0|U$ is bounded. Fix $\epsilon > 0$ and let $m \in N$ be such that $|f - f_n| < \epsilon < \epsilon v_0$ on ∂U for $n \geq m$. By [9], Lemma 2.1, we obtain $|\overline{H}_f^U - H_{f_n}^U| \leq \epsilon v_0$ and $|\underline{H}_f^U - H_{f_n}^U| \leq \epsilon v_0$. Therefore $|\overline{H}_f^U - \underline{H}_f^U| \leq 2\epsilon v_0$ and it follows that $\overline{H}_f^U = \underline{H}_f^U$. Now, f is resolutive by Proposition 5.1.

Let $x \in U$ be given. Since v_0 is bounded on U, $\{H_{f_n}^U(x)\}_{n\in N}$ is a Cauchy sequence. Immediately, $H_f^U(x) = \lim_{n\to\infty} H_{f_n}^U(x)$. Since v_0 is bounded, this limit is uniform on U.

Our quasi–linear version of the Bouligand theorem (Theorem 5.4 below) contains an additional parameter $\epsilon > 0$. This parameter is needed since multiplying a hyperharmonic function with a real number > 0 does not necessarily produce a hyperharmonic product. It remains open whether this parameter could be removed.

Definition 5.3. A point $y \in \partial U$, U open, has a barrier, if there exists a neighbourhood W of y in X such that some hyperharmonic function $w \in \mathfrak{U}_+(U)$ satisfies $\lim_{x\to y} w(x) = 0$ and $\liminf_{x\to y} w(x) > 0$ for all $y \in (\partial U \setminus W) \cup (U \cap \partial W)$.

Theorem 5.4. Let $f: \partial U \to [-\infty, +\infty)$ be an upper bounded boundary function. If a point $y_0 \in \partial U$ has a barrier, then for all sufficiently small $\epsilon > 0$,

$$\limsup_{x\in U, x\to y_0} H_{\epsilon f}^U(x) \leq \limsup_{y\in\partial U, y\to y_0} \epsilon f(y).$$

Proof. Assume $v_0 \in \mathfrak{V}(V)$, $V \supset \mathrm{cl}U$, satisfies $v_0(y_0) = 1$ and assume that $\lambda \in \mathbb{R}$ satisfies $\lambda = \lambda v_0(y_0) > \limsup_{y\in\partial U, y\to y_0} f(y)$. Take a neighbourhood W of y in X possessing a barrier w, assuming at the same time that $U \setminus W \neq \phi$ and that $f < \lambda v_0$ on $\partial U \cap W$. Since $\liminf_{x\to y} w(x) > 0$ for all $y \in \partial U \setminus W$, all sufficiently small $\epsilon > 0$ satisfy the inequality

$$\lambda \inf v_0(\partial U) + \epsilon^{-1} \inf_{\partial U\setminus W} \liminf_{x\to y} w(x) > \sup f(\partial U).$$

Consider now the hyperharmonic function $u := \lambda\epsilon v_0 + w$. For any $y \in \partial U \setminus W$, we have

$$\liminf_{x\to y} u(x) = \lambda\epsilon v_0(y) + \liminf_{x\to y} w(x) > \epsilon f(y)$$

and for any $y \in \partial U \cap W$,

$$\liminf_{x\to y} u(x) \geq \epsilon\lambda v_0(y) > \epsilon f(y).$$

Therefore $u \in \overline{\mathfrak{U}}_{\epsilon f}(U)$ and we have $\overline{H}^U_{\epsilon f} \leq u$ on U. But this implies that $\limsup_{x\in U, x\to y_0} \overline{H}^U_{\epsilon f}(x) \leq \epsilon\lambda$. Since $\lambda > \limsup_{y\in\partial U, y\to y_0} f(y)$ is arbitrary, we obtain the assertion.

<u>Remark.</u> The above proof is a slight modification of the corresponding reasoning due to Lehtola to prove Lemma 3.18 in [10]. Theorem 5.4 further suggests the following

<u>Definition 5.5.</u> A boundary point $y_0 \in \partial U$ is <u>regular</u>, if for all $f \in \mathscr{C}(\partial U)$, there exists an $\epsilon_f > 0$ such that

$$\lim_{x\to y_0} H_{\epsilon f}(x) = \epsilon f(y_0)$$

holds for all ϵ, $0 < \epsilon \leq \epsilon_f$.

An immediate consequence of Theorem 5.4 is now

<u>Corollary 5.6.</u> If a boundary point $y_0 \in \partial U$ has a barrier, then y_0 is regular.

<u>Remark.</u> The converse result seems to be more difficult. Actually, a corresponding theorem to [10], Theorem 3.20 is quite immediate. However, Axiom S of [10] is not very natural in our Bauer–Constantinesu–Cornea setting.

6. Remarks concerning global balayage

As a first step towards the converse of Corollary 5.6, we add basic results concerning the global balayage in the setting of Section 4.

<u>Definition 6.1.</u> For a numerical function $f: X \to \overline{\mathbb{R}}$, the <u>balayage</u> of f on X will be defined by

$$Rf := \inf\{u \in \mathfrak{U}(X) \,|\, u \geq f\}.$$

Remark. Immediately, $f \leq g$ implies $Rf \leq Rg$.

Proposition 6.2. Let $f:X \to (-\infty,+\infty]$ be lower semicontinuous. Then Rf is hyperharmonic. Moreover, if f has a superharmonic majorant, then Rf is superharmonic and it is finite and continuous at x, if f is finite and continuous at x.

Proof. Let $u:X \to (-\infty,+\infty]$ be defined by

$$u(x) := \liminf_{y\to x} Rf(y).$$

Clearly, u is lower semicontinuous and $\geq f$. Let now V be a resolutive, relatively compact, sufficiently small set and $v \geq f$ be hyperharmonic. Then $\underline{H}_u^V \leq \underline{H}_v^V \leq v$, hence $\underline{H}_u^V \leq Rf$ on V. Since $\underline{H}_u^V$ is lower semicontinuous by [9], Proposition 3.3, we obtain $\underline{H}_u^V \leq u$. By a standard reasoning, u is hyperharmonic and therefore $Rf = u$.

The second assertion is now trivial.

Let finally f be finite and continuous at x, and take a relatively compact, resolutive, sufficiently small neighbourhood V of x such that

$$f \leq (f(x) + \epsilon)v_o,\ Rf \geq (Rf(x) - \epsilon)v_o$$

holds on clV, v_o being a harmonic function $\in \mathfrak{W}(W)$, W open $\supset clV$, such that $v_o(x) = 1$. Define now

$$u^* := \begin{cases} Rf & \text{on } X \setminus clV \\ \inf(Rf, \underline{H}_{Rf}^V) & \text{on } clV. \end{cases}$$

By Theorem 4.3, u^* is hyperharmonic. From (6.1), it is easily seen that $f \leq \underline{H}_{Rf}^V + 2\epsilon v_o$ holds on V, and this inequality extends by lower limit to clV. Therefore $f \leq u^* + 2\epsilon v_o$ and we see that

$$Rf \leq u^* + 2\epsilon v_o \leq \underline{H}_{Rf}^V + 2\epsilon v_o$$

holds on V. Since Rf is superharmonic, $\underline{H}_{Rf}^V$ is harmonic, hence continuous on V, and we obtain

$$\limsup_{y\to x} Rf(y) \leq \lim_{y\to x} \underline{H}_{Rf}^V(y) + 2\epsilon v_o(x) \leq$$

$$\underline{H}^{V}_{Rf}(x) + 2\epsilon \leq Rf(x) + 2\epsilon \leq \liminf_{y \to x} Rf(y) + 2\epsilon,$$

giving the assertion.

7. Concluding remarks.

The reader should observe that this short excursion to axiomatic non–linear potential theories is by no means complete. Apparently there exist several different non–linear sets of axioms, having a different concrete background. We call here the reader's attention at least to the articles of K. Akô [1] and B. Calvert [3], see also a survey due to F.–Y. Maeda [12]. Moreover, a quite different set of axioms, based on the Monge–Ampère equation, has been developed by E. Bertin, see [2] and his article in the Proceedings of this Conference. The above articles as well [9] and [10] contain still further references.

References

[1] Akô, K.: On a generalization of Perron's method for solving the Dirichlet problem of second order partial differential equations. J. Fac. Soc. Univ. Tokyo 8 (1959–60), 263 – 288.

[2] Bertin, E. M. J.: Fonctions convexes et théorie du potentiel. Indag. Math. 41 (1979), 385 – 396, 397 – 409.

[3] Calvert, B.: Dirichlet problems without coercivity by the Perron – Akô – Constantinescu method. Math. Chronicle 6 (1977), 48 – 67.

[4] Constantinescu, C., Cornea, A.: Potential theory on harmonic spaces. Springer–Verlag (1972).

[5] Granlund, S., Lindqvist, P., Martio, O.: Conformally invariant variational integrals. Trans. Amer. Math. Soc. 277 (1983), 43 – 73.

[6] Granlund, S., Lindqvist, P., Martio, O.: Note on the PWB–method in the non–linear case. Pacific J. Math. 125 (1986), 381 – 395.

[7] Heinonen, J., Kilpeläinen, T.: A–superharmonic functions and supersolutions of degenerate elliptic equations. To appear in Ark. Mat.

[8] Heinonen, J., Kilpeläinen, T.: Polar sets for supersolutions of degenerate elliptic equations. Preprint.

[9] Laine, I.: Introduction to a quasi–linear potential theory. Ann. Acad. Sci. Fenn. A I 10 (1985), 339 – 348.

[10] Lehtola, P.: An axiomatic approach to non–linear potential theory. Ann.

Acad. Sci. Fenn. A I Diss. 62 (1986).

[11] Lindqvist, P.: On the definition and properties of p–superharmonic functions. J. Reine Angew. Math. 365 (1986), 67 – 79.

[12] Maeda, F.–Y.: Nonlinear harmonic space and Dirichlet problems (Japanese). Proceedings of the Conference on Potential Theory, Karuizawa (1982), 1 – 8.

[13] Martio, O.: Non–linear potential theory. In: Summer School in Potential Theory, Univ. of Joensuu Report Ser. N:o 5–M (1983), 65 – 104.

[14] **Мазья, В. Г.: О непрерывности в граничной точке решений квазилинейных эллиптических уравнений. Вестник ЛГУ** 25:13 (1970), 42 – 55.

APPLICATION OF THE POTENTIAL THEORY TO THE STUDY OF QUALITATIVE PROPERTIES OF SOLUTIONS OF THE ELLIPTIC AND PARABOLIC EQUATIONS

E.M.Landis
Dept. of Mathematics, Moscow State University
117 234, Moscow, USSR

Introduction. The problem of existence and uniqueness and of seeking solutions of different problems occupies a significant place in the theory of differential equations. The qualitative theory, however, deals with a given a priori solution whose properties need to be investigated. The potential theory is a very useful tool in these studies, yielding explicit formulae of solutions or of their estimating functions, while constructive parts of modern methods of investigation of existence, uniqueness, etc. are buried deep in depths of functional analysis. It is the qualitative questions that are treated in my report.

I limit myself to elliptic and parabolic linear equations of second order. It is mainly because of space limitation but also because, as I know, Prof. G.W. Wildenhain is reporting here on equations of higher order and Prof. I.V. Skrypnik, on non-linear equations.

It is not my aim to give any complete review of publications on the topic, for these publications are so numerous that only to name them (even the recent ones) would have made a reference list of my report; this is of course useful but disagrees with my intentions. I have chosen some of more familiar to me lines of investigation, trying to give a notion not only of the results obtained there, but also of the methods applied. I am aware that such an approach will inevitably introduce some subjective elements and distort, to a certain degree, the general picture, but, probably, it cannot be avoided: when one looks at a set of objects from within this set, the way the set presents itself to the eye depends on the observation point. I quote a rather large number of soviet mathematicians. It is my aim to attract attention to those of their results that seem of interest to me and, to my knowledge, are less-known to the outer world.

1. Non-divergence elliptic equations.

Consider an operator of the form

$$L = \sum_{i,k=1}^{n} a_{ik}(x)\frac{\partial^2}{\partial x_i \partial x_k} + \sum_{i=1}^{n} b_i(x)\frac{\partial}{\partial x_i} + c(x) \quad (1)$$

where all the coefficients are measurable functions, $a_{ik} = a_{ki}$,

$$\lambda^{-1}|\xi|^2 \le \sum a_{ij}\xi_i\xi_j \le \lambda|\xi|^2 \quad (2)$$

$$|b_i(x)| \le M, \quad c(x) \le 0 \quad (3)$$

A solution of the equation

$$Lu = 0 \quad (4)$$

in the region $G \subset R^n$ is uderstood either as a classic solution $u \in C^{2,loc}$ or $u \in W_n^{2,loc}$,satisfying the equation almost everywhere.

As is known, a solution $u \in W_n^2$ satisfies the maximum principle.

a) The regularity of boundary points. Let G R^n be a bounded region. Let Eqn.(4) be defined in G . Let $f \in C(\Omega)$ .If in a neighbourhood of any internal point there exists a solution of the Dirichlet problem,e.g. when the coefficients satisfy the Dini condition in a neighbourhood of any such point, then the generalized Wiener solution u (x) can be constructed. P. Lax [1]was first to attract attention to that.One may demand less, namely that in a neighbourhood of any internal point of G the matrix $\|a_{ik}\|$ satisfy the Cordes condition [2] (the spread of the roots of the characteristic equation be small).In this case, a generalized solution $u_f(x) \in W_n^{2,loc}(G)$ can also be constructed. Note $u \in C_{loc}^{1+\alpha}$

A boundary point $x_0 \in \partial G$ is called a regular point if, for all f belonging to $C(\partial G)$, $u_f(x) \to f(x_0)$ at $x \to x_0, x \in G$. A well known necessary and sufficient condition of regularity for the Laplace equation, Wiener's criterion, consists in the divergence (for $n > 2$) of the series

$$\sum_{k=1}^{\infty} 2^{k(n-2)} \operatorname{cap} E_k \quad (5)$$

where $E_k = (B(2^{-k},x_0) \setminus B(2^{-(k+1)},x_0)) \setminus G$, capE is the capacity of E , B(R,x) is, here and below, an open n-dimensional ball of the radius $R > 0$, centered at x, S(R,x), its boundary sphere.In the years that followed, mathematicians tried to determine what minimal demands leave the regularity conditions for (4) to be coinciding with those for $\Delta u = 0$. In 1962 M. Hervé [3] succeded in lowering the smoothness of the coefficients down to the Hölder's condition, in 1967 N.V. Krylov, to uniform fulfilment of the Dini condition [4].

The Dini condition seems to establish a boundary line for coincidence of regularity conditions for Eqn.(4) and for $\Delta u = 0$. The simple continuity does not suffice here. In 1968-69 K. Miller [5] and O.N. Zograf [6] constructed examples of equations with continuous coefficients, the condition of regularity for which did not coincide with those for the Laplace equation.An analysis of O.N. Zograf's example shows that the continuity modulus there equals $\omega(\rho)=1/|\ln\rho|$. The question still remains open, whether for a given continuity modulus $\omega(\rho)$ such that $\int_0 d\rho/\omega(\rho)$ one can construct an example of equation whose coefficient are continuous , $\omega(\rho)$ being their uniform continuity modulus, of a region $G \subset R^n$, of function $f \in C(\partial\Omega)$ so that , for a point $x_0 \in \partial G$, the regularity conditions do not coincide with those for $\Delta u=0$. A.A. Novruzov has constructed an example of non-coincidence with the coefficients having such continuity modulus at one point x_0. At the same time, A.A. Novruzov, Yu.A. Alhutov, A.I. Ibragimov [7] -[9] , for Eqn.(4) with continuous coefficients, have found necessary conditions and, separately, sufficient conditions , of regularity in terms of the continuity modulus. In general, these conditions do not come close to each other. Under some assumtions on the structure of the coefficients, however, even in the case of discontinuous coefficients one can find the conditions of the necessary and sufficient conditions coming close to each other and of their coincidence with the conditions for $\Delta u=0$.

Now we quote some of the results. Let, for the operator L , be satisfied $C(x)>-M$, in addition to (2),(3) .The function

$$n(x,y)=\sum_{i=1}^{n} a_{ii}(x)\Big/\sum_{i,k=1}^{n} a_{ik}(x_i-y_i)(x_k-y_k)\,,\quad x\in G\,,\ y\in\partial G \tag{6}$$

called the ellipticity function, is being considered.A function of this kind was introduced by Yu.A. Alhutov [7] , its form (6) was propozed by A.A. Novruzov [9]. Suppose that there exist constants s_1 and s_2 such that

$$n(x,y)-s_1 \le \varphi(|x-y|) \tag{7}$$

$$s_2 - n(x,y) \ge -\varphi(|x-y|) \tag{8}$$

where $\int_0 dt/\varphi(t)<\infty$. Consider the functions

$$G_+=\int_z^{\operatorname{diam} G} t^{1-s_1}\exp\Big[\int_t^{\operatorname{diam} G}(\varphi(\tau)/\tau)\,d\tau\Big]\,dt$$

and

$$G_-=\int_z^{\operatorname{diam} G} t^{1-s_2}\exp\Big[\int_t^{\operatorname{diam} G}(\varphi(\tau)/\tau)\,d\tau\Big]\,dt$$

Using the kernels $G_\pm$, we construct , in a usual manner, the capacities, $\gamma_\pm(E)$, of a Borel set $E \subset R^n$. Let $x_0\in\partial G$. Set up a Wiener series for these capacities: let $E_k=B(2^{-k},x_0)\setminus G$, $\gamma_k^+=\gamma_+(E_k)$, $\gamma_k^-=\gamma_-(E_k)$

The series

$$\sum_{k=1}^{\infty} 2^{k(n-2)}\gamma_k^{\pm} \tag{9}$$

is being cosidered. If inequality (7) holds and series (9) with the + (-) sign diverges (con) then x_0 is an (ir)regular point. In a particular case, when $s_1=s_2=n$, $\gamma^\pm$ are in finite ratio with the Wiener capacity of E. Thus obtained are the necessary and sufficient conditions of regularity coinciding with the corresponding conditions for the Laplace equation. A.I. Ibragimov [8] has shown that for discontinuous coefficients this condition can also be satisfied (it so happens that, nevertheless, the Cordes condition of small scattering of eigenvalues of the matrix must be satisfied in a neighbourhood of x_0).

For Eqn.(4) with continuous higher order coefficients and bounded measurable low-order coefficients P. Bauman [10] has found necessary and sufficient conditions of regularity similar in appearance to the Wiener criterion. Let $n > 2$, $0 < R' < R$. Let a fixed point $x_0 \in \overline{B(R,0) \setminus B(R',0)}$ For a compact $K \subset B(R',0)$ the capacity $cap_L K$ is defined in a following way: $cap_L K = \inf u(x_0)$ where the infimum is taken over all the supersolutions of Eqn.(4) in $B(R,0)$ (that is, functions $v \in C^2(B(R,0))$ such that $Lv \leq 0$) such that $v|_{\partial B(R,0)} = 0$, $v|_{B(R,0)} \geq 0$ and $v(x) \geq 1$ on K. The capacity $cap_L K$ depends on the choice of x_0 but the ratios of its values for different points are finite. For the operator L there exists in B(R,0) the Green function, G(x,y). Setting $\hat{G}(x,y) = G(x,y)/G(x,x_0)$, P. Bauman costructs the function G(x,y) , which she calls the "normalized Green function ".This function, though not being the Green function, has many of its properties. Now, let Ω, $\bar{\Omega} \subset B(R',0)$, $x^* \in \partial\Omega$. Let $B_k = (2^{-k}, x^*)$, $E_k = (\bar{B}_{k+1} \setminus B_{k+2}) \setminus \Omega$. The number $2^{k(n-2)}$ of (5) differs by an independent of k constant factor from the value at $x \in \partial B_k$,of the Green function for the Laplace operator for B(R,0) with the pole at x^* . Thus, for an arbitrary unit vector e the series $\sum_{k=k_0}^{\infty} G(x^*, x^* + 2^{-k}e)\, cap\, E_k$ can be considered, instead of (5), with k_0 large so that $E_k \subset B(R,0)$ for $k \geq k_0$. The divergence of the series $\sum_{k=k_0}^{\infty} \hat{G}(x^*, x^* + 2^{-k}e)\, cap_L E_k$ is the necessary and sufficient regularity condition ,as is proved by P. Bauman.

N.V. Bagozkaya (unpublish.) showed that ,in the P. Bauman theorem, the continuity of the coefficients can be replaced by the condition of small scattering of eigevalues of the matrix by Cordes (after an appropriately chosen linear transformation has been carried out).

If the Cordes condition is not satisfied but Eqn.(4) is uniformly elliptic, i.e. (2) and (3) hold, then a sufficient condition of regularity of a boundary point ,depending on λ of (2) can be formulated . The so called constant of ellipticity $e = \sup_{x \in G} \left(\sum_{i=1}^{n} a_{ii}(x)/\lambda_{min}(x)\right)$ ($\lambda_{min}(x)$ is the minimal eigenvalue of $\|a_{ik}(x)\|$) is more convenient to introduce instead of λ . (E.M. Landis [11]). Since we are interested in a small neighborhood U of x_0, we can consider $\| a_{ik}(x) \|$ in U.Carry out all linear transformations , take the ellipticity constant in the image, take the inf over all possible linear transformations and over all the possible neighbourhoods U of the point x_0. The number thus obtained will be denoted $\bar{e}_{x_0}$ and called the <u>ellipticity constant of the operator</u> L at the point x_0. Let $s > \bar{e}_{x_0} - 2$. Then there exists a neighbourhood of x_0 such that $L(1/r(x,x_0)^s) > 0$ in an appropriately chosen affine coordinate system, where $r(x,x_0)$ is the distance between x and x_0 in this system. The capacity constructed on the basis of the potential $1/r^s$ will be called s-capacity and denoted cap_s. The divergence of the series

$$\sum_{k=1}^{\infty} 2^{ks} cap_s((B(2^{-k},x_0) \setminus B(2^{-(k+1)},x_0) \setminus \Omega) \tag{10}$$

suffices for the regularity of x_0. Thus the diagram of Fig.1 is obtained.

By narrowing the class of equations to a certain degree (the widest in Fig.1) , the sufficient condition of irregularity in terms of divergence of the series (10) can be formulated. To this end, we define the constant of superellipticity $\bar{e}$ in Ω : $\bar{e} = \sup \frac{\sum a_{ii}(x)}{\lambda_{max}(x)}$ ($\lambda_{max}(x)$ is the maximal eigenvalue of the matrix $\|a_{ik}(x)\|$. The constant of superellipticity of the operator L in the point is defined in complete analogy with the ellipticity constant. $\bar{e}_{x_0} > 2$ is the condition for the equation to satisfy. Let $0 < s < \bar{e}_{x_0} - 2$. Then the convergence of (10) suffices for irregularity of x_0 .

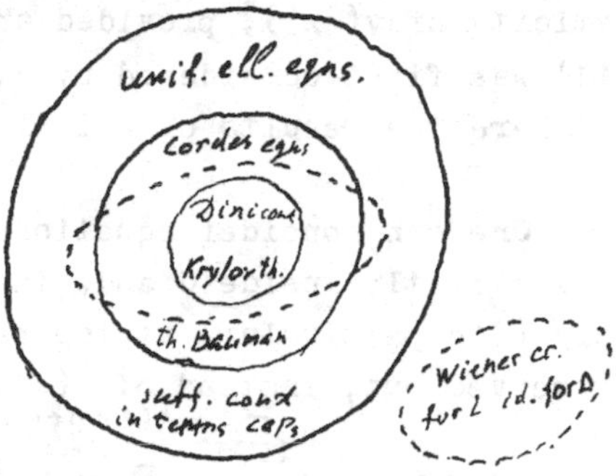

Fig. 1

How essential is the condition $\bar{e}_{x_0} > 2$? I don't know it. Note that in the known example by Gillarg and Serrin where one point retains the Dirichlet data and which will be discussed later, $\bar{e}_{x_0} < 2$ and it can be made arbitrarily close to 2 .

Consider Eqn.(4) with the sufficiently smooth coefficients in G . Let x_0 and let the boundary of the region , except x_0 = (x ,..., x), lie in the cone $\sum_{i=1}^{n-1}(x_i - x_i^0)^2 < a^2(x_n - x_n^0)$, $a \neq 0$, in a neighbourhood of x_0 . Consider the conduct of the function $v(x') = u(x', x_n^0)$ where $x' = (x_1, \ldots, x_{n-1})$ in a neighbourhood of $x_0' = (x_1^0, \ldots, x_{n-1}^0)$. If the Wiener series (5) diverges, v (x′) is obviously continuous at x_0' The relationship between the rate of divergence of the series and the continuity modulus of the function at x_0' can be established. For example , if the series diverges as an arithmetical progressio , v(x′) satisfies the Hölder condition. The slower the series diverges, the worse the continuity modulus is. These results belong to V.G. Maz'ya [12]. Finally, as soon as the series begins to converge ,the plot of function detaches itself from the boundary value at x_0. The point x_0' itself turns into a removable singular point for $v(x')$ can be additionally defined at x_0' so that it becomes continuous.

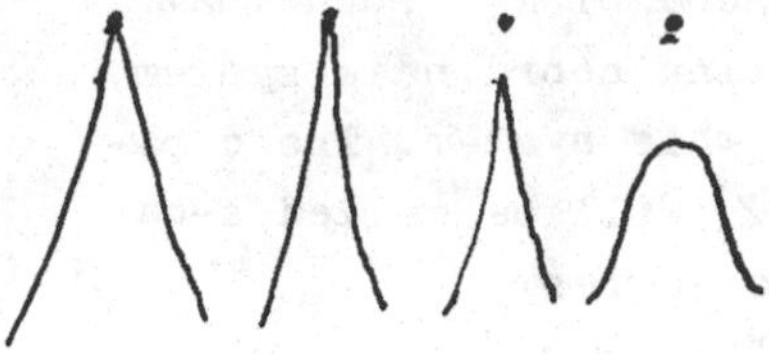

Fig.2

Further, as long as the rate of convergence of series (5) increases, the function $v(x')$ becomes more smooth. Namely, if the coefficients are sufficiently smooth and the series

$$\sum_{k=1}^{\infty} 2^{k(n-2+m+\alpha)} cap((B(2^{-k},x_0)\setminus B(2^{-k+1},x_0))\setminus G) \qquad (11)$$

converges, $m \geq 0$ an integer, $0<\alpha<1$, then $v \in C^{m+\alpha}$. If the coefficients belong to C^{∞}, it follows that the divergence of the series

$$\sum_{k=1}^{\infty} 2^{k(n-2+m_k+\alpha)} cap((B(2^{-k},x_0)\setminus B(2^{-(k+1)},x_0))\setminus G), \quad m_k \to \infty,$$

guarantees belonging of $v(x')$ to the class C^{∞}. One could coerce analyticity of $v(x')$, provided analyticity of the coefficients holds. Series (11) was first considered by E.A. Mikheeva [13] . As for me, I have reported here the results of A.I. Ibragimov [14].

One can consider equations whose coefficients have necessary smoothness strictly inside G and, in general, have discontinuities on the boundary (in particular, at the point x_0). The results on smoothness are then preserved but, instead of (11) , one must consider the series

$$\sum_{k=1}^{\infty} 2^{k(s+m+\alpha)} cap_s((B(2^{-k},x_0)\setminus B(2^{-(k+1)},x_0))\setminus G),$$

where $0<s<\bar{e}_{x_0}-2$ and $\bar{e}_{x_0}>2$ is the superellipticity constant of the operator L at the point x_0 .

b) Removable and non-essential sets. The set $E \subset G$ (for simplicity, we take it to be a compact) is called a removable set, if a solution $u(x)$ of Eqn.(4) defined and bounded on $G\setminus E$ can be continued on E so that it remains a solution. If the coefficients are the smooth functions (Hölder's, for example) then the question of removability is a trivial one: the set is removable if and only if it has the zero Wiener capacity. Now, let the coefficients be smooth (e.g. C^{∞}) outside of E but they can be discontinuous in E . What kind of set must be E in order that a solution could be smoothly continued on E or belonged to W_n^2 ? A fundamental example of non-removable set was constructed by D. Gillarg and J. Serrin [15] .E there consists of the one point, 0 , solutions being of the form z^{ε}, $0<\varepsilon<1$. Let E be a piece of a k-dimensional plane in G . There is a relationship of k,n and the scattering of the roots of characteristic equation; for the latter, there exist solutions that cannot be continued onto E as smooth functions. This problem was considered by K. Miller in [16] .

The problem of removability of a set is closely connected to the problem of its non-essentiality. Let Ω be a bounded region. The set $E \subset \Omega$ is called a non-essential set, if changing of the boundary function on it does not influence the values of the bounded solution of the Dirichlet problem. In other words, if $u|_{\partial\Omega \setminus E} = 0$ and u is bounded then $u \equiv 0$. For $G = \Omega \setminus E$, where $E \subset \Omega$ is a compact, removability of a set is equivalent to its non-essentiality for an equation with smooth coefficients. If a set is essential, it always is non-removable. That's how matters stand in the results of Gillarg and Serrin as well as K. Miller. In principle, for the equation whose coefficients are smooth functions outside of E but can have no limit while tending to the E, non-essential sets may not be also non-removable. But I do not know any corresponding example. If $\bar{e}$ is the superellipticity constant of L in G such that $\bar{e} > 2$, $0 < s < \bar{e} - 2$ and $cap_s E = 0$, the set E is non-essential (E.M. Landis [17]). The condition is only sufficient one. If E is spreaded wide enough in Ω, it can be non- essential also for $cap_s E > 0$ [17]. Landis and Bagozkaya [18],[19] give more detailed studies of conditions of non-essentiality of E depending on cap_s E and on position of E in Ω.

Another point, which I will treat now, is related to the superellipticity constant.

Let Eqn.(4) be defined in a bounded region $G \subset R^n$. For simplicity, let the boundary of the region be a smooth surface (so that the problem of irregularity of the boundary points does not arise). Let the continuous function be defined on ∂G. What should be called a "solution" of the Dirichlet problem for Eqn.(4) without supposing smoothness of the coefficients in G ? Two ways are possible. As usual, we call a smooth (of C^2 class) supersolution v such that $v|_{\partial G} \geq f$, an upper function. Set $u^+ = \inf v$. Infimum is taken over all the upper functions. u^- is defined in a similar way. Then $u^+ \geq u^-$. If $u^+ \equiv u^-$, we call this common value, naturally, a "solution". The other way is as follows. We make the coefficients of the operator L smooth (e.g., by averaging). Let L^k be the corresponding operator with smooth coefficients, let $u_k(x)$ be the solution of the Dirichlet problem: $L^k u_k = 0$, $u_k|_{\partial G} = f$
By the theorem of N.V. Krylov and M.V. Safonov 20 , 21 the sequence (as $k \to \infty$, the support of the averaging kernel tends to zero) is a precompact one in C^α for an $\alpha > 0$. If the sequence converges to a unique function u^*, it should be called also a "solution". Unfortunately, we are not sure about the uniqueness here. Let u^{**} be a partial limit of the sequence $\{u_k\}$. As easily is seen, $u^- \leq u^{**} \leq u^+$. Thus, if $u^- \equiv u^+$ then $u^- \equiv u^+ \equiv u^*$. The function u^{**} has a remarkable property (N.S. Nadirashvili [22]): for almost every point x_0, there

there exists a second degree polynomial $P_{x_0}(x)$ such that $u(x)-P_{x_0}(x)=o(|x-x_0|^2)$ and $LP_{x_0}(x_0)=0$. It is not clear what is the direction of solution of this problem, but it has a particular case when the " solution " is obtainable. Suppose that the coefficients are smooth functions outside of the compact $E \subset G$ and $cap_s E=0$ where $0<s<\bar{e}-2$, $\bar{e}$, the superellipticity constant of Eqn.(4). Then $u^- \equiv u^+$, thus $u^*=u^-=u^+$ satisfies the equation in the sense of Nadirashvili.

In a conversation N.S. Nadirashvili, N.V. Krylov , and M.V. Safonov have told me that they have a proof that if the coefficients of (4) are smooth except at an isolated point , u^* is unique independently of the size of $\bar{e}$. This fact, in particular, implies that the Gillarg-Serrin example discussed by us is a "spurious" solution, because the real solution in this case is a constant.

c). <u>On an approach to qualitative research</u>. The following lemma is very useful in qualitative investigation of conduct of solutions.

<u>The growth lemma.</u> For simplicity, assume that in the operator (1) $b_i \equiv C \equiv 0$. Denote the operator L_0 and consider the equation

$$L_0 u = 0 \tag{4'}$$

Let $R>0$, $x_0 \in R^n$ .Let $G \subset B(4R,x_0)$ be a region such that $G \cap B(R,x_0) \neq \emptyset$, $\bar{G} \cap S(4R,x_0) \neq \emptyset$. Denote $\partial G \cap B(R,x_0)=\Gamma$. Let $u(x)$ be a solution of Eqn.(4') in G , continuous in $\bar{G}$, positive in G such that $u|_\Gamma=0$. Let $s \geq e-2$ where e is the ellipticity constant of L_0 in G. Then , if we denote $E=B(R,x_0)\setminus G$,

$$M=\sup_G u \geq (1+\xi)\, cap_s E/R^s \sup_{G\cap B(R,x_0)} u \,, \tag{12}$$

where $\xi>0$ depends on s .

<u>Proof of the Lemma</u>. Let μ be the equilibrium measure of the set E corresponding to the kernel $1/r^s$ and $U(x)$, the equilibrium potential. Set $V(x)=M(1-U(x)+\mu E/(3R)^s)$. The function V is the supersolution in G , $V|_{S(4R,x_0)}>M$, $\lim V|_\Gamma \geq 0$, thus $u<V$ in G , implying that $u|_{B(R,x_0)\cap G} \leq$ $\leq M(1-cap_s E/(2R)^s+cap_s E/(3R)^s)$, that is (13) holds , with $\xi=1/2^s-1/3^s$.

<u>Remark.</u> If E contains a ball of radius ρ then $cap_s E \geq C\rho^s$ where $C>0$ depends on s , hence

$$\sup_G u \geq \left(1+\eta\left(\frac{\rho}{R}\right)^s\right) \sup_{G\cap B(R,x_0)} u \,, \tag{13}$$

where $\eta>0$ depends on s.

In the following section we will show how this Lemma works in investigation of the infinitesimal conduct of solutions of Eqn.(4) .

A theorem of the Phragmèn-Lindelöf type. Let $\varphi(t)$, where $0\le t<\infty$ be a continuously differentiable function, $\varphi(t)>0$, $|\varphi'|<\frac{1}{32}$. Denote $x'=(x_1,\dots,x_{n-1})$, $\Omega=\{x \mid |x'|<\varphi(|x_n|)\}$. Let $G\subset\Omega$ be an unbounded region. Let $u(x)$ be a solution of Eqn.(4') in G, continuous in $\bar G$ with $u|_{\partial G}\le 0$. Then either $u\le 0$ in G, or $\sup_{|x'|=z} u(x) > \exp(C\int_0^z dt/\varphi(t))$ for sufficiently large z where $C>0$ depends on the constant λ of (2).

Proof. Let $u(x_0)=a>0$, $x_0\in G$. Set $\rho_0=|x_0'|$, $z_0=16\varphi(\rho_0)$. Consider the balls $B(16z_0, x_0)$ and $B(4z_0, x)$. Let $E=B(4z_0,z_0)\setminus G$. Then E contains a ball of the radius $z_0/16$, hence, by (13), $\max_{|x'|=\rho_0+z_0}u>p\cdot a$ with $p>1$ depending on λ. Let x_1, $|x_1'|=\rho_0+z_0$ be such that $u(x_1)=\max_{|x'|=\rho_0+z_0}u(x)$. Set $\rho_1=\rho_0+z_0$ and repeat the argument assuming the radii of the balls to be $16\varphi(\rho_1)$ and $4\varphi(\rho_1)$, centering them at x_1. Repeating the process, we find a sequence of points of G, $\{x_i\}$, such that $x_i'=x_{i-1}'+16\varphi(|x_{i-1}'|)$ and $u(x_i)>p^i\cdot a$. But $\int_{\rho_{i-1}}^{\rho_i}dt/\varphi(t)<64$, thus yielding $\int_{\rho_0}^{\rho_i}dt/\varphi(t)<64i$ or $u(x)>\exp(C\int_0^{|x'|}dt/\varphi(t))$ for $|x'|$ sufficiently large, where $0<C<\ln p$.

2. Divergent elliptic second order equation.

Consider the operator

$$L=\sum_{i,k=1}^{n}\frac{\partial}{\partial x_i}\left(a_{ik}(x)\frac{\partial}{\partial x_k}\right)+\sum_{i=1}^{n}b_i(x)\frac{\partial}{\partial x_i}+c(x), \tag{14}$$

where all the coefficients are measurable bounded functions, $a_{ik}=a_{ki}$, $c\le 0$ and inequality (2) holds. A solution of the equation

$$Lu=0 \tag{15}$$

in the region G is understood as a generalized solution in the sense that the integral identity $u\in W_2^{1,loc}$ $\int_G(\sum a_{ik}u_{x_i}\varphi_{x_k}+\sum b_iu_{x_i}\varphi+c\varphi)dx=0$ $\forall\varphi\in\mathring{W}_2^1(G)$ holds. Such solutions always are the limits, in the norm in W_2^1, of classical solutions with infinitely differentiable coefficients, with the same constant λ, of (2) (Stampacchia [23]), so that investigation of the qualitative problems can be limited to investigation of equations with C^∞ coefficients. It is necessary that the obtained estimates were dependent only on λ of (2) (and on constants which bound the coefficients). We will also consider the shortened operator L_0 that is the operator L with $b_i\equiv c\equiv 0$, and the corresponding equation

$$L_0u=0. \tag{15'}$$

In comparison to non-divergent equations, it is more simple to investigate the divergent equations, because of their having the Green function, wich allows upper and lower estimates of the Laplace operator in terms of Green function. Due to this, the most of the qualitative properties of solutions are very alike to the corresponding properties of harmonic functions.

a). A two-sided estimation of the Green function. Let $G(x,y)$ be the Green function for (15) and $G\subset R^n$. Let $G'\Subset G$ and $x,y\in G'$. Then

(16)

(here and further, we assume that $n>2$). The known to me proofs of these inequalities use the Harnack inequality: if $u(x)>0$ is a solution of Eqn.(15) in B(R, x_0) , sup u / inf u < C in B(R/2, x). If one considers Eqn.(15') , C depends only on λ of (2). We limit ourselves to this case. There are different ways to prove the Harnack inequality: of J.Moser[24] who was first to prove it, the way of O.A. Ladyzenskaya and N.N. Ural'ceva [25] . I wish turn to the third way which uses the growth lemma. I give here a version of the lemma , equally suitable to both Eqn.(4') and Eqn.(15').

An universal lemma on growth. For a region $\Omega \subset R^n$ and for a family $\mathfrak{M} = \{f\}$ of functions , continuous and defined in Ω , such that $f \in \mathfrak{M}$ implies $\alpha f + \beta \in \mathfrak{M}$, the universal lemma on growth is said to be fulfilled if there exists a function $\xi(\sigma)>0$, $0<\sigma<\sigma_0$, $\xi(\sigma)\downarrow 0$ for $\sigma \to 0$ such that if $B(2R,x_0)\subset \Omega$, $G=\{x \in B(2R,x_0) \mid f(x)>0\}$, $E=B(R,x_0)\setminus G$ (it may be that $G=\emptyset$ or $E=\emptyset$) then $\sup_G f \geqslant (1+\sigma(\operatorname{mes}E/R^n)) \times \sup_{G\cap B(R,x_0)} f$.

The universal lemma on growth implies the fulfillment of the Harnack inequality with the constant C depending on $\xi(\sigma)$ for any positive function f , continuous in $B(2R,x_0)\subset\Omega$. L.V. Davydova [26] has proposed this definition and the proof of the Harnack inequality. For solutions of Eqn.(4') , N.V. Krylov and M.V. Safonov [20] ,[21] have proved this lemma earlier (they considered there a more general equation (4)).Following Moser, one can obtain the universal lemma on growth for the divergent equation (15) (a popular account of this approach is given in a paper by V.A. Kondrat'ev and E.M. Landis [27] (to be published)). Now I will mention another approach using the so-called theorem on flux ([19] , see also [28]) .It so happens that $\xi(\sigma)=\xi_0\cdot\sigma$, $\xi_0=\mathrm{const}$.

The theorem on flux. Let $x_0 \in R^n$, $0<R_1<R_2$ and let a region $\mathcal{D}$,which is situated in $B(R_2,x_0)\setminus \overline{B(R_1,x_0)}$, have accessible limit points on both of the spheres $S_1=S(R_1,x_0)$ and $S_2=S(R_2,x_0)$. Let a positively defined quadratic form $\|a_{ik}(x)\|$ be defined in $\mathcal{D}$. A smooth surface Σ is said to be separating S_1 and S_2 in $\mathcal{D}$, if any continuous curve with its ends on S_1 and S_2 respectively and whose internal points belong to $\mathcal{D}$ crosses Σ . Let $f \in C^2(\mathcal{D})$. Then there exists a separating surface Σ such that

$$\int_\Sigma \left|\frac{\partial u}{\partial \nu}\right| ds \leq C\,\frac{\operatorname{osc}_{\mathcal{D}} f \cdot \operatorname{mes}\mathcal{D}}{(R_2-R_1)^2} \tag{17}$$

($\partial/\partial\nu$ is the derivative along the co-normal). C depends on the ratio of the greatest to the lowest of the eigenvalues of $\|a_{ik}(x)\|$.

With the help of this theorem, as I told before, the universal lemma on growth and, hence, the Harnack inequality for (15') are proved (in fact, this theorem is also applicable to Eqn.(15) . But then the constant in the inequality depends on R).

Estimate (16) can also easily be obained from the theorem on flux. It suffices to consider the case when G' is a ball, y, its center. Let B_0 be a unit ball centered at the origin 0 . Since the coefficients can be regarded as belonging to C^{∞}, there exists the Green function G(x,0). Let M_k and m_k be the maximum and the minimum, respectively , of G(x,0) in the layer $z_{k+1} \leq |x| \leq z_k$,where $z_\ell = 2^{-\ell}$. By Harnack inequality $M_k/m_k < C$. There is no need of any special technique to obtain an upper estimate : $m_k < C_1/z_k^{n-2}$. If the inequality is not satisfied, one can immediately obtain a surface σ encompassing 0 such that $\int_\sigma (\partial u/\partial \nu)\, ds > 1$. More interesting is the estimate to the other direction. By applying the flux theorem to $\mathcal{D} = \{x \mid z_{k+1} < |x| < z_k\}$,we obtain $1 \leq CM_k \times z_k^n/z_k^2$ or $M_k < C_2/z_k^{n-2}$. So, for Eqn.(15) , we obtain (16) . It allows us to prove a more exact lemma on growth in terms of capacity.

b). <u>Lemma on growth for a divergent equation.</u> Let $x_0 \in R^n$, $0 < R < R_0$ Let $G \subset B(2R,x_0)$, $G \cap B(R,x_0) \neq \emptyset$,$\partial G \cap S(2R,x_0) \neq \emptyset$, $\partial G \cap B(R,x_0) = \Gamma$, $B(R,x_0) \setminus \mathcal{D} = E$. Let a solution u(x) of Eqn.(15) be defined in G, such that it is continuous in $\overline{G}$, $u > 0$ in G , $u|_\Gamma = 0$. Then $\sup_G u \geq (1 + \xi \,\mathrm{cap}\, E/R^{n-2}) \sup_{G \cap B(R,x_0)} u$, where $\xi > 0$ depends on λ of (2), that is on the constant bounding lower-order terms and on R . (On λ only, for Eqn.(15')). Here $\mathrm{cap}\, E$ is the common Wiener capacity of the set E . The growth lemma in this form was first proved by V.G. Maz'ya [29] for $\Delta u = 0$.

This lemma enables one to prove an essentially finer theorem of the Phragmén-Lindelöf type. For simplicity, we limit ourselves to Eqn.(15').

The region G is said to have an <u>internal diameter smaller than R up to the capacity γ</u> , if $\forall x_0 \in R^n$ $\mathrm{cap}(B(R,x_0) \setminus G) > \gamma R^{n-2}$ .This definition was proposed by V. A. Kondrat'yev. Let G be an unbounded region. We say that $u \in W_2^{1,loc}$ satisfies the zero Dirichlet conditions on ∂G , if $\forall \varphi \in C_0^{\infty}$,$\mathcal{D} = \{\varphi(x) \neq 0\}$ implies $u\varphi \in \mathring{W}_2^1(G \cap \mathcal{D})$.

<u>A theorem of the Phragmén - Lindelöf type</u>. Let G be an unbounded region with the internal diameter smaller than R up to the capacity γ. Let $u \in W_2^{1,loc}$ be a solution of (15') satisfying the zero Dirichlet conditions. Then if $u \not\equiv 0$, $\sup_{|x|=z} |u(x)| > \exp(\frac{\alpha}{R} z)$ for large z , where α depends on R , γ ,and λ .

In order to show that this theorem is finer than the Phragmén - Lindelöf type theorem of Sec.1, we give an example. Consider an integer lattice in R^n. We place at every lattice point the center of an n-1 dimensional disc of radius not smaller than r, $0<r<\frac{1}{2}$. Every disc can be oriented arbitrarily. By E, we denote the union of the (closed discs). Set $G = R^n \setminus E$.

c). Regularity of boundary points for the Dirichlet problem. By inequality (16), conditions of regularity of boundary points for non-degenerate divergence equations coincide with regularity conditions for the Laplace equation. These are the well-known results by W. Littman, and G. Stampacchia, and H.F. Weinberger [30] (for Eqn. (15') and Stampacchia [23] (for Eqn. (15)). There is a number of papers, e.g. [31], dealing with different kinds of degeneration. I will not mention them.

d). The Neumann condition. Trichotomy. Let $G \subset R^{n+1} = \{(x_1,\ldots,x_{n-1})\}$ and $\mathcal{D}^+ = G \times \times (0 < x_n < \infty)$. If a harmonic function is defined in $\mathcal{D}^+$, if on its side surface S the Neumann condition holds (∂G is supposed to be smooth) then, as $x_n \to \infty$, the function either 1) exponentially tends to a constant or 2) exponentially approaches a linear function, or 3) $\lim_{t\to\infty} (\ln M(t)/t) > 1$, where $M(t) = \sup_{|x_n|=t} |u(x)|$ Such division of all the solutions into three classes is called trichotomy. It also is true for the uniformly elliptical divergence equations. For simplicity we again restrict ourselves to Eqn. (15').

Let $\Omega \subset R^n$ be a region and $\Gamma \subset \partial\Omega$, an open in $\partial\Omega$ set. We say that a generalized solution u of Eqn. (15') satisfies the zero Neumann conditions on Γ if $\forall \varphi \in C_0^\infty$, $\operatorname{supp}\varphi \cap \partial G \subset \Gamma$, $\int_\Omega a_{ik} u_{x_i} \varphi_{x_k}\, dx = 0$.

Let $\mathcal{D}^+$ and S be the same as above. S.S. Lahturov [32] has proved that the same trichotomy theorem holds for a generalized solution of Eqn. (15') in $\mathcal{D}^+$, satisfying the zero Neumann conditions on S (a purely linear function there is replaced by a solution of linear growth). N.S. Nadirashvili [33] has shown non-emptiness of all the three classes.

The solutions of the class 2) differ from the solutions of the class 3) also in preserving of the sign of the solutions of the former class, while the solutions of the class 3) change sign necssarily. This feature allows one to transfer the notion of trichotomy on very general region described in terms of isoperimertical inequalities. Isoperimetrical inequalities are, in turn, closely related to the notion of relative capacity of sets: Let G, E_1, E_2, be an open set and closed sets, respectively, in R^n, $E_1 \cap G \neq \emptyset$, $E_2 \cap G \neq \emptyset$, $E_1 \cap E_2 = \emptyset$. Then

$cap(G, E_1, E_2) = \inf_G \int \sum a_{ik} v_{x_i} v_{x_k} dx$, where infimum is taken over all the functions $v \in C^\infty$ equalling unity on E_1 and zero, on E_2. These are the results by A.I. Ibragimov [34].

We give here a simplified formulation: Let $\varphi(t)$, $0 \leq t < 1$, be a continuously differetiable function, $\varphi(0) = 0$, $\varphi(t) > 0$ for $t > 0$, $|\varphi'(t)| < 1$. Let $\Omega = \{x \in R^n \mid 0 < x_n < 1, (\sum_{i=1}^{n-1} x_i^2)^{1/2} \leq \varphi(x_n)\}$. Let a solution of Eqn.(15') satisfying the zero Neumann condition on S = $= \{0 < x_n < 1, (\sum_{i=1}^{n-1} x_i^2)^{1/2} = \varphi(x_n)\}$ be defined on Ω. Let $M(t) = \max_{x_n = t} |u(x)|$. Then one of three options holds : 1) there exists a constant M_0 such that $u(x) = M_0 + O(\exp(-a\int_{x_n}^1 d\tau/(\varphi(\tau))^{n-1}$ where $a > 0$ is a constant depending on λ and n ; 2) there exist constants l_1, l_2, $0 < l_1 < l_2$ such that, for sufficiently small x_n $l_1 \int_{x_n}^1 d\tau/(\varphi(\tau))^{n-1} \leq u(x) \leq l_2 \int_{x_n}^1 d\tau/(\varphi(\tau))^{n-2}$ or $-l_2 \int_{x_n}^1 d\tau/(\varphi(\tau))^{n-1} d\tau \leq u(\tau) \leq -l_1 \int_{x_n}^1 d\tau(\varphi(\tau))^{n-1}$, not changing sign; 3) $\overline{\lim_{t \to \infty}} (\ln M(t) / \int_t^1 d\tau/(\varphi(\tau))^{n-1}) > 0$, changes its sign.

The flux theorem plays a decisive rôle in proving of these statements. There are several generalizations of a such theorem [35].

e) Elliptic equations on manifolds. When investigating global properties of solutions of a divergence equation, it is natural to consider it not in R^n but on a Riemann manifold. Further we will consider the Laplace-Beltrami equation, $\Delta u = 0$, on a smooth connected Riemann manifold M. In case, the manifold has an edge, we will consider only solutions, which satisfy the Neumann condition on the edge : $\frac{\partial u}{\partial \nu}\big|_{\partial M} = 0$, where ν is a normale to ∂M.

One of the first Liouville type theorem on a manifold was proved by Cheng and Yan [36]. It consists in the following : If M is a complete manifold and if the volume V_R of a geodesic ball of radius R with a fixed center satisfies the inequality $V_{R_k} \leq C R_k^2$ for a sequence $R_k \to \infty$, then any superharmonic on M function is a constant.

It is obtained, in a number of papers, the one-side Liouville theorem on harmonic functions on certain manifolds (see [37], [38], [39] etc.)

Let us dwell upon the Liouville theorem on harmonic functions with the finite Dirichlet integral

$$\int_M |\nabla u|^2 < \infty \qquad \text{(D)}$$

Definition. An open set $\Omega \subset M$ is called the massive set, if there exists a harmonic function $u \in C(\bar{\Omega}) \cap C^1(\Omega)$ such that $0 \leq u \leq 1$, $u \not\equiv 0$,

$u|_{\partial\Omega} = 0$ and $\int_\Omega |\nabla u|^2 < \infty$.

The Liouville theorem with condition (D). A harmonic on a manifold M function satisfying the condition (D) is a constant if and only if any two massive sets intersect.

Definition . Any triple of sets $(F_0, F_1; \Omega)$ in M such that F_0 , F_1 are closed , Ω , an open set , is called a condenser. The capacity of a condenser is the number $cap(F_0, F_1; \Omega) = \inf_\varphi \int_\Omega |\nabla\varphi|^2$, where infimum is taken over the functions $\varphi \in C^\infty$ such that $\varphi|_{F_0} = 0$, $\varphi|_{F_1} = 1$ (cf. to the definition of relative capacity in heading d)). If K is a compact in M , $\{F_k\}$, a decreasing sequence of closed sets such that $M \setminus F_k$ is precompact , and $\bigcap_{k=1}^{\infty} F_k = \emptyset$, there exists a limit $\lim_{k\to\infty} cap(K, F_k, \Omega) \equiv cap(K, \infty; \Omega)$ independent of $\{F_k\}$.

Definition. An open set $\Omega \subset M$ is called a parabolic set , if, for any compact $K \subset M$, $cap(K, \infty; \Omega) = 0$, and , a hyperbolic set, otherwise.

A criterion of massivity. An open Ω is a massive set, if and only if when there is a hyperbolic set $\Omega_1 \subset \Omega$ such that $cap(\Omega_1, M \setminus \bar{\Omega}; M) < \infty$.

Corollary 1. The validity of the Liouville theorem on harmonic functions satisfying condition (D) is preserved under quasi-isometric transformations of a manifold (that is , the diffeomorphisms which change the Riemann metrics no more than by a finite factor).

Corollary 2. If M is a region with smooth boundary in R^n which is a region of rotation of a subgraph of a function around a ray, the Liouville theorem with condition (D) holds in M (as in a manifold with an edge).

The massivity criterion allows also constructing examples ofregions in R^n for which the Liouville theorem with condition (D) is not valid and which are diffeomorph to a semi-space.

The Liouville theorem with condition (D) as well as the massivity criterion have been proved by Grigor'yan [40], [41]. Applying them, he uses estimates of capacity which he proved in [42].

3. Non-divergence parabolic equations.

Let

$$L=\sum_{i,k=1}^{n} a_{ik}(x,t)\frac{\partial^2}{\partial x_i \partial x_k}+\sum_{i=1}^{n} b_i(x,t)\frac{\partial}{\partial x_i}+c(x,t) \tag{18}$$

be an uniformly elliptic operator defined in a region $G=R^{n+1}=R^n_x \times R^1_t$. In this section we consider parabolic equations

$$Lu-u_t=0. \tag{19}$$

A function $u\in C^{2,1}$, which turns (21) into an identity, is called a solution of Eqn.(21) in G. We call $v\in C^{2,1}$, $Lv-v_t\le 0$ a super-, and $w\in C^{2,1}$, $Lw-w_t\ge 0$ a subparabolic function. Further on, the cylinder $\{(x,t)\in R^{n+1} \mid |x-x_0|<r,\ t_0<t<t_1\}$ will be denoted by $C^{t_0,t_1}_{x_0,r}$. The set $\gamma(G)\subset\partial G$ is called the upper lid of G, if $\forall (x,t)\in\gamma(G)$ there exists $\varepsilon>0$ such that $C^{t,t+\varepsilon}_{x,\varepsilon}\cap G=\emptyset$ and $C^{t-\varepsilon,t}_{x,\varepsilon}\subset G$. The set $\Gamma(G)=\partial G\setminus\gamma(G)$ is called the parabolic boundary of G.

a). Sub- and superparabolic functions of the potential type. A uniqueness theorem for solutions of Cauchy problem. Consider a function in R^{n+1}:

$$E_{s,\beta}(x,t)=\begin{cases}\frac{1}{t^s}e^{\frac{-|x|^2}{4\beta t}} & \text{for } t>0\\ 0 & \text{for } t\le 0\end{cases}, \text{ except } x=0,\ t=0.$$

$$s=\text{const}>0,\quad \beta=\text{const}>0.$$

Let $M_1=\sup\sum_{i=1}^n a_{ii}(x,t)$, $M_2=\inf\sum_{i=1}^n a_{ii}(x,t)$, $\alpha_1=\inf\min_{|\xi|=1}\sum_{i,k=1}^n a_{ik}(x,t)\xi_i\xi_k$, $\alpha_2=\sup\max_{|\xi|=1}\sum_{i,k=1}^n a_{ik}(x,t)\xi_i\xi_k$, with sup and inf taken over all $(x,t)\in G$.

Direct calculations yield:

L.1. If

$$|b_i(x,t)|\le K_1(|x|+1),\ i=1,\dots,n,\quad |c(x,t)|\le K_2(|x|^2+1), \tag{20}$$

$K_1, K_2=\text{const}>0$ and if

$$\beta>\alpha_2\ (\beta<\alpha_1),\quad s<M_2/2\beta\ (s>M_1/2\beta) \tag{21}$$

holds then there exists $\eta>0$ such that $E_{s,\beta}(x,t)$ is a superparabolic (subparabolic) function for $x\in R^n$, $0\le t\le\eta$, except x=0, t=0.

L.2. If

$$\sum_{i=1}^n b_i(x,t)\cdot x_i\ge 0\ \Big(\sum_{i=1}^n b_i(x,t)\cdot x_i\le 0\Big),\ c(x,t)\le 0\ (c(x,t)\ge 0) \tag{22}$$

and (21) holds then $E_{s,\beta}$ is a superparabolic (subparabolic) function for all x,t, except the origin.

Let $E\subset R^{n+1}$ be a Borél set and μ, a measure on it, $\mu<\infty$. Consider a function $U(x,t)=\int_E E_{s,\beta}(x-\xi,t-\tau)\,d\mu(\xi,\tau)$ outside of $\overline{E}$. L.1 and L.2 both imply that $LU-U_t\le 0\ (LU-U_t\ge 0)$ under conditions (21) and (20) or under (21) and (22). The function $E_{s,\beta}$ is a very convenient threshold for different qualitative theorems. As an example, I will show how it can be used in a simple proof of uniqueness of solution of the Cauchy problem for Eqn.(20) with lower-order coefficients which grow according to inequalities (2) in the Tikhonov class of functions.

Let $u(x,t)$ be a subparabolic function of the operator $L-\frac{\partial}{\partial t}$ in the layer $t_0\le t\le T$ and $u|_{t=t_0}\le 0$. Let $u(x,t)\le C_1 e^{C_2|x|^2}$ for some $C_1>0$ and $C_2>0$. Show that $u\le 0$. It suffices to consider the case when $T-t_0<\varepsilon$ where $\varepsilon>0$ is sufficiently small. Set $\varepsilon=\min(T-t_0, 1/64 C_2\beta, \frac{1}{2})$ and, for an arbitrary $R>1$ consider the function $v_R=Me^{C_2R^2}\int_{|\xi|=R}E_{s,\beta}(t-t_0+\varepsilon, x-\xi)\,ds_\xi$. Choose $M>0$ so that $v_R>C_1e^{C_2R^2}$ on the side surface of $\{|x|<R, t_0<t<t_0+\varepsilon\}$. In order to do that, we set $M=C_1(2\varepsilon)^s/a$ where a is a constant such that $\int_{|\xi|=R} e^{-\frac{|x-\xi|^2}{4\beta\varepsilon}}ds>a$, $R>1$. Applying the maximum principle, we obtain, for $x<R/2$, $t_0<t<t_0+\varepsilon$, that $\le\frac{\omega_n}{\varepsilon^s}e^{-C_2R^2}R^{n-1}$. It remains to tend R to infinity.

Slightly different proof of the theorem was given by L. I. Kamyhin and N.B. Himchenko [43]. Conditions (22) make it possible to obtain the uniqueness theorem without restrictions put on the lower-order coefficients growth, but, of course, under assumption that these conditions are satisfied, since, generally speaking, the growth of lower-order coefficients influences the uniqueness classes. Many papers deal with this subject, but I will not mention them.

Using the kernel $E_{s,\beta}$, we can, in a usual way, construct the capacity $\mathrm{cap}_{s,\beta}E$. Note, if $s=n/2$ and $\beta=1$ then $\mathrm{cap}_{s,\beta}E=\mathrm{cap}_pE$, where cap_pE is the commonly known heat capacity (see,e.g., [44]). The capacity enables one to prove a lemma on growth, which is similar to the corresponding lemma of Sec.1.

b) <u>A parabolic lemma on growth.</u> Let τ, $0<\tau<1$, and $(x_0,t_0)\in R^{n+1}$ be given. Consider the three cylinders $C_1=C_{x_0,\tau}^{t_0,t_0+\tau^2}$, $C_2=C_{x_0,\tau/2}^{t_0+3/4\cdot\tau^2,t_0+\tau^2}$, $C_3=C_{x_0+\tau/2}^{t_0+\tau^2/4,t_0+\tau^2/2}$. Let $\mathcal{D}\subset C_1$ be an open set. Let $\mathcal{D}_1\cap C_2\ne\emptyset$ and $\Gamma=\partial\mathcal{D}\cap C_1$, $C_3\setminus\mathcal{D}=E$. Let a solution $u(x,t)$ of Eqn.(20) be defined in $\mathcal{D}$ such that $u(x,t)>0$ and $u|_\Gamma=0$. Then $\sup_{\mathcal{D}}u>(1+\xi\,\mathrm{cap}_{s\beta}E\times R^{-2s})\sup_{\partial\cap C_2}u$, where $\xi>0$ depends on α_1, α_2, M_1, M_2, as well as on constant which bound the lower-order terms of L. This Lemma imply, for example, following properties of solutions of equations.

<u>An analogue of the Liouville theorem.</u> As is known, for parabolic equations, there exists no one-side Liouville theorem: e^{x+t} is a positive solution of the equation $u_{xx}=u_t$. However, the following

propositions, which could be regarded as some analogues of the one-side Liouville theorem, are valid.

Let $u(x,t)$ be a solution of Eqn.(19) (with $b_i \equiv c \equiv 0$), non-negative and defined in the semi-space $t \leq 0$. Let $u(x,0) < M$. Then $u \equiv$ const.

Let $u(x,t)$ be a solution of the same equation in the semi-space $t \leq 0$ and let $\inf_{t\leq 0} u = m > -\infty$ ($\sup_{t \leq 0} u = M < +\infty$). Then $u(x,t) \to m$ ($u(x,t) \to M$) as $t \to -\infty$, for every x.

These two last theorems have been proved by R. Ya. Glagoleva [45].

c). First boundary-value problem. Regularity of the boundary points.
Let $G \subset R^{n+1}$ be a bounded region and let a solution of Eqn. (19) be defined in G. Suppose that the coefficients are smooth and they may be discontinued at boundary points. Let a continuous function f be defined on the parabolic boundary Γ of the region G. Then a generalized solution u_f of the equation (by the method of upper and lower functions of Perron or, by the Wiener method, by approximation the region from within by step functions ; as was shown by J. Lukes and I. Netuka[46], the same function u_f was obtained there). The question of regularity of a boundary point arises : $(x_0,t_0) \in \Gamma$ is a regular point, if $u_f(x,t) \to f(x_0,t_0)$ as $(x,t) \to (x_0,t_0)$, $(x,t) \in G$.
In contrast to elliptic equations, here even for the simplest equations $u_t = a^2 u_{xx}$ and $u_t = b^2 u_{xx}$ the regularity conditions differ if $a^2 \neq b^2$. We thus begin from the pure heat equation :

$$u_t = \Delta u. \tag{23}$$

Let $(x_0,t_0) \in \Gamma$. We will try to find a regularity condition, which is analoguous to the Wiener criterion. To do so, we turn to the Wiener criterion for the equation $\Delta u = 0$. Let $\mathcal{D} \subset R^n$. Consider a boundary point $x_0 \in \partial\mathcal{D}$, constructing for it known series (5) :

$$\sum_{k=1}^{\infty} 2^{k(n-2)} \operatorname{cap}((B(2^{-k},x_0) \setminus B(2^{-(k+1)},x_0)) \setminus \mathcal{D}).$$

There can be different interpretations of the series terms.

1. In a ring $C_k = B(2^{-k},x_0) \setminus B(2^{-(k+1)},x_0)$, the fundamental solution of the Laplace equation differs from $2^{k(n-2)}$ by a common for all k constant factor. Thus, a term of the series can be considered as a product of capacity on the value of the fundamental solution in the ring.

2. Let $U_k(x)$ be the equilibrium potential for $E_k = C_k \setminus G$. Then terms of the series (5) by a common for all k constant factor differ

from $U_k(x_0)$, thus the Wiener series can be rewritten in the form $\sum_{k=1}^{\infty} \tilde{U}_k(x_0)$. Note that here in the second case the decrease of the radii of balls according to geometrical progression is not obligatory . The radii can decrease more rapidly (but not slower) than the terms of the progression. It is even not necessary that it were balls. Let O_k $O_{k+1} \subset O_k$ be a sequence of neighbourhoods of x_0 contracting to x_0 such that dist $(O_{k+1}, \partial O_k) > q^k$, $0<q<1$, and let $H_k=(O_k \setminus O_{k+1}) \setminus \mathcal{D}$, $V_k(x)$ be the equilibrium potential of the set $H_k=(O_k \setminus O_{k+1}) \setminus \mathcal{D}$. Then the divergence of the series $\sum_{k=1}^{\infty} V_k(x_0)$ is a necessary and sufficient condition of regularity. All the same, it is reasonable to take the balls and use the geometrical progression, since the construction of balls is the simplest and it is easier to extract different geometrical criteria of regularity or irregularity. When proceeding to parabolic equation there are advantages in generalizations of either of the approaches.

By $G(x,\xi;t,\tau)$ we denote the Green function for the equation conjugated to the heat equation and let K_k be a ring between equipotential surfaces of this function : $K_k=\{(x,t) \mid 2^k \le G(x,x_0;t,t_0) \le 2^{k+1}\}$. Let $E_k=K_k \setminus \mathcal{D}$. Then divergence of the series $\sum_{k=1}^{\infty} \mathrm{cap}_p E_k \cdot 2^k$ is a necessary and sufficient condition of regularity of the point (x_0,t_0) . The necessity was proved by E. Lanconelli [47] , while the sufficiency , by L. Evans and R.F. Gariepi [48] .

The equipotential surfaces of the Green function have a rather complicated structure. One could use the second approach by replacing rings with plane layers: let $\{t_k\}$ be an increasing sequence , tending to t_0 . Set $H_k=\{(x,t) \mid x \in R^n, t_k<t \le t_{k+1}$. Let $U_k(x,t)$ be the equilibrium potential of the set H_k . Then , if t_k tends sufficiently rapidly to , divergence of the series $\sum_{k=1}^{\infty} U_k(x_0,t_0)$ is a necessary and sufficient condition of regularity of (x_0,t_0) [44] ([49] deals with the subject in more detail). Finally , we give a result by A.A. Novruzov [50] :

If the coefficients of (19) are variable , and $a_{ik}(x_0,t_0)=\delta_{ik} a_{ik}$ satisfying Dini's condition, regularity conditions coincide with those for(23).

References

1. Courant R.: Partial differential equations, New York,London,1962.
2. Cordes H.O.: Die erste Randwertaufgabe bei Differentialgleichungen in mehr als zwei Variabeln. Math. Ann., Bd. 131, Heft 3(156), 278-318.
3. Hervé R.M.: Recherche axiomatique sur la théorie des fonctions surharmoniques et du potentiel. Ann. Inst. Fourier (Grenoble) 12 (1962), 415-571.
4. Krylov N.V.: On the first boundary-value problem for the elliptic equations. Diff. uravnenija 3(1967), No 2, 315-325.
5. Miller K.: Exceptional boundary points for the nondivergence equation which are regular for the Laplace equation and vice-versa. Ann. Sc. norm. sup. di Pisa, s.III, XXII, F.2, 1968,315-330.
6. Zograf O.N.: An example of second order elliptic equations with continuous coefficients for which the regularity conditions of boundary point are different of similar conditions for the Laplace equation (Russian). Vestnik Mosk. Univ. 72(1969), 30-40.
7. Alkhutov Yu.A.: On the regularity of boundary points with respect to the Dirichlet problem for second-order elliptic equations (Russian). Mat. Zametki 30(1981), No 3, 333-342, 461.
8. Ibragimov A.I.: On some qualitative properties of solutions of elliptic equations with continuous coefficients (Russian). Mat. Sbornik 121(1983), No 4, 454-468.
9. Novruzov A.A.: An approach to the investigation of qualitative properties of solutions of second order elliptic equations not in divergence form (Russian). Mat. Sb. (N.S.) 122(164)(1983), No 3, 360-387.
10. Bauman P.: The Wiener test for second order elliptic equations with continuous coefficients. Indiana Univ. J. 34(1985), No 4, 825-844.
11. Landis E.M.: Second order equations of elliptic and parabolic types. "Nauka", Moskva 1971.
12. Maz'ja V.G.: On the modulus of continuity of the solution of the Dirichlet problem near the nonregular boundary. Probl. Math. Anal., Ed. LGU, 1966, 45-58.
13. Mihuva E.A.: On the behavior of the solution of second order elliptic equation in neighborhood of the nonregular point (Russian). Mat. Sbornik 80(1969), No 3, 503-512.
14. Ibragimov A.I.: On some qualitative properties of solutions of elliptic equations with continuous coefficients (Russian). Mat. Sbornik 121(1983), No 4, 454-468.
15. Gilbarg D., Serrin J.: Uniqueness of axially symmetric subsonic flow past a finite body. J. Rational Mech. Anal. 4, 1955, 169-175.

16. Miller K.M.: Nonequivalence of regular boundary points for the Laplace and nondivergence equations even with continuous coefficients. Ann.Sc. norm. sup. di Pisa 3(24),1970, 159-163.

17. Landis E.M.: Uniqueness theorems for the solutions of the Dirichlet problem for second order elliptic equations. Trudy Moskov. Mat. Ob. 42, 1981, 50-63.

18. Landis E.M.: On the structure of sets nonessential relative to the Dirichlet problem for second order elliptic operators. Trudy Moskov. Mat. Ob. 46, 1983, 122-135.

19. Bagozkaya N.V.: On the structure of nonessential sets for the Dirichlet problem. Trudy Moskov. Mat. Ob. 46, 1983, 136-144.

20. Krylov N.V., Safonov M.V.: On the propriety of solution of parabolic equation with measurable coefficients (Russian). Izv. AN SSSR, ser. mat. 44(1980), No 1, 161-175.

21. Safonov M.V.: On the Harnack inequality for the nondivergence elliptic equations. Zap. nauchn. sem. LOMI 96, 1980, 262-287.

22. Nadirashvili N.S.: Some differential properties of the solution of elliptic equations with measurable coefficients (Russian). Izv. AN SSSR, ser. mat. 49(1985), No 6, 1329-1335.

23. Stampacchia G.: Le problème de Dirichlet pour les équations elliptiques du second ordre à coefficients discontinus. Ann. Inst. Fourier (Grenoble) 15, 1965, 189-259.

24. Moser J.: On Harnack's theorem for elliptic differential equations. Comm. Pure and Appl. Math. 14(1961), No 3, 577-591.

25. Ladyzenskaja O.A., Ural'ceva N.N.: Linear and quasilinear equations of elliptic type. "Nauka", Moskva, 1966.

26. Davydova L.V.: The Harnack inequality for quasilinear elliptic and parabolic equations (Russian). Mat. Sbornik 125(1984),No 6, 332-346.

27. Kondrat'jev V.A., Landis E.M.: The qualitative theory for linear second order partial differential equations (to appear). Itogi Nauki, Moskva, 1988.

28. Landis E.M.: The integral theorem on the flux (Russian). Mat. Zametki 42(1987), No 3, 73-78.

29. Maz'ja V.G.: On regularity on the boundary of solutions of elliptic equations and conformal mappings. Doklady Akad. Nauk SSSR 152(1963), No 6, 1297-1300.

30. Littman W., Stampacchia G., Weinberger H.F.: Regular points for elliptic equations with discontinuous coefficients. Ann. Sc. norm. sup. di Pisa 17, 1963, 43-77.

31. Fabes E., Kenig C., Serapioni R.: The local regularity of solutions of degenerate elliptic equations. Comm. Partial Differential Equations 7(1), 1982, 77-116.

32. Lahturov S.S.: Asymptotic behavior of the solutions of the second boundary value problem in unbounded domains (Russian). Uspekhi Mat. Nauk 35(1980), No 4, 195-196.

33. Nadirashvili N.S.: On one differential property of the solution of elliptic equations of the second order (Russian). Uspekhi Mat. Nauk 40(1985), No 2, 167.

34. Ibragimov A.I.: Some qualitative properties of the solutions of a mixed problem for elliptic equations in nonsmooth domains (Russian). Doklady Akad. Nauk SSSR 265(1982), No 1, 27-31.

35. Ibragimov A.I., Landis E.M.: On the trichotomy of solutions of the Neumann problem for Laplace equation (Russian). Doklady Akad. Nauk SSSR 275(1984), No 4, 137-152.

36. Cheng S.Y., Yau S.T.: Differential equations on Riemannian manifolds and their geometric applications. Comm. Pure and Appl. Math. 28(1975), No 3, 333-354.

37. Bombieri E., Giusti E.: Harnack's inequality for elliptic differential equations on minimal surfaces. Invent. Math. 15(1972), 24-46.

38. Yau S.T.: Harmonic functions on complete Riemannian manifolds. Comm. Pure and Appl. Math. 28(1975), No 3, 201-228.

39. Grigorian A.A.: A Liouville theorem on a manifold (Russian). Uspekhi Mat. Nauk 37(1982), No 3, 181-182.

40. Grigorian A.A.: A criterion for the existence on a Riemannian manifold of a nontrivial bounded harmonic function with the Dirichlet finite integral (Russian). Doklady Akad. Nauk SSSR 293 (1987), No 3, 529-531.

41. Grigorian A.A.: On the D-Liouville theorem for Laplace-Beltrami equation on Riemann manifolds (Russian). Mat. Sbornik 132(1987), No 4, 496-516.

42. Grigorian A.A.: On existence of the positive fundamental solution of the Laplace equation on a Riemannian manifold (Russian). Mat. Sbornik 128(1985), No 3, 354-363.

43. Kamynin L.I., Khimchenko N.B.: The Tikhonov-Petrovskij problem for second-order parabolic equations (Russian). Sibirsk. Mat. Zh. 22(1981), No 5, 78-109, 222.

44. Watson N.A.: The thermal capacity. Proc. London Math. Soc. 37 (1978), No 3, 342-362.

45. Glagoleva R.: The Liouville type theorems for parabolic equations (Russian). Mat. Zametki 5(1969), No 5, 599-606.

46. Lukeš J., Netuka I.: The Wiener type solution of the Dirichlet problem in potential theory. Math. Ann. 244(1976), No 2, 173-178.

47. Lanconelli E.: Sul problema di Dirichlet per l'equazione del calor. Ann. Mat. Pure ed Appl. 97(1973), ser. 4, 83-117.

48. Evans L., Gariepi R.F.: The Wiener test for heat equation. Arch. Rat. Mech. Anal. 72(1982), No 4, 293-314.

49. Landis E.M.: The necessary and sufficient conditions of regularity of boundary point for Dirichlet problem for heat equation (Russian). Doklady Akad. Nauk SSSR 165(1969), No 3, 507-510.

50. Novruzov A.A.: On the regularity of boundary point for parabolic equations (Russian). Doklady Azerb. SSR, ser. math. 1970, No 10, 128-131.

WEIGHTED EXTREMAL LENGTH AND BEPPO LEVI FUNCTIONS

Makoto Ohtsuka
Department of Mathematics, Gakushuin University
1-5-1 Mejiro, Toshima-ku, Tokyo, 171 Japan

§1. The notion of extremal length was first found by Beurling around 1950 and was developed by him and Ahlfors. It has been very useful in the theory of functions.

First we recall the definition of weighted extremal length of a family Γ of non-point locally rectifiable curves in the Euclidean sapce R^n, $n \geq 2$. We shall say that a Borel measurable function $\rho \geq 0$ in R^n is Γ-admissible if $\int_\gamma \rho ds \geq 1$ for every $\gamma \in \Gamma$. A measurable function $w \geq 0$ in R^n will be called a weight. We define the weighted module of order $p > 0$ of Γ by

$$M_p(\Gamma;w) = \inf\{\int \rho^p w dx;\ \rho \text{ is } \Gamma\text{-admissible}\}$$

and the weighted extremal length by $\lambda_p(\Gamma;w) = 1/M_p(\Gamma;w)$. This was defined on Riemann surfaces in 1955 ([10, p.194]) for some quasi-conformal mappings. For simplicity we consider here a twice continuously differentiable homeomorphism f of a plane domain D onto a plane domain and let $q = q(z)$ be the dilatation quotient of f. Then, for any family Γ of curves in D we obtain $\lambda_2(\Gamma;1) \leq \lambda_2(f(\Gamma);1/q)$.

The module $M_p(\Gamma;w)$ is in a way similar to a measure and a family Γ of curves with $M_p(\Gamma;w) = 0$ corresponds to a set of measure zero. So we say that a property holds for (p,w)-a.e. curve if the exceptional family Γ of curves has $M_p(\Gamma;w) = 0$ or $\lambda_p(\Gamma;w) = \infty$.

The definition of weighted extremal length was greatly generalized by Fuglede in his Acta Math. paper [5, p.176] in 1957. Instead of a curve family Γ a class M of measures $\mu \geq 0$ in R^n is considered and a general basic measure $m \geq 0$ is taken. We define

$$M_p(M) = \inf\{\int \rho^p dm;\ \int \rho d\mu \geq 1 \text{ for all } \mu \in M\}$$

and $\lambda_p(M) = 1/M_p(M)$. The weighted extremal length stated above of a curve family Γ is obtained if the arc-length of $\gamma \in \Gamma$ is taken as μ and wdx is taken as dm. Likewise Fuglede's definition includes the weighted extremal length $\lambda_p(\Gamma;w)$ of a family Γ of surfaces which is defined as the reciprocal of

$$M_p(\Gamma,w) = \inf\{\int \rho^p w dx;\ \int_S \rho d\sigma \geq 1 \text{ for all } S \in \Gamma\},$$

where σ is the area on S.

We shall say that a weight w satisfies the Muckenhoupt A_p condition ([9]) if

$$(A_p) \qquad \sup_Q \frac{1}{|Q|}\int_Q w dx \left(\frac{1}{|Q|}\int_Q w^{1/(1-p)} dx\right)^{p-1} < \infty,$$

where Q is a cube with sides parallel to the axes and $|Q|$ stands for the volume of Q. Then we write $w \in A_p$.

We denote by $L^{p,w}$ the family of measurable functions f in R^n with

$$\|f\|_{p,w} = \left(\int |f|^p w dx\right)^{1/p} < \infty.$$

A Riesz potential is defined by

$$U_\alpha^f(x) = \int \frac{1}{|x-y|^{n-\alpha}} f(y)dy$$

when this has a meaning. We call a set $X \subset R^n$ (α,p,w)-polar if there is $f \in L^{p,w}$, $f \geq 0$, such that $U_\alpha^f = \infty$ on X but $U_\alpha^f \not\equiv \infty$. We say that a property holds (α,p,w)-q.e. if the exceptional set is (α,p,w)-polar. We denote by $\Lambda_m(X)$ the family of curves or even m-dimensional Lipschitz surfaces each of which meets a set X. Fuglede proved that $\lambda_p(\Lambda_m(X);1) = \infty$ if and only if X is $(m,p,1)$-polar in case $p > 1$ and $mp \leq n$. One can show for curve families that if $w \in A_p$ and $p > 1$, then

$$\lambda_p(\Lambda_1(X);w) = \infty \rightleftarrows X \text{ is } (1,p,w)\text{-polar}$$

and

$$\lambda_p(\Lambda_1(\infty);w) = \infty \rightleftarrows \int_{|x|\geq 1} |x|^{-(n-1)p'} w^{1/(1-p)} dx = \infty,$$

where $1/p + 1/p' = 1$ and $\Lambda_1(\infty)$ denotes the family of curves tending to the point at infinity. We point out that the origin as a point set is (α,p,w)-polar if and only if $\int_{|x|<1} |x|^{-(n-\alpha)p'} w^{1/(1-p)} dx = \infty$, not necessarily under the A_p condition but under the condition $\int_{|x|<1} w^{1/(1-p)} dx < \infty$. The higher dimensional case is not yet considered. Hereafter we assume $w \in A_p$. In case $w \equiv 1$ we shall omit w.

§2. Next we are concerned with precise functions. Fuglede [5, p.218] gave the following definition: He writes $f \in BL^p(R^n)$ if f is absolutely continuous along p-a.e. curve and $\int |\text{grad } f|^p dx < \infty$. We may call it p-precise with Ziemer [12], and call f (p,w)-precise if f is absolutely continuous along (p,w)-a.e. curve and $\int |\text{grad } f|^p w dx < \infty$. The following result gives a relation between the notions of extremal length and precise functions. Any (p,w)-precise function f has a finite limit along (p,w)-a.e. curve and tends to a finite unique value if $|x| \to \infty$ along (p,w)-a.e. curve of $\Lambda_1(\infty)$. To prove the latter fact we need an integral representation of f. Denote by $f(\infty)$ the limit at the point at infinity. We can show that any (p,w)-precise function f in R^n is approximated in such a way that, given a compact set K and $\varepsilon > 0$, there exists $\varphi \in C_0^\infty$ satisfying $\int_K |f-f(\infty)-\varphi|^p w dx < \varepsilon$ and $\|\text{grad}(f-\varphi)\|_{p,w} < \varepsilon$. In 1970's I discussed integral representations of p-precise functions in lectures.

§3. In what follows we are mainly concerned with integral representations of Beppo Levi functions of general order. Let $m \geq 1$, $\alpha = (\alpha_1,\ldots,\alpha_n)$, $|\alpha| = \alpha_1+\cdots+\alpha_n$, $x^\alpha = x_1^{\alpha_1}\cdots x_n^{\alpha_n}$, $D^\alpha = \partial^{|\alpha|}/\partial x_1^{\alpha_1}\cdots\partial x_n^{\alpha_n}$. A locally integrable function f in R^n is called a Beppo Levi function of order m with weight w by Aikawa [1, §1] if $D^\alpha f$ defined in the distribution sense may be regarded as a function in $L^{p,w}$ for every multiindex α with $|\alpha| = m$. We shall write then $f \in BL_{m,p,w}$. In case $m > 1$ we do not know how the notion of extremal length is used in the definition of (precise) Beppo Levi functions. We shall discuss, therefore, Beppo Levi functions without using extremal length. We limit ourselves to the case when $p > 1$ and $n \geq 3$ unless otherwise stated.

There are several kernels with respect to which we express $f \in BL_{m,p,w}$ in integrals. First we consider the general Riesz kernels

$$k_m(x) = \begin{cases} |x|^{m-n} & \text{if } m < n \text{ or if } m > n \text{ and } m-n \text{ is odd} \\ -|x|^{m-n}\log|x| & \text{if } m \geq n \text{ and } m-n \text{ is even.} \end{cases}$$

It is well-known that $\Delta_m k_{2m} = c_{n,m}\delta$ with certain constant $c_{n,m}$, where Δ_m is the Laplace operator iterated m times. Hereafter we assume that m is a positive integer less than n for simplicity. Let $|\alpha| = m$. As in Mizuta [8, p.388] we use

$$K_{m,\alpha,q}(x,y) = \begin{cases} D_x^\alpha k_{2m}(x-y) - \sum_{|\beta|\leq q-1} \frac{x^\beta}{\beta!} D_x^\beta D_x^\alpha k_{2m}(x-y)\Big|_{x=0} & \text{if } |y| \geq 1 \\ D_x^\alpha k_{2m}(x-y) & \text{if } |y| < 1; \end{cases}$$

we set $K_{m,\alpha,0}(x,y) = D_x^\alpha k_{2m}(x-y)$. If $2|x| \leq |y|$ and $|y| \geq 1$, then $|K_{m,\alpha,q}(x,y)| \leq \text{const.}|x|^q|x-y|^{m-n-q}$. By an elementary computation we can show

$$\iint |K_{m,\alpha,m}(x,y)f(y)\varphi(x)|dydx < \infty$$

if $f \in L^{p,w}$, $\varphi \in C_0^\infty$ and $|\alpha| = m$, and then that $\int K_{m,\alpha,m}(x,y)f(y)dy$ belongs to $BL_{m,p,w}$. This leads to the representation

$$f(x) = \sum_{|\alpha|=m} a_\alpha \int K_{m,\alpha,m}(x,y)D^\alpha f(y)dy + P(x) \quad \text{a.e.}$$

and naturally to

$$f(x) = \int \sum_{|\alpha|=m} a_\alpha K_{m,\alpha,m}(x,y)D^\alpha f(y)dy + P(x) \quad \text{a.e.}$$

for any $f \in BL_{m,p,w}$, where $a_\alpha = m!/(c_{n,m}\alpha!)$ and P belongs to the family P_{m-1} of polynomials of degree at most $m-1$. As to the kernels $K_{m,\alpha,q}$ we can prove that $f \in BL_{m,p,w}$ is expressed as

$$f(x) = \sum_{|\alpha|=m} a_\alpha \int K_{m,\alpha,q}(x,y)D^\alpha f(y)dy + P(x) \quad \text{a.e.}$$

if and only if

$$(1) \quad \int_{|y|\geq 1} |D^\beta D^\alpha k_{2m}(-y)D^\alpha f(y)|dy < \infty$$

for every α with $|\alpha| = m$ and β, $q \leq |\beta| \leq m-1$. Even if each of $\int K_{m,\alpha,q}(x,y)D^\alpha f(y)dy$ does not exist, it may happen that $\int \sum_{|\alpha|=m} a_\alpha K_{m,\alpha,q}(x,y)D^\alpha f(y)dy$ exists. Actually we have

$$f(x) = \int \sum_{|\alpha|=m} a_\alpha K_{m,\alpha,q}(x,y)D^\alpha f(y)dy + P(x) \quad \text{a.e.}$$

if and only if

$$\int_{|y|\geq 1} \Big| \sum_{|\alpha|=m} D^\beta D^\alpha k_{2m}(-y)D^\alpha f(y)\Big|dy < \infty$$

for every β, $q \leq |\beta| \leq m-1$. We can show that condition (1) is equivalent to

$$\int_{|y|\geq 1} |y|^{m-n-q}|D^{\alpha}f(y)|dy < \infty$$

required for every α with $|\alpha| = m$. It is easy to see that this condition follows from

$$\int_{|y|\geq 1} |y|^{(m-n-q)p'}w(y)^{1/(1-p)}dy < \infty.$$

In general, Aikawa [1, §4] writes $w \in A_{p,k}$ when w satisfies

$$\int_{|y|\geq 1} |y|^{(k-n)p'}w(y)^{1/(1-p)}dy < \infty,$$

and investigated this condition in various ways. We recall that $\lambda_p(\Lambda_1(\infty); w) < \infty \not\rightleftarrows w \in A_{p,1}$.

The uniqueness of the above representation is stated as follows: Let $0 \leq q \leq m$, $f \in BL_{m,p,w}$ and assume that $\{f_\alpha\}_{|\alpha|=m}$ in $L^{p,w}$ satisfy $\partial f_\alpha/\partial x_j = \partial f_\beta/\partial x_k$ in the distribution sense whenever $\alpha+e_j = \beta+e_k$ (cf. [6, §5]), where the components of e_j vanish except for the jth which is equal to 1. If

$$f(x) = \int \sum_{|\alpha|=m} a_\alpha K_{m,\alpha,q}(x,y)f_\alpha(y)dy + P'(x) \qquad \text{a.e.}$$

with $P' \in P_{m-1}$, then $f_\alpha = D^\alpha f$ a.e. for each α and $P' = P$.

From the integral representation in terms of $K_{m,\alpha,m}$ it follows that there exist $\{\varphi_j\}$ in C_0^∞ such that $\|D^\alpha f - D^\alpha \varphi_j\|_{p,w} \to 0$ as $j \to \infty$ for each α with $|\alpha| = m$ although this was already proved by Sobolev [11] in 1963 in the case when $w \equiv 1$.

Next we consider the kernels

$$\kappa_{\alpha,q}(x,y) = \begin{cases} \dfrac{(x-y)^\alpha}{|x-y|^n} - \displaystyle\sum_{|\beta|\leq q-1} \frac{x^\beta}{\beta!} D^\beta \left.\frac{(x-y)^\alpha}{|x-y|^n}\right|_{x=0} & \text{if } |y| \geq 1 \\ \dfrac{(x-y)^\alpha}{|x-y|^n} & \text{if } |y| < 1 \end{cases}$$

in case $q \geq 1$ and $\kappa_{\alpha,0}(x,y) = (x-y)^\alpha/|x-y|^n$. We can discuss the integral representations

$$f(x) = \sum_{|\alpha|=m} b_\alpha \int \kappa_{\alpha,q}(x,y)D^\alpha f(y)dy + P(x) \qquad \text{a.e.}$$

and

$$f(x) = \int \sum_{|\alpha|=m} b_\alpha \kappa_{\alpha,q}(x,y) D^\alpha f(y)dy + P(x) \quad \text{a.e.}$$

with $b_\alpha = (-1)^m m/(\alpha!\sigma_n)$ and $P \in P_{m-1}$ similarly, where σ_n is the area of $\{|x| = 1\}$. In the case where $q = 0$ and $m = 1$ both integrals

$$\int \sum_{|\alpha|=1} a_\alpha K_{1,\alpha,0}(x,y) D^\alpha f(y)dy \quad \text{and} \quad \int \sum_{|\alpha|=1} b_\alpha \kappa_{\alpha,0}(x,y) D^\alpha f(y)dy$$

are equal to

$$\frac{1}{(n-2)\sigma_n} \int \text{grad}_x |x-y|^{2-n} \cdot \text{grad } f(y)dy.$$

The representation of $f \in BL_{1,p}$ in this integral plus a constant was called canonical by Deny and Lions [3, Chap.II.6].

Lastly we treat the Riesz kernel itself. Set

$$\gamma(m) = \pi^{n/2} 2^m \Gamma(\tfrac{m}{2})/\Gamma(\tfrac{n-m}{2})$$

and $h_m = k_m/\gamma(m)$. Set also

$$h_{m,q}(x,y) = \begin{cases} h_m(x-y) - \sum_{|\beta| \le q-1} \frac{x^\beta}{\beta!} D^\beta h_m(-y) & \text{if } |y| \ge 1 \\ h_m(x-y) & \text{if } |y| < 1; \end{cases}$$

set $h_{m,0}(x,y) = h_m(x-y)$. For $g \in L^{p,w}$ and α with $|\alpha| = m$ it was proved by Coifman and Fefferman [2, p.244] that $(D^\alpha h_m)*g$, defined in the distribution sense, may be regarded as a function of $L^{p,w}$ and $\|(D^\alpha h_m)*g\|_{p,w} \le \text{const.}\|g\|_{p,w}$. For $f \in BL_{m,p,w}$ we set

$$g_{f,m} = \sum_{|\alpha|=m} c_\alpha (D^\alpha h_m) * D^\alpha f$$

with $c_\alpha = (-1)^m m!/\alpha!$. Let $0 \le q \le m$. Then one can prove that

$$f(x) = \int h_{m,q}(x,y) g_{f,m}(y)dy + P(x) \quad \text{a.e.}$$

with $P \in P_{m-1}$ if and only if

$$\int_{|y|\ge 1} |y|^{m-n-q} |g_{f,m}(y)| dy < \infty.$$

As Aikawa [1, Theorem 1] we can give a uniqueness theorem.

Aikawa established an approximation theorem [1, Theorem 2] similar to the one stated at the end of §2.

We call a function f quasicontinuous or simply q.c. in R^n if, given $\varepsilon > 0$, there is an open set U such that the weighted Bessel capacity $B^w_{m,p}(U)$ (see [1, §6]) of U of order m is less than ε and the restriction of f to $R^n - U$ is continuous. Let us say that f is an (m,p,w)-precise function if $f \in BL_{m,p,w}$ and f is q.c. in R^n. A property is said to hold q.e. if the exceptional set has $B^w_{m,p}$-capacity zero. Aikawa [1, §6] proved that f is represented by an integral with modified Riesz kernel $h_{m,q}$ q.e. if so is a.e. The same holds for the other two kernels too.

We have some results as to relations between canonical and Riesz potential representations in case m = 1 and $w \equiv 1$. There exists a precise function which admits a canonical representation but not a Riesz potential representation, and vice versa, although an example for the latter is found only in case n = 2. The condition $\int_{|y|\geq 1} |y|^{1-n}|\mathrm{grad}\, f|dy < \infty$ is evidently sufficient for the canonical representation, but it is not necessary. It is not a necessary condition for the Riesz representation. However, we do not know whether it is sufficient.

Still in the case when m = 1 and $w \equiv 1$ the integrand in the canonical representation is of the form $\mathrm{grad}|x-y|^{2-n} \cdot \mathrm{grad}\, f$ and the kernel in the Riesz representation is $|x|^{1-n}$. One may ask if one can represent p-precise functions in terms of $\mathrm{grad}|x|^{s-n}$ and $|x|^{s-n}$ with s in (0,n). Actually, given s, $0 < s \leq 2$, one can prove under some condition that a p-precise function f is written in the form

$$f(x) = \int \mathrm{grad}|x-y|^{s-n} \cdot h(y)dy + \text{const.} \qquad \text{q.e.}$$

with a vector-valued function h, and given s, $0 < s \leq 1$, in the form

$$f(x) = U^g_s + \text{const.} \qquad \text{q.e.}$$

We can show that for any $s \in (2,n)$ (resp. $s \in (1,n)$) there exists a p-precise function which does not admit any representation in the first (resp. second) form. Counter-examples are provided by p-precise Riesz potentials of measures without density.

Finally we are concerned with applications of integral representations. Replacing exceptional sets of measure zero by those of capacity zero, Mizuta generalized results by Landkof [7, Theorem 1.21] and Fefferman [4] who proved the existence of limits along a.e. lines. For example, he proved (see [8, Theorem 4]) that if mp < n and f is an (m,p,1)-precise function (a precise (m,p,1)-function in our terminology), then there exist $P \in P_{m-1}$ and a set E with vanishing $B_{m,p}(E)$

such that $f(x',t) - P(x',t) \to 0$ as $t \to \infty$ for every $(x',0) \in R^{n-1}\times\{0\} - E$. Kurokawa [6, §7] applies integral representation to an embedding theorem in the case when $w \equiv 1$.

References

[1] H. Aikawa: On weighted Beppo Levi functions, in preparation.

[2] R. R. Coifman and C. Fefferman: Weighted norm inequalities for maximal functions and singular integrals, Studia Math. 51 (1974), 241-250.

[3] J. Deny and J. L. Lions: Les espaces du type de Beppo Levi, Ann. Inst. Fourier 5 (1955), 305-370.

[4] C. Fefferman: Convergence on almost every line for functions with gradient in $L^p(R^n)$, Ann. Inst. Fourier 24 (1974), no.3, 159-164.

[5] B. Fuglede: Extremal length and functional completion, Acta Math. 98 (1957), 171-219.

[6] T. Kurokawa: Riesz potentials, higher Riesz transforms and Beppo Levi spaces, in preparation.

[7] N. S. Landkof: Foundations of modern potential theory, Grundlehren Math. Wiss. 180, Springer, 1972.

[8] Y. Mizuta: On the existence of limits along lines of Beppo Levi functions, Hiroshima Math. J. 16 (1986), 387-404.

[9] B. Muckenhoupt: Weighted norm inequalities for the Hardy maximal function, Trans. Amer. Math. Soc. 165 (1972), 207-226.

[10] M. Ohtsuka: Sur un théorème étoilé de Gross, Nagoya Math. J. 9 (1955), 191-207.

[11] S. L. Sobolev: The density of finite functions in the $L_p^{(m)}$ space, Soviet Math. Dokl. 4 (1963), 313-316.

[12] W. P. Ziemer: Extremal length as a capacity, Mich. Math. J. 17 (1970), 117-128.

AN INTRODUCTION TO ITERATIVE TECHNIQUES FOR POTENTIAL PROBLEMS

G. F. Roach
Department of Mathematics, University of Strathclyde
26 Richmond Street, Glasgow G1 1XH

In this paper we examine means of obtaining approximate solutions to potential problems in such a way that the approximations can be successively improved and the rate of convergence to the required exact solution conveniently estimated.

There are a number of ways of approaching this general question; for example, variational methods, Galerkin procedures or finite difference schemes. Here we shall adopt an integral equation approach and confine attention to the development of iteration schemes.

Iteration methods are recognised as one of more immediate ways of obtaining approximate solutions to operator equations. Many methods have been developed and in this connection we would cite as source references the work of Petryshyn [37] and the survey monograph of Patterson [36]. However, most of the work on iterative methods has been concerned with solving linear equations which are uniquely solvable and which involve a self adjoint operator. In our case the integral equations we have to deal with are not necessarily uniquely solvable and moreover involve a non self adjoint operator. Difficulties arising from the lack of uniqueness can be overcome by using a technique originating in the work of Wiarda [49]. To resolve problems centred on the lack of unique solvability we shall use either an eigenvalue shifting technique [30] or a modified Greens Function approach [18], [25], [26].

The paper is organised as follows. In the next section the main notations to be used are introduced together with those classes of potential problems which are of immediate physical interest. In §3 an operator formulation of these problems is given together with an indication of the prerequisites for such representations. An integral equation approach is developed in §4 and boundary integral equations are obtained which are equivalent to the given potential problem. It will be seen that the operator equations derived in §3 and §4 have the same generic form. In §5 we recall a number of basic results from the general theory of linear operators centred mainly on those questions of solvability and well posedness which have to be addressed later. In §6 we outline some of the main iteration techniques which have been developed and in §7 we indicate how these can be made available to deal with the potential problems in which we are interested.

2. Notions and statements of problems

Let $\Omega_+ \subset \mathbb{R}^n$, $n \geq 2$ be the unbounded region exterior to a closed, bounded surface $\Gamma \subset \mathbb{R}^{n-1}$. Let $R = R(p,q)$ denote the distance between two typical points

$p,q \in \mathbb{R}^n$ and $\hat{n}_p$ the unit normally erected at $p \in \Gamma$ directed into Ω_+. We denote by $\Omega_- := \mathbb{R}^n \setminus \{\Omega_+ \cup \Gamma\} = \mathbb{R}^n \setminus \bar{\Omega}_+$ the interior of Γ. When there is no immediate need to distinguish between Ω_+ and Ω_- we shall simple suppress the subscripts.

The problems with which we shall be concerned have the following typical form

$$\Delta u(p) = f(p,u), \quad p \in \Omega \tag{2.1}$$

$$u(p) \in (bc) \quad , \quad p \in \Gamma . \tag{2.2}$$

Such problems can be broadly classified as

Nonlinear

when f given as a function of position and of the required unknown.

Linear

when f given as a function of position only.

Interior

when the problem is posed in Ω_-.

Exterior

when the problem is posed in Ω_+.

The expression labelled (2.2) denotes that solutions of (2.1) must also satisfy certain imposed conditions on Γ denoted typically by (bc); these will be made specific later.

We should also be more precise about the smoothness assumptions we may adopt for u,f and other quantities which might appear in (bc), and if $\Omega = \Omega_+$ we shall also want to impose conditions on their behaviour at infinity.

However, for the time being we shall work on a formal level and assume that sufficient smoothness is always available and that for exterior problems the required solution satisfies a suitable set of conditions at infinity [22]; these latter we shall simply refer to as Kellogg conditions.

Some comment should also be made regarding the smoothness of the boundary Γ. For the purposes of this article we shall be mainly concerned with smooth boundaries, certainly no worse than $C^{1,1}$. For problems involving non smooth boundaries we would refer to the excellent monograph [11] and the references cited there.

In specifying (bc) we define a particular boundary value problem. The two most important problems, and to which we will devote most of our attention, are the following.

Dirichlet problem:

Given f on Ω and g on Γ find a quantity u(p), $p \in \bar{\Omega}$ satisfying

$$\Delta u(p) = f(p), \ p \in \Omega \tag{2.3}$$

$$u(p) = g(p), \ p \in \Gamma . \tag{2.4}$$

Neumann problem:

Given f on Ω and g on Γ find a quantity u(p), $p \in \bar{\Omega}$ satisfying

$$\Delta u(p) = f(p), \ p \in \Omega \tag{2.5}$$

$$\frac{\partial u}{\partial n_p}(p) = g(p), \ p \in \Gamma \tag{2.6}$$

where $\partial/\partial n_p$ denotes the directional derivative in the direction of $\hat{n}_p$.

Other problems of interest are for example the Robin problem, in which a linear combination of the solution and its normal derivative is required to satisfy a given condition on Γ, and the mixed boundary value problems in which the solution is required to satisfy different conditions on different parts of the boundary. In this connection we would make reference to [27], [8], [48].

In this paper we shall concentrate on linear problems and adopt an integral equation approach to their solution as it lends itself well to the development of iteration techniques. By doing this a reasonably unified approach can be developed.

3. An operator formulation

In analysing a problem of the form

$$\Delta u(p) = f(p), \quad p \in \Omega \tag{3.1}$$

$$Bu(p) = g(p), \quad p \in \Gamma, \tag{3.2}$$

where B is a linear differential expression defined on Γ, we introduce spaces of functions $X(\Omega)$, $Y(\Omega)$ and $Z(\Omega)$ which ensure that we can properly define operators L and γ according to

$$\begin{aligned} &L : X(\Omega) \to Y(\Omega) \\ &Lu = \Delta u, \quad u \in D(L) \\ &D(L) = \{u \in X(\Omega) : \Delta u \in Y(\Omega)\} \end{aligned} \tag{3.3}$$

and

$$\begin{aligned} &\gamma : X(\Omega) \to Z(\Gamma) \\ &\gamma u = Bu, \quad u \in D(\gamma) \\ &D(\gamma) = \{u \in X(\Omega) : Bu \in Z(\Gamma)\}. \end{aligned} \tag{3.4}$$

Inherent in this definition of L is the so-called Shift Theorem governing the smoothness relationship between the unknown and the data functions. We remark that its availability for problems in non-smooth domains often requires quite delicate investigation [11]. In the definition of γ we have an embodiment of the Trace Theorem [31], [4]. Again, in general, the availability of this theorem must be carefully checked.

Once the definitions (3.3), (3.4) can be made then the boundary value problem (3.1), (3.2) can be given the following operator realisation:

Given $f \in Y(\Omega)$ and $g \in Z(\Omega)$ determine $u \in X(\Omega)$ satisfying

$$Lu = f, \quad u \in D(L) \tag{3.5}$$

$$\gamma u = g, \quad u \in D(\gamma) \tag{3.6}$$

In a more compact form this can be written

$$[L,\gamma]u = [f,g], \quad u \in D(L) \cap D(\gamma) \tag{3.7}$$

which in turn suggests the introduction of an operator $\mathcal{a}$ defined by

$$\begin{aligned} &\mathcal{a} : X(\Omega) \to Y(\Omega)\times Z(\Gamma) \\ &u = [L,\gamma]u \quad u \in D(\mathcal{a}) \\ &D(\mathcal{a}) = \{u \in D(L) \cap D(\gamma) : [L,\gamma]u \in Y(\Omega)\times Z(\Gamma)\}. \end{aligned} \tag{3.8}$$

Setting

$$F = [f,g]$$

The problem (3.1), (3.2) then has the operator realisation

$$\mathcal{Q}u = F. \tag{3.9}$$

Thus the task of solving (3.1), (3.2) has been reduced to that of determining $\mathcal{Q}^{-1}$. Therefore we must examine

(i) the solvability of (3.9)

(ii) the uniqueness of solutions to (3.9)

(iii) the continuous dependence of the solution on the given data; that is we must examine the well posedness of (3.9).

We remark that (iii) implies that $\mathcal{Q}^{-1}$ is required to be bounded; a property which is also of great importance in most approximation procedures.

The determination of $\mathcal{Q}^{-1}$ is, in general, very difficult if indeed at all possible in a constructive sense. In an abstract treatment $\mathcal{Q}^{-1}$ is the Green's operator, an integral operator the kernel of which is the Green's function for the given problem. However, it is only in the simplest cases that the exact Green's function can be determined and, in general, recourse must be made to some approximate procedure [12], [13].

If approximation methods are to be developed then it is often felt that representing the given problem in terms of an equivalent integral equation offers good prospects. This we do in the next section.

4. An integral equation approach

One of the principal tools at our disposal for this approach is the fundamental solution, $\gamma(p,q)$, for the Laplacian Δ defined by [34]

$$\gamma(p,q) = \begin{cases} \dfrac{R^{2-n}}{(2-n)\omega_n}, & n > 2 \\[2ex] \dfrac{1}{2\pi}\log R, & n = 2 \end{cases} \tag{4.1}$$

where $R = |p-q|$, $p,q \in \mathbb{R}^n$ and

$$\omega_n = 2\pi^{n/2}/\Gamma(n/2). \tag{4.2}$$

For ease of presentation we restrict attention to problems in $\mathbb{R}^3$ in which case the appropriate fundamental solution has the form

$$\gamma(p,q) = -\frac{1}{2\pi R}, \quad R = |p-q| . \tag{4.3}$$

If $w \in C(\Gamma)$ then we introduce integral operators S and D which define single and double layer distributions of density w as follows

$$(Sw)(p) := \int_\Gamma \gamma(p,q)w(q)dS_q, \quad p \in \mathbb{R}^3\backslash\Gamma \tag{4.4}$$

$$(Dw)(p) := \int_\Gamma \frac{\partial\gamma}{\partial n_p}(p,q)w(q)dS_q, \quad p \in \mathbb{R}^3\backslash\Gamma . \tag{4.5}$$

A general investigation of these operators particularly as concerns their mapping properties can be found in the monographs [28], [34].

It is convenient to introduce an integral operator K defined by

$$(Kw)(p) := \int_\Gamma \frac{\partial\gamma}{\partial n_p}(p,q)w(q)dS_q, \qquad p \in \Gamma \tag{4.6}$$

and its $L_2(\Gamma)$ adjoint

$$(K^*w)(p) := \int_\Gamma \frac{\partial\gamma}{\partial n_q}(p,q)q(q)dS_p, \qquad p \in \Gamma \tag{4.7}$$

The jump conditions satisfied by the single and double layer distributions [22], [34] can now be conveniently written in the form [24]

$$\frac{\partial}{\partial n_p^\pm}\left\{Sw\right\}(p) = (\pm I+K)w(p), \qquad p \in \Gamma \tag{4.8}$$

$$\lim_{p\to p^\pm}\{Dw(p)\} = (\mp I+K^*)w(p), \qquad p \in \Gamma \tag{4.9}$$

where +(-) denotes that p tends to Γ from the exterior (interior) of Γ.

With this notation appropriate forms of Green's Theorem for interior and exterior problems are

$$(Du_-)(p)-(S\partial u_-)(p) = \begin{cases} 2u_-(p) \, . & p \in \Omega_- \\ u_-(p) \, , & p \in \Gamma \\ 0 \, , & p \in \Omega_+ \end{cases} \tag{4.10}$$

$$(S\partial u_+)(p)-(Du_+)(p) = \begin{cases} 2u_+(p) \, , & p \in \Omega_+ \\ u_+(p) \, , & p \in \Gamma \\ 0 \, , & p \in \Omega_- \end{cases} \tag{4.11}$$

where $u_-(u_+)$ is the required solution to the interior (exterior) potential problem and

$$\partial u := \partial u/\partial n.$$

If now either a Green's Theorem Method or a Layer Theoretic Method is used to solve boundary value problems of the form (2.1) to (3.3) then we obtain the following representations of solutions to interior problems and their associated boundary integral equations. Similar results can be obtained for exterior problems [24], [28].

Boundary	Representation of Solutions	Boundary Integral Equations
Dirichlet	$u = \frac{1}{2}(Df-Sw)$	$(I-K)w = D_n f$
$u=f$	$u = Dw$	$(I-K^*)w = f$
	$u = Sw$	$Sw = f$
Neumann	$u = \frac{1}{2}(Dw-Sf)$	$(I+K^*)w = -Sf$
$\frac{\partial u}{\partial n} = f$	$u = -Sw$	$(I+K)w = f$
	$u = Dw$	$D_n w = f$

Table 1

We also notice that the unknown quantity in the representations (4.10) and (4.11) can always be obtained as a solution to the first kind integral equation generated by the third component of (4.10) and (4.11).

Thus in all these approaches we are faced with the problem of solving an equation of the form

$$Aw = f \tag{4.12}$$

which depending on the form of A could be either a differential equation or an integral equation or a boundary integral equation of either first or second kind. In the next section we gather together some results from linear operator theory which must be borne in mind when constructing solutions to equations of the form (4.12).

5. Concerning linear operator equations

Linear operator equations of the first and second kind may or may not be solvable in the classical sense and when they are solvable they might not be uniquely solvable. Physical situations certainly exist which lead to such situations. Perhaps the simplest example is that of a boundary value problem which can be reduced to an equivalent integral equation which has an associated eigenvalue problem. In this case if the data appearing in the integral equation satisfies a compatibility condition of the Fredholm Alternative type then an infinity of solutions exist: otherwise we have no solution.

In those cases when the equations do not have solutions in the classical sense then solutions of minimal norm or least square solutions are sought mainly by means of the techniques of generalised inverses [19].

Most of the work directed towards developing iteration methods for linear equations has been devoted to equations which have unique solutions [39], [37], [32], [33]. In practical problems this is sometimes a luxury; the restoration of unique solvability is a matter which we shall discuss in later sections when we return to the particular case of developing constructive methods for potential problems.

We also remark that the analysis of first kind equations leads to difficulties since solutions to such equations do not, in general, depend continuously on the data. The required continuous dependence can be achieved by a regularisation process [12], [19].

Let H_1, H_2 be two Hilbert spaces over the same field, and let T be a bounded linear operator on H_1 into H_2, whose range R(T) is not necessarily closed. We denote the null space of T by N(T) and the adjoint of T by T^*. For any subspace S of H_k, k = 1,2 we denote by $S^\perp$ the orthogonal complement of S and by $\bar{S}$ the closure of S. Then the following relations hold [50], [21]

$$H_1 = N(T) \oplus N(T)^\perp$$
$$H_2 = N(T^*) \oplus N(T^*)^\perp$$

$$\overline{R(T)} = N(T^*)^{\perp}, \qquad \overline{F(t^*)} = N(T)^{\perp} \tag{5.1}$$

and $T|_{N(T)^{\perp}}$ the restriction of T to $N(T)^{\perp}$ has an inverse which is not necessarily continuous.

Now consider the linear operator equations

$$(T-\lambda I)x = y \tag{5.2}$$

$$Tx = y \tag{5.3}$$

where y is a given element in a Hilbert space H, I is the identity operator on H and T is a compact linear operator on H into H. For any $\lambda \neq 0$, $R(T-\lambda I)$ is closed [46] and we have from (4.1)

$$H = R(T-\lambda I) \oplus N(T^*-\bar{\lambda}I) = R(T^*-\bar{\lambda}I) \oplus N(T-\lambda I). \tag{5.4}$$

The criteria for solvability of (4.2) can be determined by means of the Fredholm-Riesz theory which are based on (4.4) and the following relations [50], [51]

$$\dim N(T-\lambda I) = \dim N(T^*-\bar{\lambda}I) < \infty$$

$$\dim R(T-\lambda I) = \dim R(T^*-\bar{\lambda}I). \tag{5.5}$$

Consequently we can deduce that for all $\lambda \neq 0$

(i) (4.2) has a solution for all $y \in H$ if and only if $N(T-\lambda I) = \{0\}$

(ii) (4.2) has a solution for given $y \in H$ if and only if y is orthogonal to $N(T^*-\bar{\lambda}I)$

(iii) if $\lambda \neq 0$ is not an eigenvalue of T then $(T-\lambda I)^{-1}$ exists, is bounded and $R(T-\lambda I) = H$

(iv) if $\lambda \neq 0$ is an eigenvalue of T then $R(T-\lambda I)$ is a closed proper subspace of H.

The solvability of the first kind equation (4.3) is not so readily settled. For a given $y \in H$ the condition $y \in N(T^*)^{\perp}$ is necessary but not sufficient since R(T) is not closed unless it is finite dimensional [46]. Furthermore, the Alternative Theorem does not hold and we cannot obtain a decomposition of H in terms of R(T) and $N(T^*)$. However, it can be shown that (4.3) is solvable provided [6], [45]

(i) $y \in N(T^*)^{\perp}$

(ii) $\Sigma \frac{1}{\mu_n} |(y,\vartheta_n)|^2 < \infty$

where $\{\mu_n\}$ are the eigenvalues and $\{\vartheta_n\}$ the associated orthonormal eigenvectors of the operator TT^*.

6. Iteration techniques

6.1 Introduction

Germane to the development of approximation procedures for equations of the form

$$Aw = f \tag{6.1}$$

is a consideration of equations of the form

$$Tx = x \tag{6.2}$$

where $T : B \to B$ = Banach space need not necessarily be a linear operator. When (6.1) is an equation of the second kind then the associated equation (6.2) is readily obtained. This is also the case for first kind equations as we shall see later.

Introduce a sequence $\{x_n\}$ defined by

$$x_n = Tx_{n-1}, \quad n = 1,2,... \tag{6.3}$$

where x_o is an initial element which is assumed to satisfy $x_o \in D(T^n)$ for all $n = 1,2,...$.

If (i) $x_n \to x^*$ in B (ii) T is continuous at x^*, then x^* is a solution of (6.2). Therefore a necessary condition for the convergence of (6.3) is that (6.2) should be solvable, and in this case (6.3) can be regarded as a source of increasingly better approximations to the solutions of (6.2).

A fundamental result in this area is

Theorem 6.1 (Contraction Mapping Principle) [46], [20].

Let

(1) $M \subset B$ = Banach space be a closed set

(ii) $A : M \to M$ into

(iii) A be a contraction operator

That is, A satisfies

$$||Ax-Ay|| \leq q||x-y|| \quad \text{for all } x,y \in M$$

with $q < 1$.

Then for any initial element $x_o \in M$ the iteration scheme defined by

$$x_n = Ax_{n-1}, \quad n = 1,2,... \tag{6.4}$$

converges to the unique solution x^* of the equation $Ax = x$.

The rate of convergence of the process (6.4) can be expressed by either

$$||x_n-x^*|| \leq q^n||x_o-x^*||, \quad n = 1,2... \tag{6.5}$$

or, rather more conveniently by

$$||x_n-x^*|| \leq \frac{q^n}{1-q}\,||x_o-Ax_o||, \quad n = 1,2,... \tag{6.6}$$

6.2 Linear equations

In general the results (6.5), (6.6) cannot be improved; simply take the equation $Ax \equiv qx = x$. However, additional assumptions may guarantee more rapid convergence. For instance consider the linear equation

$$x = Lx + f \tag{6.7}$$

then the following result can be obtained.

Theorem 6.2 [29]

If $\rho(L)$, the spectral radius of the linear operator L, satisfies $\rho(L) < 1$ then the iteration scheme

$$x_n = Lx_{n-1}+f \quad n = 1,2,... \tag{6.8}$$

converges to a solution x^* of (6.7) and for any ε satisfying $0 < \varepsilon < (1-\rho(L))$ the following error estimate holds

$$||x-x^*|| \le \frac{\{\rho(L)+\varepsilon\}^n}{1-\rho(L)-\varepsilon} ||x_o-Lx_o-f|| \; . \tag{6.9}$$

We note that uniqueness is not guaranteed here. This centres on the uncertainty of closedness required in Theorem 6.1.

6.3 Non linear equations

As an indication of the results which can be obtained in this case we have

Theorem 6.3 [29]

Let

(1) A be a non linear operator in a Banach space B.

(ii) A be Fréchet differentiable at a point $x^* \in B$ which is a solution of the equation

$$Ax = x.$$

(iii) ρ_o denote the spectral radius of the linear operator $A'(x^*)$.

(iv) $\rho_o < 1$.

Then the iteration scheme

$$X_n = Ax_{n-1} \qquad n = 1,2...$$

converges to x^* provided the initial approximation x_o is sufficiently close to x^*, in which case

$$||x_n-x^*|| \le c(x_o,\varepsilon)(\rho_o+\varepsilon)^n$$

for ε an arbitrary positive scalar.

For further details in this connection we refer to [29], [13], [35] and the references cited there.

6.4 Transformation of linear equations

Given a linear equation of the form

$$Lx = b \tag{6.10}$$

it would be convenient to be able to convert this into an equivalent equation of the form

$$x = Sx + f \tag{6.11}$$

where (i) S is a linear operator (ii) $\rho(S) < 1$. Once the form (6.11) has been achieved then it can be solved by the iteration scheme indicated earlier.

To obtain the form (6.11) we make the following assumptions

(1) L has a continuous inverse

(ii) there is a linear operator L_1 such that $(L+L_1)$ has a bounded inverse.

As a consequence (6.10) can now be expressed in the equivalent form

$$x = (L+L_1)^{-1}L_1x+(L+L)^{-1}b =: Sx+f \; . \tag{6.12}$$

If L_1 can be chosen to ensure

$$\rho((L+L_1)^{-1}L_1) < 1$$

then solutions of (6.10) can be obtained by an iteration scheme developed for (6.11)

The choice of L_1 is a complicated matter in general. The main requirements are that L_1 and $(L+L_1)^{-1}$ are easy enough to compute. This approach has been pursued in [30], [23]. We shall return to it in a later section.

6.5 Equations with compact operators

Assume that (6.10) has the particular form

$$x = Ax+f \tag{6.13}$$

where $A: B \to B$ = Banach space is a compact operator. Assume that there is a linear operator A_1 such that $(I-A_1)^{-1}$ exists and is continuous. Then (6.13) can be expressed in the form

$$x = (I-A_1)^{-1}(A-A_1)x+(I-A_1)^{-1}f =: Sx+f. \tag{6.14}$$

Again we want to choose A_1 such that $\rho(S) < 1$ so as to ensure that (6.14) has a unique solution which can be obtained by iteration. With this in mind introduce a projection operator P which projects B onto a finite dimensional subspace of B. Possible forms for A_1 are either

$$A_1 = PA \quad \text{or} \quad A_1 = AP, \tag{6.15}$$

In the former case (6.14) becomes

$$x = (I-PA)^{-1}(I-P)Ax+(I-PA)^{-1}f \tag{6.16}$$

and in the latter

$$x = (I-AP)^{-1}A(I-P)x+(I-AP)^{-1}f. \tag{6.17}$$

In the case when A is compact and B has a basis then A may be approximated arbitrarily well by the operators of finite rank PA (or AP). If $\lambda = 1$ is not an eigenvalue of A then the operator $(I-PA)^{-1}$ exists and approximates $(I-A)^{-1}$. Consequently if PA is close to A in the sense described here then the operator $(I-PA)^{-1}(A-PA)$ will have a small norm and an iteration scheme based on (6.16) will be efficient. Similar remarks can be made with regard to (6.17).

So far there has been no real restriction placed on P other than that it should be a projection. In practice a possible way to proceed is as follows. Denote by $\{\lambda_n\}$, $\{u_n\}$ the eigenvalues and eigenvectors respectively of the operator A. Further, let $\{v_n\}$ denote the eigenvectors of A^*. Now let

$$\lambda_o = \max_n |\lambda_n|$$

and take ℓ_o to be an eigenvector of A^* corresponding to the same eigenvalue λ_o as u_o. Finally denote by ρ_o the radius of a disc containing the entire spectrum of A with the exception of λ_o. A possible choice for P is defined by

$$Px = \ell_o(x)u_o, \qquad x \in B \tag{6.18}$$

If $\rho_o < 1$ then a suitable projection is given by (6.18).

Similar arguments hold when several eigenvalues lie outside a disc of radius $\rho_o < 1$. Here P should be taken to project B onto a subspace whose dimension is at least the sum of the multiplicities of the eigenvalues of A outside the disc $|\lambda| \leq \rho_o$. This approach is developed in the monographs [10], [36].

6.6 α-processes

In this section we consider equations of the form

$$Bx = b \tag{6.19}$$

where B is a bounded, positive, self adjoint operator in a Hilbert space H. Consequently, the spectral theorem indicates that B can be represented in the form

$$Bx = \int_m^M \lambda dE_\lambda x, \qquad 0 < m < M < \infty \tag{6.20}$$

where $\lambda \in \sigma(B)$ the spectrum of B and E_λ are the usual spectral projectors.

Let x^* be a solution of (6.19) and x_o be an approximation to x^*. Define

(i) Residual $=: \Delta_o = Bx_o - b$

(ii) Error $=: \delta_o = x_o - x^*$

then clearly $\Delta_o = B\delta_o$.

A new approximation, x_1, to the required solution can be generated by

$$x_1 = x_o - c_1 \Delta_o$$

where the coefficient c_1 is chosen as the solution of some extremum problem (e.g. minimising of the residual and or error). Continuing we obtain the iteration scheme

$$x_{n+1} = x_n - c_{n+1}\Delta_n \tag{6.21}$$

where

$$\Delta_n = Bx_n - b = B\delta_n$$

$$\delta_n = x_n - x^*.$$

A promising line to adopt now is to set

$$c_n = k\ , \quad n = 1,2,\dots\ .$$

In this case (6.21) reduces to

$$x_{n+1} = (1-kB)x_n + kb, \quad n = 0,1,\dots$$

and k is to be chosen to minimise the norm of the operator (1-kB).

Rather more generally, let $\alpha \in \mathbb{R}$ be fixed and define a sequence $\{x_n^{(\alpha)}\}$ in H by

$$x_{n+1}^{(\alpha)} = x_n^{(\alpha)} - \frac{(B^\alpha \Delta_n, \Delta_n)}{(B^{\alpha+1}\Delta_n, \Delta_n)}\ \Delta_n, \qquad n = 0,1,\dots \tag{6.22}$$

where $x_o^{(\alpha)}$ is any element of H and

$$\Delta_n := Bx_m^{(\alpha)} - b\ , \qquad n = 0,1,2,\dots\ .$$

The iteration scheme (6.22) is known as an α-process. It embodies as special cases a number of well known approximaiton procedures. For example

$\alpha = 1$ yields the method of minimal residuals

$\alpha = 0$ yields the method of steepest descent

$\alpha = -1$ yields the method of minimal error.

We remark that an $\alpha = -1$ process is only feasible if B^{-1} is known to exist.

The above particular methods have been extensively studies and in this connection we would simply refer to [29], [20] and the references cited in these monographs.

Two results for α-processes of particular interest are

Theorem 6.3

Every α-process is convergent with a rate of convergence equal to that of a geometric progression with ratio $[\frac{M-m}{M+m}]$.

Theorem 6.4

For $\alpha \geq -1$ all α-processes are monotone in the sense that

$$||x_{n+1}^{(\alpha)} - x^*|| \leq ||x_n^{(\alpha)} - x^*||$$

6.7 The case of non-self adjoint operators

Many iteration schemes rely quite heavily on the given operators being self-adjoint. This situation can always be achieved by a method introduced by Wiarda [49] and extended by Bialy [5]. Specifically, given an equation of the form

$$Ax = a, \quad a,x \in H \tag{6.23}$$

where A and A^{-1} are continuous but A is not necessarily self-adjoint it can be shown [5], [36] that (6.23) is equivalent to

$$Bx := A^*Ax = A^*a. \tag{6.24}$$

The operator B is a self-adjoint positive definite operator and (6.24) can be solved by the α-process introduced above. In particular for $\alpha = -1$ we obtain the iteration scheme

$$x_{n+1} = x_n - \frac{(Ax_n-a, Ax_n-a)}{(A^*(Ax_n-a), A^*(Ax_n-a))} A^*(Ax_n-a)$$

6.8 Concerning regularisation

The ill posed nature of first kind equations can be overcome by a process of regularisation introduced by Tikhonov [1], [12]. In this method an equation of the first kind

$$Ku = f \tag{6.25}$$

is replaced by the equation

$$K_\alpha u_\alpha = f. \tag{6.26}$$

The approximate solutions u_α are defined variationally by

$$\underset{u \in W}{\text{minimise}} \{||Ku-f||^2 + \alpha\Omega(u)\} \tag{6.27}$$

where

W = space of smooth functions

α = a positive constant; the regularisation parameter

Ω = a non-negative stabilizing functional controlling the stability of the regularised solutions u_α to perturbations of the data f.

The aim is to show that $u_\alpha \to u^*$ a solution of (6.25).

Tikhonov regularisation is very common and indeed useful. However generalisations of the method exist and regularisation with differential operators is more in keeping with the spirit of this paper. In this approach regularisation is effected by

$$\underset{u \in W}{\text{minimise}} \{||Ku-f||^2+\alpha||Tu||^2\} \tag{6.28}$$

where $T: W \to L_2$ is a linear differential operator of the form

$$Tu = \sum_{i=0}^{m} w_i u^{(i)}$$

$$w_i \in C^i[a,b], \qquad w_m^{-1} \in C[a,b] \tag{6.29}$$

The following result can be obtained

Theorem 6.5 [1], [12]

Given (6.25) and (6.29) the problem (6.28) has a unique solution if $N(K) \cap N(T) = \{0\}$.

Furthermore it can be shown [10] that the problem (6.28) is equivalent to that of solving

$$[(K^*K-\alpha I)+\alpha(T^*T+I)]u = K^*f \tag{6.30}$$

which in turn is equivalent to

$$G(K^*K-\alpha I)u+\alpha u = GK^*f \tag{6.31}$$

where $G: L_2 \to D(T^*T)$ is the bounded inverse of (T^*T+I).

Thus the first kind equation (6.25) has been reduced to an equation of the second kind which is well posed and for which questions regarding the existence and uniqueness of solution can be settled in the usual way.

7. Solution methods for potential problems

In this section we confine attention almost entirely to second kind equations. The rationale for doing this is, as we have indicated, that when dealing with first kind equations regularisation processes tend towards developing an equivalent equation of the second kind. A dicsussion of first kind equations in their own right as they appear in potential theory can be found in [14], [15], [16], [17], [43].

As we see from Table 1 the second kind equations with which we are concerned here have the typical form

$$(I \pm A)u = f \tag{7.1}$$

where A is either K or K^* the $L_2(\Gamma)$ adjoint of K.

In order to develop iteration schemes for (7.1) we require details of A and the solvability of (7.1). These aspects are dealt with quite fully in [30]. In summary we find that compactness of K and K^* follows from the weak singularity of their kernels. Further, as these operators are adjoints of each other the Fredholm Alternative indicates that they have the same spectrum. Finally it is shown in [30] that the spectrum $\sigma(K) = \sigma(K^*) \subset \mathbb{R}$ and that all eigenvalues of K and K^* have absolute value less than or equal to unity.

The ideal situation for constructing iteration schemes is when the given equation is uniquely solvable and the operator involved is compact, self adjoint with norm less than unity.

In our case the operators are compact and thus the Fredholm Alternative is readily available for discussions of solvability; uniqueness however is not always present. Self-adjointness can be achieved by Bialy's method, uniqueness can be obtained by an eigenvalue shifting technique whilst an operator with an adequately small norm can often be introduced by means of a modified Green's function approach. We shall briefly outline these various techniques referring to original sources for details.

7.1 Bialy's method [49], [5]

This has been developed to handle nonself-adjoint operators in the following manner.

Given a Hilbert space H and a not necessarily self-adjoint operator $K: H \to H$ we require to construct a solution of the equation

$$Lw := (I-K)w = f. \tag{7.2}$$

We rewrite (7.2) in the form

$$w = w-\alpha L^*Lw+\alpha L^*f, \qquad \alpha \in \mathbb{R}. \tag{7.3}$$

The operator L^*L is clearly self-adjoint and we now try to choose the parameter α so that the operator

$$B := (I-L^*L)$$

has a norm which, ideally, is less than unity. When this is the case then (7.3) can be solved by any of the iteration techniques described earlier.

7.2 An eigenvalue shifting technique

Straightforward iteration of (7.2) yields the scheme

$$w_n = Kw_n+f. \tag{7.4}$$

However, as we have already indicated $\rho(K)$ the spectral radius of K is unity and therefore, in general, the iteration scheme (7.4) does not converge. To overcome this we rewrite (7.2) in the form

$$(I-T)w := [I-(cI+(1-c)K)]w = (1-c)f. \tag{7.5}$$

The intention now is to choose c to ensure $\rho(T) < 1$. In [30] the authors set $c = \frac{1}{3}$ and it was then shown that $\rho(T)$ satisfies

$$\frac{1}{3} \leq \rho(T) < 1 \tag{7.6}$$

specifically it was shown

<u>Theorem 7.1</u> [16]

Starting with an arbitrary $w_0 \in C(\Gamma)$ the iteration

$$w_{n+1} = \frac{1}{3} w_n + \frac{2}{3} Kw_n + \frac{2}{3} f \tag{7.7}$$

converges in $C(\Gamma)$ to the unique solution of the integral equation (7.2).

A more general treatment of this technique is given in [23].

7.3 A modified Green's function approach

This technique was introduced by Jones [48] to study the boundary integral equations which arose in acoustic scattering problem. Essentially these equations were not uniquely solvable and involved an operator which was inherently nonself-adjoint. This technique was extended in [46], [47] under the generic heading of modified Green's functions and it was shown there that apart form ensuring unique solvability a number of other criteria, such as minimising a norm, could also be satisfied simultaneously. Whilst the approach was developed to deal with problems centred on the Helmholtz equation it can also be employed to advantage with potential problems. In essence the approach can be described as follows. We recall first that an exact Green's function $G(p,q)$ for the class of problems we are dealing with can be expressed in terms of a fundamental solution $\gamma_0(p,q)$ in the form

$$G(p,q) = \gamma_0(p,q)+g(p,q). \tag{7.8}$$

The determination of γ_0 is relatively easy but g can really only be determined in the simplest cases [42], [41], [44]. Nevertheless the form of the right hand side of (7.8) can be used to generate another fundamental solution which can have many advantages over γ_0. The resulting modified Green's function, denoted by γ_1, is then used instead of γ_0 in the development given in §4. Repeating the arguments given there we arrive again at boundary integral equations of the form

$$(I \pm A_j)w = f, \quad j = 0,1 \tag{7.9}$$

where A_j is either K_j or K_j^* and the subscript indicates which particular fundamental solution was used in generating the various boundary integral operators.

In choosing the modification $g(p,q)$ in (7.8) we are only restricted initially by the requirement that it should be harmonic with respect to both p and q. For potential problems a suitable form for the modified Green's function is

$$\gamma_1(p,q) = \gamma_0(p,q) + \frac{c_0}{r_p r_q}$$

where r_p and r_q are the radial coordinates of the points p and q respectively and c_0 is a coefficient at our disposal. The intention now is to choose c_0 so that $\rho(K_1)$, the spectral radius of the associated modified boundary integral operator K_1, satisfies

$$\rho(K_1) < 1.$$

It can be shown [2] that for problems involving a sphere c_0 can be chosen to ensure $\rho(K_1) = \frac{1}{3}$. Choosing c_0 in this way therefore ensures both unique solvability and small norm and consequently iteration methods developed for the modified equation, (7.9) with $j = 1$, will converge well to the unique solution of the problem.

8. Concluding remarks

As the title of this paper suggests it is virtually impossible to offer an exhaustive survey of this field. Here we have simply adopted an integral equation approach and concentrated on the second rather than first kind equations which arise.

Unlike a generalised inverse approach we have chosen to settle problems centred on unique solvability and difficulties near eigenvalues before developing an iteration scheme. For an account of the generalised inverse approach we would make particular reference to [19] and the extensive bibliography which it contains. Throughout our intention has been directed towards ensuring that the operators generating an approximation method have a spectral radius less than unity. A recent paper [7] has indicated how this particular requirement can be relaxed in certain cases.

For approximation processes involving first kind equations, in addition to the works already cited, we would refer to the survey articles appearing in [1].

The area of nonlinear problems has a vast associated literature which frequently becomes very specialised when dealing with constructive aspects. This is largely responsible for our considering here, apart form one brief comment, only linear equations.

Finally, we remark that a comparison of a number of the methods outlined here is to be found in [2].

References

1. R. S. Anderssen et al: The application and numerical solution of integral equations.
   Sijthoff and Noordhoff, The Netherlands, 1980.

2. T. S. Angell, R. E. Kleinman, G. F. Roach: Iterative methods for potential problems.
   (This volume), 1987.

3. K. E. Atkinson: The solution of non-unique linear integral equations.
   Numer. Math. 10, 117-124, 1967.

4. Ju. M. Berezanskii: Expansions in eigenfunctions of self adjoint operators.
   A.M.S. Translations Math. Mono., 17, Providence, 1968.

5. H. Bialy: Iterative behandlung linearer Funktrangleighungen.
   Arch. Rat. Mech. Anal. 4, 166-176, 1959.

6. R. Courant, D. Hilbert: Methods of mathematical physics.
   Interscience, New York, 1953.

7. H. W. Engl: A successive approximaiton method for solving equations of second kind with arbitrary spectral radius.
   Jour. Int. Equ. 8, 239-247, 1985.

8. G. I. Eskin: Boundary value problems for elliptic pseudo-differential equations.
A.M.S. Translations of Math. Mono. 52, Providence, 1981.

9. P. R. Garabedian: Partial Differential Equations.
Wiley, New York, (341-344), 1964.

10. I. C. Gohberg, I. A. Fel'dman: Convolution equations and projection methods for their solution.
A.M.S.Translations of Math. Mono. 41, Providence, 1974.

11. P.Grisvard: Elliptic problems in non-smooth domains.
Monographs and studies in mathematics 24, Pitmans, London, 1985

12. C. W. Groetsch: Theory of Tikhonov regularisation for Fredholm equations of the first kind.
Research Notes in Mathematics, 105, Pitmans, London, 1984.

13. S. Heikkila, G. F. Roach: On equivalent norms and the contraction mapping principle.
Jour. Nonlinear Analysis TMA 8, No.10, 1185-1193, 1985.

14. J. L. Howland: Symmeterising kernels and the integral equations of the first kind of classical potential theory.
Proc. Amer. Math. Soc. 19, 1-7, 1968.

15. G. C. Hsaio, R. C. McCamy: Solution of boundary value problems by integral equations of the first kind.
SIAM. Rev. 15, 687-705, 1973.

16. G. C. Hsaio, W. Wendland; On Galerkin's method for a class of integral equations of the first kind.
Applicable Anal. 6, 155-157, 1977.

17. G. C. Hsaio, G. F. Roach: On the relationship between boundary value problems.
Jour. Math. Anal. Appl. 68, (2), 557-566, 1979.

18. D. S. Jones: Integral equations for the exterior acoustic problem.
Q. Jour. Mech. Appl. Math. XXVII, 1, 129-142, 1974.

19. W. J. Kammerer, M. Z. Nashed: Iterative methods for the best approximate solutions of linear integral equations of the first and second kinds.
Jour. Math. Anal. and Applic. 40, (3), 547-573, 1972.

20. L. V. Kantorovich, G. F. Akilov: Functional Analysis in Normed Spaces.
Pergamon, London, 1964.

21. T. Kato: Perturbation theory for linear operators.
Springer Verlag, Berlin, 1966.

22. O. D. Kellogg: Fundations of Potential Theory.
Dover Publications, New York, 1953.

23. R. E. Kleinman: Iterative solutions of boundary value problems. Function Theoretic Methods for Partial Differential Equations.
Springer LNM 561, 198-313, 1976.

24. R. E. Kleinman, G. F. Roach: Boundary integral equations for the three dimensional Helmholtz equation.
SIAM Reviews 16, (2), 214-236, 1974.

25. R. E. Kleinman, G. F. Roach: On modified Green's functions in exterior problems for the Helmholtz equation.
Proc. Roy. Soc. Lond. A 383, 313-332, 1982.

26. R. E. Kleinman, G.F. Roach: Operators of minimal norm via modified Green's functions.
Proc. Roy. Soc. Edin. 94 A, 163-178, 1983.

27. V. A. Kondratiev: Boundary value problems for elliptic equations in domains with conical and angular points.
Trudy Moskovkogo Mat. Obschetsva 16, 209-292.
Transactions Moscow Mat. Soc. 16, 227-313, 1967.

28. J. Kral: Integral operators in potential theory. Lecture Notes in Mathematics 823.
Springer Verlag, 1980.

29. M. A. Krasnosel'skii et al: Approximate solution of operator equations.
Wolters-Noordhoff, Groningen, 1969.

30. R. Kress, G. F. Roach: On the convergence of successive approximations for an integral equation in a Green's function approach to the Dirichlet problem.
Jour. Math. Anal. and Applic. 55, (1), 102-111, 1976.

31. J. L. Lions, E. Magenes: Non homogeneous boundary value problems and applications.
Springer Verlag, Berlin, 1972.

32. A. T. Lonseth: Approximate solution of Fredholm-type integral equations.
Bull.Amer. Math. Soc. 60, 415-430, 1954.

33. S. G. Mikhlin, K. L. Smolitskiy: Approximate Methods for Solution of Differential and Integral Equations.
American Elsevier, New York, 1967.

34. C. Miranda: Partial differential equations of elliptic type (2nd Ed.).
Springer Verlag, Berlin, 1970.

35. J. Mooney, G. F. Roach: Iterative bounds for the stable solutions of convex nonlinear boundary value problems.
Proc. Roy. Soc. Edin. 76A, 81-94, 1976.

36. W. M. Patterson: Iterative methods for the solution of a linear operator equation in Hilbert space. Lecture Notes in Mathematics No. 394.
Springer Verlag, Berlin, 1974.

37. W. V. Petryshyn: On a general iterative method for the approximate solution of linear operator equations.
Maths. Comp. 17, 1-10, 1963.

38. W. V. Petryshyn: On generalised inverses and on the uniform convergence of $(I-\beta K)^n$ with application to iterative methods.
Jour. Math. Anal. Appl. 18, 417-439, 1967.

39. L. B. Rall: Error bounds for iterative solutions of Fredholm integral equations.
Pacific J. Maths. 5, 977-986, 1955.

40. G. F. Roach: On the approximate solution of elliptic self adjoint boundary value problems.
Arch. Rat. Mech. Anal. 17, (3), 243-254, 1967.

41. G. F. Roach: Approximate Green's functions and the solution of related integral equations.
Arch. Rat. Mech. Anal. 36. (1), 79-88, 1970.

42. G. F. Roach: Green's functions (2nd Ed.).
Cambridge University Press, Cambridge, 1970.

43. G. F. Roach: On the commutative properties of boundary integral operators.
Proc. Amer. Math. Soc. 73, (2), 219-227, 1979.

44. G. F. Roach: Weighted norms and the contraction mapping principle.
Univ. Göttingen, NAM Bericht No. 10, 1-11, 1974.

45. I. Stakgold: Boundary value problems of mathematical physics.
Macmillan, New York, 1966.

46. A. E. Taylor: Introduction to Functional Analysis.
Wiley, New York, 1958.

47. A. Walther, B. Dejon: General report on the numerical treatment of integral and integro-differential equations.
Symposium on the Numerical Treatment of ODE, Int. Equations, Integro-differential equations (645-671), Birkhauser, Basle, 1960.

48. W. L. Wendland, E. Stephan, G. Hsaio: On the integral equation method for the plane mixed boundary value problem of the Laplacian.
Math. Method in Applied Sci. 1, 265-321, 1979.

49. G. Wiarda: Integralgleichungen.
Teubner, Leipzig, 1930.

50. K. Yosida: Functional Analysis.
Springer Verlag, Berlin, 1968.

51. A. C. Zaanen: Linear Analysis.
North Holland, Amsterdam, 1964.

POTENTIAL THEORY METHODS FOR HIGHER ORDER ELLIPTIC EQUATIONS

G. Wildenhain
Wilhelm-Pieck-Universität Rostock, Sektion Mathematik
Universitätsplatz 1, Rostock, DDR-2500

1. Preliminary remarks

One of the motivating subjects for my lecture is the classical Dirichlet problem for elliptic equations. For that reason it is useful to start with some concise remarks describing the situation in the case of the Laplace equation.

Let $\Omega \subset \mathbb{R}^n$ be a bounded domain and $g \in C(\partial\Omega)$ a continuous function on the boundary $\partial\Omega$. Concerning the Dirichlet problem there is a decomposition $\partial\Omega = \partial_r\Omega \cup \partial_i\Omega$ of the boundary $\partial\Omega$ ($\partial_r\Omega$ regular, $\partial_i\Omega$ irregular boundary points) and an unique bounded harmonic function u_g in Ω with

$$\lim_{\substack{x \to y \\ x \in \Omega}} u_g(x) = g(y) \quad \text{for every } y \in \partial_r\Omega \text{ and every } g \in C(\partial\Omega).$$

Moreover, there exists a family $(\mu_x)_{x\in\Omega}$ of harmonic measures, such that

$$u_g(x) = \int g(y)\, d\mu_x(y).$$

The function u_g is the so-called generalized solution of the Dirichlet problem (see e.g. L. L. Helms /18/). N. Wiener /25/ constructed this solution in the following way. Let $(\Omega_k)_{k=1,2,\ldots}$ be a sequence of smooth domains with $\Omega_k \subset \overline{\Omega}_k \subset \Omega_{k+1} \subset \Omega$ $(k=1,2,\ldots)$,

$\Omega = \bigcup_{k=1}^{\infty} \Omega_k$, $f \in C(\mathbb{R}^n)$ an extension of $g \in C(\partial\Omega)(u|_{\partial\Omega} = g)$ and $u_{k,f}$ the sequence of the solutions of the Dirichlet problems

$$\Delta u_{k,f} = 0 \text{ in } \Omega_k,\ u_{k,f}|_{\partial\Omega_k} = f.$$

Then the generalized solution u_g is the uniform limit of $u_{k,f}$ in Ω (i.e., u_g is independent of the approximation of the domain and of the special extension f).

2. The Dirichlet problem for higher order elliptic equations

In the case of elliptic equations of higher order we do not have such definitive results as in the case of equations of second order. In the follqwing we shall try to characterize the actual situation. Let

$$L(x,D) := \sum_{|\alpha|\leq 2m} a_\alpha(x) D^\alpha$$

be a properly elliptic differential operator with smooth coefficients.

A continuous vectorfunction $\tilde{g} = (g_\alpha)_{|\alpha|\leq 1}$ on a compact set $K \subset \mathbb{R}^n$ is called a Whitney-Taylorfield of order 1, if there exists a function $g \in C^1(\mathbb{R}^n)$ such that $D^\alpha g|_K = g_\alpha$ $(|\alpha| \leq 1)$. Let $W^1(K)$ be the vector space of all this Whitney-Taylorfields.

$$\|\tilde{g}\| = \sum_{|\alpha|\leq 1} \|g_\alpha\|_{C(K)}$$

is a norm in $W^1(K)$.

Using $W^{m-1}(\partial\Omega)$ we are able to formulate the Dirichlet problem in the following natural way.

Dirichlet problem (classical formulation):

Let $\Omega \subset \mathbb{R}^n$ be a given bounded domain and $\tilde{g} \in W^{m-1}(\partial\Omega)$. We are looking for a function $u \in C^{2m}(\Omega) \cap C^{m-1}(\bar{\Omega})$ with $Lu = 0$ in Ω and $D^\alpha u|_{\partial\Omega} = g_\alpha$ $(|\alpha| \leq m-1)$.

In the case ord $L > 2$ this problem is far from a definitive solution. Usually the Dirichlet problem is studied in the so-called weak form.

Dirichlet problem (weak formulation):

For given $g \in W_2^m(\Omega)$ we are looking for a weak solution of the equation $Lu = 0$ (i.e. $\int u\, L^*\varphi\, dx = 0$ for every $\varphi \in C_0^\infty(\Omega)$) with $u-g \in \mathring{W}_2^m(\Omega)$.

Here $W_2^m(\Omega)$ denotes the classical Sobolev space, $\mathring{W}_2^m(\Omega)$ the closure of $C_0^\infty(\Omega)$ in $W_2^m(\Omega)$ and L^* the adjoint differential operator.

For a sufficiently smooth boundary $\partial\Omega$ the solution of the weak problem is a classical solution (see e.g. S. Agmon, A. Douglis, L. Nirenberg /3/), but also for non-smooth domains a lot of results concerning the behaviour of the weak solution of the Dirichlet problem near the boundary were proved. There is an extensive literature in this field (see e.g. J. Nečas /22/ and the references given there). In the following we shall summarize some results, which are in connection with the basic notions of nonlinear and fine potential theory (see e.g. B. Fuglede /10/, V. P. Havin; V. G. Mazja /13/, N. G. Meyers /21/, D. R. Adams; N. G. Meyers /1/, L. I. Hedberg /16/).

Let G_s ($s \geq 1$ an integer) be the Bessel kernel of order s. Any function $f \in W_p^s(\mathbb{R}^n)$ $(1 < p < \infty)$ can be represented almost everywhere (in the Lebesgue sense) as a Bessel potential $G_s * h$ with $h \in L^p(\mathbb{R}^n)$. There are constants $C_1 > 0$, $C_2 > 0$ such that

$$C_1 \|h\|_{L^p(\mathbb{R}^n)} \leq \|f\|_{s,p,\mathbb{R}^n} \leq C_2 \|h\|_{L^p(\mathbb{R}^n)},$$

where $\| \|_{s,p,\mathbb{R}^n}$ denotes the classical Sobolev norm.

For a set $E \subset \mathbb{R}^n$ the (s,p)-capacity is defined by

$$C_{s,p}(E) = \inf \left\{ \|h\|^p_{L^p(\mathbb{R}^n)} : h \in L^p(\mathbb{R}^n),\ h \geq 0,\ G_s * h \geq 1 \text{ on } E \right\}.$$

(s,p)-almost everywhere (a.e.) means "with exception of a set of (s,p)-capacity zero". Because of the Sobolev imbedding theorem for $sp > n$ any $f \in W_p^s(\mathbb{R}^n)$ contains a continuous representant. In the case $sp \leq n$ the following holds: The Bessel potential $G_s * h$ is (s,p)-a.e. defined and (s,p)-quasi continuous.

Quasi-continuity means: For every $\varepsilon > 0$ there exists an open set Ω_ε with $C_{s,p}(\Omega_\varepsilon) < \varepsilon$ such that $(G_s * h)|_{\mathbb{R}^n \setminus \Omega_\varepsilon}$ is continuous on $\mathbb{R}^n \setminus \Omega_\varepsilon$.

Hence the restriction $f|_E$ of a function $f \in W_p^s(\mathbb{R}^n)$ on E can be defined as the trace of a (s,p)-quasi-continuous representant of f (which is defined (s,p)-a.e.).

From $f \in W_p^s(\mathbb{R}^n)$ follows $D^\alpha f \in W_p^{s-|\alpha|}(\mathbb{R}^n)$, i.e. $D^\alpha f|_E$ is defined $(s-|\alpha|,p)$-a.e. for $|\alpha| \leq s-1$.

There is an interesting connection with notions of nonlinear potential theory (see V. G. Mazja and V. P. Havin /13/, L. I. Hedberg /16/, L. I. Hedberg and T. H. Wolff /17/ and references given there).

For $sp \leq n$ the set $E \subset \mathbb{R}^n$ is called (s,p)-thin in x iff

$$\int_0^1 \left(\frac{C_{s,p}(E \cap B_x^\delta)}{\delta^{n-sp}}\right)^{q-1} \frac{d\delta}{\delta} < \infty$$

($\frac{1}{p}+\frac{1}{q}=1$, B_x^δ the open ball with centre x and radius δ).
E is (s,p)-thick in x iff the integral diverges.
For $sp > n$ we say that E is (s,p)-thin ($(s,p$-thick) in x iff $x \notin \overline{E}$ ($x \in \overline{E}$). The sets $\{\complement E : E\ (s,p)\text{-thin in } x\}$ form the base of neighbourhoods of x of the so-called (s,p)-fine topology with the corresponding notions (s,p)-lim fine and (s,p)-fine continuity.
B. Fuglede /9/ proved: If the function f is (s,p)-quasi-continuous, then f is (s,p)-a.e. (s,p)-finely continuous.
The preceding remarks motivate the so-called fine Dirichlet problem, which is a version of the Dirichlet problem, adapted to the notions of fine topology and first formulated by B. Fuglede (see /24/).

<u>Dirichlet problem (fine formulation):</u>

1. <u>Variant:</u> Given $g \in W_2^m(\mathbb{R}^n)$ we are looking for $u \in W_2^m(\mathbb{R}^n)$ with $Lu = 0$ in Ω and $D^\alpha(u-g)|_{\partial\Omega} = 0$ $(|\alpha| \le m-1)$.
2. <u>Variant:</u> Given $g \in W_2^m(\Omega)$ we are looking for $u \in W_2^m(\Omega)$ with $Lu = 0$ in Ω and $\underset{\substack{x \to y \in \partial\Omega \\ x \in \Omega}}{(m-|\alpha|,2)\text{-lim fine}}\ D^\alpha(u(x)-g(x)) = 0$
$(m-|\alpha|,2)$-a.e. on $\partial\Omega$ for $|\alpha| \le m-1$.

<u>Theorem 1:</u> For any bounded domain $\Omega \subset \mathbb{R}^n$ the fine Dirichlet problem has an unique solution.

The existence is easy to prove (see /24/). Concerning the uniqueness, there is an interesting connection with the so-called spectral synthesis. We say that a closed set $E \subset \mathbb{R}^n$ admits the (s,p)-spectral synthesis, if any function $f \in W_p^s(\mathbb{R}^n)$ with $D^\alpha f|_E = 0$ $(|\alpha| \le s-1)$ is contained in the closure of $C_0^\infty(\mathbb{R}^n \setminus E)$ with respect to the Sobolev norm $\|\ \|_{s,p,\mathbb{R}^n}$.
B. Fuglede showed that the solution (in the first variant) is unique iff $\complement\Omega$ admits the $(m,2)$-spectral synthesis. L. I. Hedberg /14,15/ proved that for $p > 2-\frac{1}{n}$ the (s,p)-spectral synthesis holds for any closed set and in 1983 L. I. Hedberg and T. H. Wolff /17/ have given the same result for every $p > 1$.
The uniqueness of the solution of the Dirichlet problem in the second variant has been proved by T. Kolsrud /20/.
The classical results for the Laplace equation, mentioned in the first chapter, and the Theorem 1 give rise to the following open questions.

1. May be replaced the fine limit by the usual limit?
2. What can be proved about the exceptional sets on the boundary?

The conjecture is that the investigation of this problems should lead to a classification of the boundary points as in the classical case, but the resulting regularity notion probably depends on the order of derivation on the boundary (see also /29/).

3. Applications of the Agmon-Miranda-inequality, representation of solutions

The classical maximum principle can be considered as an essential background for the theory of harmonic functions - moreover for elliptic equations of second order. This - on the other hand - does not hold for higher order elliptic equations. But in some ways we have the so-called Agmon-Miranda-inequality as a compensation in this case, proved by C. Miranda and S. Agmon (see /2/) for smooth domains.

Let $\Omega \subset \mathbb{R}^n$ be a bounded domain with a sufficiently smooth boundary $\partial\Omega$ and let L be an uniformly elliptic operator of order $2m$. Moreover we suppose that the homogeneous Dirichlet problem for $Lu = 0$ in Ω only has the solution $u \equiv 0$. Then for the solutions $u \in C^{2m}(\Omega) \cap C^{m-1}(\overline{\Omega})$ the inequality

$$\sum_{|\alpha| \le m-1} \sup_{x \in \Omega} |D^\alpha u(x)| \le C \cdot \sum_{|\alpha| \le m-1} \sup_{x \in \partial\Omega} |D^\alpha u(x)| \qquad (1)$$

holds, where $C > 0$ denotes a constant, not depending on u.
The Agmon-Miranda-inequality (1) is not only true for smooth domains. A.-M. Sändig /23/ and G. Albinus /4/ proved (1) for special plane domains with corners.
There are two fundamental open questions.

1. Under which conditions with respect to the domain the inequality (1) holds?
   (For arbitrary simply connected domains?)
2. What can be proved with respect to the dependence of the constant C in (1) from the domain?
   (In a sense continuous dependence?)

In the work of G. Anger, B.-W. Schulze and the author (see for instance /5/, /26/, /24/) the inequality (1) is taken as a starting point for the development of a potential theory for higher order elliptic equations. We shall summarize some aspects of this theory.

It is supposed in the following that for the domains $\Omega \subset \mathbb{R}^n$ and the operator L in consideration the inequality (1) is true.
We consider the space of vector functions

$$D(\bar{\Omega}) = \left\{ h = (h_\alpha)_{|\alpha| \le m-1} : h_\alpha = D^\alpha u,\ u \in C^{2m}(\Omega) \cap C^{m-1}(\bar{\Omega}),\ Lu = 0 \text{ in } \Omega \right\}$$

as a subspace of $W^{m-1}(\bar{\Omega})$ (space of Whitney-Taylorfields).
$D(\partial\Omega) \subset W^{m-1}(\partial\Omega)$ denotes the space of restrictions on $\partial\Omega$.

Theorem 2: For any given vector measure $\nu = (\nu^\alpha)_{|\alpha| \le m-1}$ with $\operatorname{supp} \nu^\alpha \subseteq \bar{\Omega}$ there exists a vector measure $\mu = (\mu^\alpha)_{|\alpha| \le m-1}$ with $\operatorname{supp} \mu^\alpha \subseteq \partial\Omega$ such that

$$\sum_{|\alpha| \le m-1} \int h_\alpha \, d\mu^\alpha = \sum_{|\alpha| \le m-1} \int h_\alpha \, d\nu^\alpha \quad \text{for every } h \in D(\bar{\Omega}).$$

By $\Pi \nu := \mu$ we define the so-called balayage operator (choosing a suitable μ for given ν).
For $x \in \bar{\Omega}$ and $|\alpha_0| \le m-1$ let $\delta_{x_0,\alpha_0} := (\nu^\alpha)_{|\alpha| \le m-1}$ be the special measure, given by

$$\nu^\alpha = \begin{cases} 0 & \text{for } \alpha \neq \alpha_0 \\ \delta_x & \text{for } \alpha = \alpha_0 \end{cases}$$

(δ_x Dirac measure supported in x). Then

$$\Pi \delta_{x,\alpha_0} =: \mu_{x,\alpha_0} = (\mu^\alpha_{x,\alpha_0})_{|\alpha| \le m-1}$$

is the "generalized harmonic measure" with respect to the domain Ω and the operator L. It is characterized by the representability

$$D^{\alpha_0} u(x) = \sum_{|\alpha| \le m-1} \int D^\alpha u(y) \, d\mu^\alpha_{x,\alpha_0}(y)$$

of every $u \in C^{2m}(\Omega) \cap C^{m-1}(\bar{\Omega})$ with $Lu = 0$ in Ω, $|\alpha_0| \le m-1$.
In order to give a potential theoretic interpretation of the balayage we suppose the existence of a global fundamental solution $\Phi(x,z)$ of the operator L. Writing $\Phi^\beta_\alpha(x,z) = D^\alpha_x D^\beta_z \Phi(x,z)$ ($|\alpha| \le m-1$, $|\beta| \le m-1$), the potential (resp. the adjoint potential) of a given vector measure $\mu = (\mu^\alpha)_{|\alpha| \le m-1}$ is defined by the vector function

$$\Phi\mu(x) = ((\Phi\mu)_\alpha(x))_{|\alpha|\leq m-1} \text{ with } (\Phi\mu)_\alpha(x) = \sum_{|\beta|\leq m-1}\int \Phi_\alpha^\beta(x,z)d\mu^\beta(z)$$

$$(\text{resp. } \Phi^*\mu(z) = ((\Phi^*\mu)_\beta(z))_{|\beta|\leq m-1} \text{ with } (\Phi^*\mu)_\beta(z) = \sum_{|\alpha|\leq m-1}\int \Phi_\alpha^\beta(y,z)d\mu^\alpha(y)).$$

Let J_λ be the system of sets of λ-measure zero and

$$F_1^+ = \left\{\lambda \geq 0 : \int (\Phi_\alpha^\beta(x,z))^\pm \, d\lambda(z) \text{ continuous for } |\alpha|\leq m-1,\ |\beta|\leq m-1\right\}.$$

Defining $J = \bigcap_{\lambda\in F_1^+} J_\lambda$ we get a suitable system of "sets of capacity zero". J-almost everywhere (J-a.e.) means "with exception of a set of J". It follows

<u>Theorem 3:</u> $\prod \nu = \mu$ holds iff

$$\Phi^*\mu(z) = \Phi^*\nu(z) \quad \text{for every } z \in \complement\overline{\Omega} \text{ and J-a.e. on } \partial\Omega.$$

Especially for $|\alpha_0|\leq m-1$, $|\beta|\leq m-1$ we have the relation

$$\Phi_{\alpha_0}^\beta(x,z) = \sum_{|\alpha|\leq m-1}\int \Phi_\alpha^\beta(y,z)d\mu_{x,\alpha_0}^\alpha(y) \tag{2}$$

for $z \in \complement\overline{\Omega}$ and J-a.e. on $\partial\Omega$.

Denote $F_1 = F_1^+ - F_1^+ = \left\{\lambda_1 - \lambda_2 : \lambda_1 \in F_1^+,\ \lambda_2 \in F_2^+\right\}$,

$$F = \prod_{|\alpha|\leq m-1} F_1,\quad F(\complement\Omega) = \left\{\lambda \in F : \operatorname{supp}\lambda \subseteq \complement\Omega\right\},$$

$$S(\overline{\Omega}) = \left\{\Phi\lambda|_{\overline{\Omega}} : \lambda \in F(\complement\Omega)\right\},$$

$$S(\partial\Omega) = \left\{\Phi\lambda|_{\partial\Omega} : \lambda \in F(\complement\Omega)\right\}.$$

The vector measure $\mu = (\mu^\alpha)_{|\alpha|\leq m-1}$ is said to be J-absolutely continuous iff $J \subseteq J_\mu \equiv \bigcap_{|\alpha|\leq m-1} J_{\mu^\alpha}$.

Now we are able to formulate some assertions, which are equivalent in the classical case of the Laplace-equation. It can be conjectured that the equivalence also is true in the general case. The assertions are

(I) The classical Dirichlet problem has an unique solution with the representation

$$D^{\alpha}u(x) = \sum_{|\beta|\le m-1} \int g_{\beta}(y)\,d\mu^{\beta}_{x,\alpha}(y) \quad (x\in\Omega,\ |\alpha|\le m-1)$$

(II) $\overline{S(\partial\Omega)} = W^{m-1}(\partial\Omega)$

(III) There exist $\mathcal{J}$-absolutely continuous generalized harmonic measures such that in (2) the equality holds in every point on $\partial\Omega$.

(IV) There exist $\mathcal{J}$-absolutely continuous generalized harmonic measures with $\|\mu_{x,\alpha}\| \le C$ for every $x\in\Omega$ and $|\alpha|\le m-1$ such that $\mu_{x,\alpha} \to \delta_{y,\alpha}$ (weak convergence) for $x\to y$ and every $y\in\partial\Omega$, $|\alpha|\le m-1$.

The following implications are proved up to the present.

Theorem 4: (II) $\Rightarrow$ (I), (IV) $\Rightarrow$ (I), (III) $\Rightarrow$ (II).

For smooth domains the generalized harmonic measures are given by the Poisson kernels. We shall describe this notion for more general elliptic boundary conditions (see e.g. /8/, /27/). Let

$$Lu = f \ \text{ in } \ \Omega,\ B_j u|_{\partial\Omega} = \varphi_j \quad (j=1,\dots,m) \tag{3}$$

be an elliptic boundary value problem. If we suppose

$$(f,\varphi_1,\dots,\varphi_m) \in W_2^{s-2m}(\Omega) \times \prod_{j=1}^{m} W_2^{s-m_j-\frac{1}{2}}(\partial\Omega)$$

($s>\frac{n}{2}$, m_j = ord $B_j \le 2m-1$) and if $N=\{0\}$ holds for the kernel of the problem (3), then (3) has an unique solution, given by

$$u(x) = (f,\Gamma_x^{(B)}) + \sum_{j=1}^{m} (\varphi_j,\ C_j'\Gamma_x^{(B)}). \tag{4}$$

Here the scalar product means the duality between $W_2^{s-2m}(\Omega)$ and $W_2^{2m-s}(\Omega)$ resp. between $W_2^{s-m_j-\frac{1}{2}}(\partial\Omega)$ and $W_2^{-s+m_j+\frac{1}{2}}(\partial\Omega)$, C_j' are boundary conditions, which complet the adjoint boundary conditions (B_j') of (B_j) to a Dirichlet system, $\Gamma_x^{(B)}(y) = \Gamma^{(B)}(x,y) \in W_2^{2m-s}(\Omega)$ is the Green function, associated with the homogeneous boundary conditions $B_j u = 0$ and $\Lambda_j(x,y) := C_j'\Gamma_x^{(B)}$ are the Poisson kernels of the boundary value problem. If the integer s is sufficiently large, (4) can be written in the following form.

$$u(x) = \int_{\Omega} f(y)\cdot \Gamma^{(B)}(x,y)dy + \sum_{j=1}^{m} \int_{\partial\Omega} \varphi_j(y)\cdot\Lambda_j(x,y)d\sigma(y). \qquad (5)$$

For the case of the Dirichlet problem one get a representation of the generalized harmonic measures.
Now we consider the following question: Under which conditions a given solution of $Lu = f$ in a bounded smooth domain Ω can be represented in the form (5) with suitable boundary functions φ_j ?
Define $\Omega_{\varepsilon} = \{x \in \Omega : \text{dist}\,(x, \partial\Omega) > \varepsilon\}$. In order to give an answer to the question we mention the classical Agmon-Douglis-Nirenberg inequality

$$\|u\|_{C^{l+\sigma}(\overline{\Omega}_{\varepsilon})} \leq C_{\varepsilon}\cdot\Big\{\|g\|_{C^{l-2m+\sigma}(\overline{\Omega}_{\varepsilon})} + \sum_{j=1}^{m} \|B_j' u\|_{C^{l-m_j'+\sigma}(\partial\Omega_{\varepsilon})}\Big\},$$

which holds for solutions of $L^* u = g$ ($m_j' = \text{ord}\, B_j'$, $0 < \sigma < 1$) in Ω_{ε}.

<u>Theorem 5:</u> Suppose $f \in W_2^{s-2m}(\Omega)$, $s > \frac{n}{2}$ and $C_{\varepsilon} \leq C$ for $0 \leq \varepsilon \leq \varepsilon_0$. Let u be a solution of $Lu = f$ in Ω which fulfils the condition $\left\| B_j u |_{\partial\Omega_{\varepsilon}} \right\|_{s-m_j-\frac{1}{2}} \leq C_1$ $(j=1,\dots,m;\ 0 \leq \varepsilon \leq \varepsilon_0)$.

Then there exist functions $\varphi_j \in W_2^{s-m_j-\frac{1}{2}}(\partial\Omega)$ such that

$$u(x) = (f, \Gamma_x^{(B)}) + \sum_{j=1}^{m} (\varphi_j, \Lambda_j).$$

With respect to the proof see /27/. Theorem 5 can be generalized for the L^p-case (see /11/).

<u>4. Approximation properties and stability of domains</u>

In this chapter we want to summarize some generalizations of classical approximation results for the Laplace equation resp. elliptic equations of second order to higher order elliptic equations. There are connections with stability notions. First we remind of the notion of stability of a domain with respect to the Dirichlet problem of the Laplace equation, defined by M.V. Keldyš /19/ in 1941.
We suppose $\partial\Omega = \partial\overline{\Omega}$ and consider a sequence of smooth domains with

$$\overline{\Omega} \subset \Omega^{k+1} \subset \overline{\Omega}^{k+1} \subset \Omega^{k}, \quad \overline{\Omega} = \bigcap_{k=1}^{\infty} \Omega^{k}.$$

Let be given any function $g \in C(\partial\Omega)$ and an extension $f \in C(\mathbb{R}^n)$, i.e. $f|_{\partial\Omega} = g$. Then the classical solutions u_f^k of the Dirichlet problems

$$\Delta u_f^k = 0 \quad \text{in} \quad \Omega^k, \; u_f^k|_{\partial\Omega^k} = f$$

exist. Keldyš proved that (in analogy to the Wiener procedure) the sequence (u_f^k) uniformly converges to a harmonic function u^g, the so-called outer solution of the Dirichlet problem, which does not depend on the approximating sequence (Ω^k) and the special extension f. Now the domain is said to be stable with respect to the Dirichlet problem iff $u_g = u^g$ in Ω for every $g \in C(\partial\Omega)$.
There is a generalization of this notion by I. Babuška /6/ and L. I. Hedberg /14/, using Sobolev spaces. Let $\overline{C_o^\infty(\Omega)}^{\| \|_{s,p,\mathbb{R}^n}}$ be the closure of $C_o^\infty(\Omega)$ in the Sobolev space $W_p^s(\mathbb{R}^n)$ and

$$(W_p^s(\mathbb{R}^n))_{\overline{\Omega}} = \left\{ u \in W_p^s(\mathbb{R}^n) \; : \; \text{supp } u \subseteq \overline{\Omega} \right\}$$

($s \geq 0$ an integer). The domain $\Omega \subset \mathbb{R}^n$ is said to be (s,p)-stable iff

$$(W_p^s(\mathbb{R}^n))_{\overline{\Omega}} = \overline{C_o^\infty(\Omega)}^{\| \|_{s,p,\mathbb{R}^n}}.$$

The notion was introduced by I. Babuška in the case $p = 2$ first.
Now let L be an elliptic operator of second order and $K \subset \mathbb{R}^n$ a compact set,

$H(K) = \{u : Lu = 0$ in a neighbourhood of K, depending on $u\}$,

$S(\mathring{K}) = \{v \in C(K) : Lv = 0$ in $\mathring{K}\}$ ($\mathring{K}$ the interior of K),

$A_p(\mathring{K}) = \{v \in L^p(K) : Lv = 0$ in $\mathring{K}\}$ $(1 < p < \infty)$.

There are many papers, where conditions for the density $\overline{H(K)}|_K = S(\mathring{K})$ resp. $\overline{H(K)}|_K = A_p(\mathring{K})$ are given. (see e.g. C. Runge (1885), S. N. Mergeljan (1952), A. G. Vituškin (1967), V. P. Havin (1968), L. I. Hedberg (1972), T. Bagby (1972) for the Cauchy-Riemann operator and by M. V. Keldyš (1941), M. Brelot (1945), J. Deny (1945), A. A. Gončar (1963-65) for the Laplace-case).
A typical result is the following : $\overline{H(K)}|_K = S(\mathring{K})$ holds iff $C_{1,2}(w\backslash K) = C_{1,2}(w\backslash\mathring{K})$ ($w \subset \mathbb{R}^n$ arbitrary open set) or iff $\mathbb{R}^n\backslash K$ has only regular points.
The following results, given by U. Hamann /12/, are corresponding results for higher order elliptic equations. There is a connection with the stability notion on the one hand and with notions of fine potential theory on the other hand.

Condition 1: Let $\Omega \subset \mathbb{R}^n$ be a bounded domain, $\partial\Omega \in C^\infty$, $\Omega_1 \subset \bar{\Omega}_1 \subset \Omega$ and $U \subset \bar{U} \subset \Omega\setminus\bar{\Omega}_1$ subdomains.

Condition 2: Let L be a properly elliptic operator of order 2m and $B = (B_j)_{j=1}^m$ a system of boundary operators such that the boundary value problem (L,B) is elliptic in the usual sense (see /24/).

For $1 < p < \infty$, $\frac{1}{p} + \frac{1}{q} = 1$ we use the following notations.

$A_p^k(\Omega_1) = \{v \in W_p^k(\Omega_1) : Lv = 0 \text{ in } \Omega_1\}$ $(-\infty < k < \infty)$.

$W_p^k(\bar{\Omega}_1) = \{\tilde{v} = (v_\alpha)_{|\alpha|\le k} : \exists v \in W_p^k(\mathbb{R}^n),\ D^\alpha v|_{\bar{\Omega}_1} = v_\alpha,\ |\alpha| \le k\}$ $(0 \le k < \infty)$.

$A_p^k(\bar{\Omega}_1) = \{\tilde{v} = (v_\alpha)_{|\alpha|\le k} \in W_p^k(\bar{\Omega}_1) : Lv_0 = 0 \text{ in } \Omega_1\}$ $(0 \le k < \infty)$.

$M_U(\Omega) = \{u \in C^\infty(\bar{\Omega}) : \text{supp } Lu \subset U,\ Bu|_{\partial\Omega} = 0\}$.

$R_k' M_U(\Omega) = \{(D^\alpha u|_{\bar{\Omega}_1})_{|\alpha|\le k} : u \in M_U(\Omega)\}$ $(k \ge 0)$.

Condition 3: For the adjoint operator L^* in $\Omega\setminus\bar{\Omega}_1$ the unique continuation property holds, i.e. if $L^*u = 0$ in $\Omega\setminus\bar{\Omega}_1$ and $u \equiv 0$ in an open subset $\Omega_0 \subset \Omega\setminus\bar{\Omega}_1$, then $u \equiv 0$ in $\Omega\setminus\bar{\Omega}_1$ follows.

<u>Theorem 6 (U. Hamann):</u> If we suppose that the conditions 1-3 hold and that $\Omega\setminus\bar{\Omega}_1$ is connected, and if Ω_1 is $(2m-k,q)$-stable, then

$$\overline{M_U(\Omega)|_{\Omega_1}}^{\|\ \|_{k,p,\Omega_1}} = A_p^k(\Omega_1) \quad \text{for } k < 0,$$

$$\overline{R_k' M_U(\Omega)}^{\|\ \|_{k,p,\bar{\Omega}_1}} = A_p^k(\bar{\Omega}_1) \quad \text{for } k \ge 0.$$

On the left hand side is meant the closure with respect to the corresponding norm. The stability condition can be replaced by potential theoretic conditions on the boundary $\partial\Omega_1$.

<u>Theorem 7 (U. Hamann):</u> We suppose that the conditions 1-3 hold and that $\Omega\setminus\bar{\Omega}_1$ is connected.

(i) If $\mathbb{R}^n\setminus\bar{\Omega}_1$ is (j,q)-a.e. (j,q)-thick on $\partial\Omega_1$ for $j = 1,\dots,2m-k$, then

$$\overline{M_U(\Omega)|_{\Omega_1}} = A_p^k(\Omega_1) \quad \text{for } k < 0$$

and $\overline{R_k' M_U(\Omega)} = A_p^k(\bar{\Omega}_1)$ for $0 \le k < 2m$.

(ii) If Ω_1 is (j,p)-a.e. (j,p)-thick on $\partial\Omega_1$ for $j = 1,\dots,k-2m$, then

$$\overline{R_k' M_U(\Omega)} = A_p^k(\bar{\Omega}_1) \quad \text{for } k > 2m.$$

(iii) If $m_n(\partial\Omega_1) = 0$ (m_n Lebesgue measure), then

$$\overline{R'_{2m}M_U(\Omega)} = A_p^{2m}(\overline{\Omega}_1)$$

Next we generalize an approximation result by H. Beckert /7/.

Let $\Omega \subset \mathbb{R}^n$ be a bounded domain with a smooth boundary $\partial\Omega$, $\Gamma \subset \Omega$ a smooth surface with $\dim \Gamma = n-1$, such that $\Omega \setminus \overline{\Gamma}$ is connected, $V \subset \partial\Omega$ a fixed (small) open part of the boundary and $g_1 \in W_2^1(\Gamma)$, $g_2 \in L^2(\Gamma)$ given functions on Γ. Then for any $\varepsilon > 0$ there exists a solution of the Dirichlet problem $\Delta u = 0$ in Ω, $u|_{\partial\Omega\setminus V} = 0$ such that

$$\|g_1 - u|_\Gamma\|_{W_2^1(\Gamma)} + \|g_2 - \frac{\partial u}{\partial n}|_\Gamma\|_{L^2(\Gamma)} < \varepsilon \quad .$$

This result has been generalized by A. Göpfert, G. Anger, B.-W. Schulze, G. Wildenhain, G. Wanka, U. Hamann and K. Beyer in different directions. Using methods of potential theory, mentioned above, results can be given, which solve this approximation problem in a definitive sense.

First we formulate our conditions.

Condition 1': Let $\Omega \subset \mathbb{R}^n$ be a bounded domain, $\partial\Omega \in C^\infty$, $\Gamma \in C^\infty$, $\overline{\Gamma} \subset \Omega$, $\dim \Gamma = n-1$, U an open set with $U \subset \overline{U} \subset \Omega \setminus \overline{\Gamma}$, $V \subseteq \partial\Omega$ a relatively open set on the boundary.

$$\text{Let } R_{2m-1}M_U(\Omega) = \left\{R_{2m-1}u = \left(\frac{\partial^{j-1}u}{\partial n^{j-1}}\Big|_\Gamma\right)_{j=1}^{2m} : u \in M_U(\Omega)\right\},$$

$$N_V(\Omega) = \{u \in C^\infty(\overline{\Omega}) : Lu = 0 \text{ in } \Omega \text{ , } Bu|_{\partial\Omega\setminus V} = 0\}.$$

$R_{2m-1}N_V(\Omega)$ is defined in the corresponding way.

Let $D_j(\Gamma)$ ($j=1,\dots,2m$) be function spaces on Γ, for which we suppose

Condition 4: $\overline{C^\infty(\overline{\Gamma})} = D_j(\Gamma)$ ($j=1,\dots,2m$). There exists an integer $s_0 \geq 0$ such that $C^{s_0}(\overline{\Gamma})$ is continuously imbedded in $D_j(\Gamma)$ ($j=1,\dots,2m$).

One can choose e.g. the spaces $C^k(\overline{\Gamma})$, $W_p^t(\Gamma)$ ($0 \leq t < \infty$, $1 < p < \infty$) or the dual spaces $(W_p^t(\Gamma))'$.

<u>Theorem 8 (U. Hamann):</u> Let $\Omega \setminus \overline{\Gamma}$ be connected. We suppose further that the conditions 1', 2, 4 hold and that condition 3 is fulfilled in $\Omega \setminus \overline{\Gamma}$. Then the density

$$\overline{R_{2m-1}M_U(\Omega)} = \overline{R_{2m-1}N_V(\Omega)} = \prod_{j=1}^{2m} D_j(\Gamma) \qquad \text{holds.}$$

The result does not hold, if we replace R_{2m-1} by R_{2m}. In this sense the result is best possible.
If we suppose

Condition 5: $\Omega \setminus \Gamma$ is non connected, $\Omega = \Omega_i \cup \Omega_a \cup \Gamma$, i.e. Ω is divided into two open parts, bounded by Γ resp. $\partial\Omega \cup \Gamma$,

then we have a much more complicated situation. In this case, roughly speaking, Theorem 8 holds for R_{m-1}, where R_{m-1} cannot be replaced by R_m.
In order to approximate in spaces of Whitney-Taylorfields, we need further conditions.

Condition 6: $W^{m-1}(\Gamma)$ is a Banach space.

Condition 7: The Dirichlet problem

$$Lu = 0 \quad \text{in} \quad \Omega_i, \quad D^\alpha u|_\Gamma = g_\alpha \quad (|\alpha| \leq m-1, \; g \in W^{m-1}(\Gamma))$$

has an unique solution.

Condition 8: There exists a global fundamental solution for L in Ω.

With the further notations $M_U(\Gamma) = M_U(\Omega)|_\Gamma$, $N_V(\Gamma) = N_V(\Omega)|_\Gamma$ we have the following

Theorem 9: We suppose that the conditions 1',2 ,5-8 hold and that condition 3 is fulfilled in Ω_a. Then the density

$$\overline{M_U(\Gamma)} = \overline{N_V(\Gamma)} = W^{m-1}(\Gamma) \quad \text{holds.}$$

Also this result is best possible in the sense mentioned above.
For the proof of Theorem 9 see /28/.
In case of the Laplace equation the corresponding assertion can be proved also for a non-smooth surface Γ.

Theorem 10: Let $L = \Delta$. If condition 5 holds, then $\overline{N_V(\Gamma)} = C(\Gamma)$ iff $\partial\Omega_i$ has only regular points.

Concluding the paper, we mention that a suitable condition of outer stability also gives a sufficient condition for the assertion of Theorem 9.

Additional Remark: V.G. Mazja and S.A. Nazarov /30/ constructed for $n \geq 8$ counter-examples for an operator of fourth order, where the Agmon-Miranda-inequality does not hold.

References

1. Adams, D.R.; Meyers, N.G.: Thinness and Wiener criteria for nonlinear potentials. Indiana Univ. Math. J. 22, 169-197 (1972).
2. Agmon, S.: Maximum theorems for solutions of higher order elliptic equations. Bull. Amer. Math. Soc. 66, 77-80 (1960).
3. Agmon, S.; Douglis, A.; Nirenberg, L.: Estimates near the boundary for solutions of elliptic partial differential equations satisfying general boundary conditions I. Comm. Pure Appl. Math. 12, 623-727 (1959).
4. Albinus, G.: Estimates of Agmon-Miranda type for solutions of the Dirichlet problem for linear elliptic differential operators of order 2m in plane domains with corners. Preprint. Akademie der Wiss. der DDR, Inst. f. Math. Berlin 1981.
5. Anger, G.: Funktionalanalytische Betrachtungen bei Differentialgleichungen unter Verwendung von Methoden der Potentialtheorie I. Berlin: Akademie-Verlag 1967.
6. Babuška, I.: Stability of the domain with respect to the fundamental problems in the theory of partial differential equations, mainly in connection with the theory of elasticity I,II. (Russian). Czechoslovak Math. J. 11(86), 76-105, 165-203 (1961).
7. Beckert, H.: Eine bemerkenswerte Eigenschaft der Lösungen des Dirichletschen Problems bei linearen elliptischen Differentialgleichungen. Math. Ann. 139, 255-264 (1960).
8. Berezanskij, Ju. M.; Rojtberg, Ja.A.: Homeomorphism-theorems and Green functions for general elliptic boundary value problems. (Russian). Ukr. mat. J. 19, 3-32 (1967).
9. Fuglede, B.: Quasi topology and fine topology. Séminaire Brelot-Choquet-Deny, 10e année 1965-1966.
10. Fuglede, B.: Applications du thèoréme minimax à l'étude de diverses capacités. C.R. Acad. Sci. Paris, Ser. A 266, 921-923 (1968).
11. González, L.A.; Wildenhain, G.: Representatión de soluciones de ecuaciones diferenciales lineales elipticas (Parte I). Revista Ciencias Matematicas, vol.VI, No. 1, 72-82 (1985).
12. Hamann, U.: Approximation durch Lösungen allgemeiner elliptischer Randwertprobleme bei Gleichungen beliebiger Ordnung. Dissertation Wilhelm-Pieck-Universität Rostock 1986.
13. Havin, V.P.; Mazja, V.G.: Nonlinear potential theory. (Russian). Usp. Mat. Nauk 27:6, 67-138 (1972).
14. Hedberg, L.I.: Spectral synthesis and stability in Sobolev spaces. In Euclidean harmonic analysis (Proc. Univ. of Maryland 1979), Lecture Notes in Math. 779, 73-103, Springer Verlag 1980.
15. Hedberg, L.I.: Spectral synthesis in Sobolev spaces, and uniqueness of solutions of the Dirichlet problem. Acta Math. 147, 237-264 (1981).
16. Hedberg, L.I.: Nonlinear potential theory and Sobolev spaces. Proc. Spring School Lytomysl 1986.
17. Hedberg, L.I.; Wolff, T.H.: Thin sets in nonlinear potential theory. Ann. Inst. Fourier, Grenoble 33,4, 161-187 (1983).
18. Helms, L.L.: Introduction to potential theory. New York- London- Sydney- Toronto: Wiley- Interscience Series in Pure and Appl. Math. 22, 1969.
19. Keldyš, M.V.: On the solubility and stability of the Dirichlet problem. Usp. Mat. Nauk 8, 171-231 (1941) (Russian). Amer. Math. Soc. Translations (2) 51, 1-73 (1966).
20. Kolsrud, T.: A uniqueness theorem for higher order elliptic partial differential equations. Dept. of Mathematics. Univ. of Stockholm, Report No. 9 (1981).

21. Meyers, N.G.: A theory of capacities for functions in Lebesgue classes. Math. Scand. 26, 255-292 (1970).
22. Nečas, J.: Les méthods directes en théorie des equations elliptiques. Prag: Academia-Verlag 1967.
23. Sändig, A.-M.: Das Maximum-Prinzip vom Miranda-Agmon-Typ für Lösungen der biharmonischen Gleichung in einem Rechteck. Math. Nachr. 96, 49-51 (1980).
24. Schulze, B.-W.; Wildenhain, G.: Methoden der Potentialtheorie für elliptische Differentialgleichungen beliebiger Ordnung. Berlin: Akademie-Verlag 1977, Basel-Stuttgart: Birkhäuser-Verlag 1977.
25. Wiener, N.: The Dirichlet Problem. J. Math. Physics 3, 127-146 (1924)
26. Wildenhain, G.: Potentialtheorie linearer elliptischer Differentialgleichungen beliebiger Ordnung. Berlin: Akademie-Verlag 1968.
27. Wildenhain, G.: Darstellung von Lösungen linearer elliptischer Differentialgleichungen. Berlin: Akademie-Verlag, Math. Forschung, Band 8, 1981.
28. Wildenhain, G.: Uniform approximation by solutions of general boundary value problems for elliptic equations of arbitrary order I,II. Z. Anal. Anw. 2(6), 511-521 (1983), Math. Nachr. 113, 225-235 (1983).
29. Wildenhain, G.: Das Dirichlet-Problem für lineare elliptische Differentialgleichungen höherer Ordnung. Jahrbuch Überblicke Mathematik, 137-162 (1983).
30. Mazja, V.G.; Nazarov, S.A.: Conical points can be irregular in the sense of Wiener for elliptic equations of fourth order. Mat. zametki 39, 24-28 (1986) (Russian).

COLLECTION OF PROBLEMS

"What does mathematics really consist of ? Axioms (such as the parallel postulate)? Theorems (such as the fundamental theorem of algebra)? Proofs (such as Gödel's proof of undecidability)? Concepts (such as sets and classes)? Definitions (such as the Menger definition of dimension)? Theories (such as category theory)? Formulas (such as Cauchy's integral formula)? Methods (such as the methods of successive approximations)?

Mathematics could surely not exist without these ingredients; they are essential. It is nevertheless a tenable point of view that none of them is at the heart of the subject, that the mathematician's main reason for existence is to solve problems, and that, therefore, what mathematics really consists of is problems and solutions".

(From P. R. Halmos : The heart of mathematics, Amer. Math. Monthly 87(1980), 519-524.)

On the occasion of the Conference on Potential Theory, Praha, 19.-24. July, 1987, a Collection of Problems from Potential Theory was compiled.

The first part of the Collection consists of those problems sent to the Organizing Committee until 15 May, 1987.

In the second part, problems formulated on the occasion of Copenhagen's Colloquium on Potential Theory, 1979, are reproduced (with kind permission of Professor B. Fuglede).

The third part includes selected problems (related to potential theory) from the collection "Research problems in complex analysis". (The problems are reproduced here with kind permission of Professor W. K. Hayman.)

We would appreciate if solutions to (or comments on) problems proposed here as well as any relevant information could be sent to

Ivan Netuka, MFF UK, Sokolovská 83, 18600 Praha 8, Czechoslovakia.

We hope that this Collection will stimulate further research in Potential Theory.

The Organizing Committee

PROBLEMS ON DISTORTION UNDER CONFORMAL MAPPINGS

José L. FERNANDEZ, College Park, USA

(1) W. Hayman and J. M. Wu have shown that if Ω is a simply connected domain in the plane and f is a conformal mapping from Ω onto the unit disk Δ then for every circle L one has that $f(\Omega \cap L)$ has finite length and moreover is a regular curve in Ahlfors' sense.

When L is contained in Ω, O. Martio has shown that $f(L)$ is a quasicircle. In the general case, is it true that each component of $f(\Omega \cap L)$ is a quasiline ? .

(2) The Hayman-Wu theorem shows that $f' \in L^1(\Omega \cap L)$; it is natural to conjecture that $f' \in L^p$ for each $p < 2$, or even further that

$$\text{length}\big(f(E)\big) \leq c \,\text{length}(E)^{1/2}$$

if $E \subset L$. (This is a conjecture of A. Baernstein.) It is true at least when E is an interval.

(3) The following is a stronger version of the so-called Brennan's conjecture : if $A \subset \Omega$ then

$$\text{area}\big(f(A)\big) \leq c \,\text{area}(A)^{1/2} .$$

This is the case when A is any disk.

(4) A domain G in the plane has uniformly perfect boundary if the following capacitary condition is satisfied :

$$\text{cap}\big(\{z : |z - a| < r\} \cap G\big) \geq c \cdot r$$

for any $a \in \partial G$, $0 < r < \text{diam}(G)$. Is the Hayman-Wu theorem valid for this domain when f^{-1} is replaced by the universal covering map ? .

ON THE RIESZ REPRESENTATION OF FINELY SUPERHARMONIC FUNCTIONS

Bent FUGLEDE, Copenhagen, Denmark

In view of recent results on integral representation of fine potentials (see [5], [6] for the classical situation and [1], [2] in the frame of standard H-cones), it seems justified to call attention to a problem which has remained open since the very beginning of the study of finely superharmonic functions. This problem bars the way towards a complete understanding of the fine topology version of the Riesz representation of a superharmonic function as the sum of a potential plus a harmonic function. After formulating the problem we single out, as a particular case, the question whether an irregular boundary point for a usual domain in $\mathbb{R}^n$ can give rise to more than one point of the minimal Martin boundary.

Let U denote a *fine domain*, e.g. in $\mathbb{R}^n$, and $G^U : U \times U \to]0,+\infty]$ the Green kernel for U. (If $n = 2$ we thus assume that CU is not inner polar.) Let $\mathcal{S}$ denote the convex cone of all *finely superharmonic* functions ≥ 0 on U. Let $\mathcal{P}$ denote the band in $\mathcal{S}$ formed by all *fine potentials* on U, i.e. those functions $p \in \mathcal{S}$ for which 0 is the only non-negative finely *subharmonic* minorant of p (in U). Equivalently [6], [1], $\mathcal{P}$ consists of all G^U-potentials $p = G^U\mu (\not\equiv +\infty)$ of positive Borel measures μ on U:

$$p(x) = G^U\mu(x) = \int_U G^U(x,y)\, d\mu(y), \quad x \in U .$$

Let $\mathcal{H}$ denote the band in $\mathcal{S}$ generated by all finely harmonic functions $h \geq 0$. Clearly, $\mathcal{P} \cap \mathcal{H} = \{0\}$.

<u>Question :</u> Is $\mathcal{S} = \mathcal{P} + \mathcal{H}$? In other words, *if a finely superharmonic function* $s \geq 0$ *on* U *has no finely harmonic minorants* ≥ 0 *other than* 0, *does in then follow that* s *is a fine poten-*

tial ?

This question was raised in [3, p. 105]. It is known that the answer is affirmative if s is finite valued (hence also if s is representable as the pointwise sum of a sequence of finite valued finely superharmonic functions ≥ 0). If the question (for general s) has an affirmative answer for every *regular* fine domain U , then likewise for arbitrary fine domain. (This can easily be shown, using e.g. [3, Theorem 9.14] and [5, proof of Proposition 2.8].)

In the case of a regular fine domain U it follows from [6] together with [5, §§ 2.5, 4.4], that the *orthogonal band* $\mathcal{P}^{\perp}$ to $\mathcal{P}$ within $\mathcal{S}$ consists of all $s \in \mathcal{S}$ which admit a countable covering $\mathcal{V}$ of U by finely open sets V of fine closure contained in U such that

$$s(x) = \int s \, d\varepsilon_x^{CV} \quad (\leq +\infty) \quad \text{for every} \quad x \in V \in \mathcal{V} .$$

The stated question therefore amounts to whether $\mathcal{P}^{\perp} = \mathcal{H}$. In other words : *is every finely superharmonic function* $s \geq 0$ *on* U *which admits a covering* $\mathcal{V}$ *as stated, representable as a countable sum of non-negative finely harmonic functions ?*

In a particular case our question takes the following form :

Let Ω denote a *usual* domain in $\mathbb{R}^n$ with a Green kernel G^{Ω} , and let x_0 be an irregular point of the boundary of Ω . Consider a minimal positive harmonic function u on Ω such that

$$\operatorname*{fine\,lim}_{x \to x_0} u(x) = + \infty .$$

Does it follow that there is a constant c such that

$$u(x) = c \operatorname*{fine\,lim}_{y \to x_0} G^{\Omega}(x,y) , \quad x \in \Omega \,?$$

It is well known that the two fine limits exist. This latter question is subsumed in the former, taking $U = \Omega \cup \{x_0\}$ (a fine domain) and s = the extension of u from Ω to U by fine continuity at x_0 , cf. [3, Theorem 9.14]. Any finely harmonic minorant $h \geq 0$ of s in U leads, by restriction to Ω , to a usual harmonic minorant of u , cf. [3, Theorem 10.16]. By minimality of u there is a constant $\alpha \geq 0$ such that $h = \alpha u$, and here $\alpha = 0$ because otherwise $h(x_0) = \alpha s(x_0) = +\infty$. Consequently, s is a fine potential (if our general question has an affirmative answer). This fine potential s is finely harmonic ($= u$) in $U \setminus \{x_0\}$, and s must therefore be proportional to the fine Green function $G^U(\cdot,x_0)$ according to [4] :

$$s(x) = cG^U(x,x_0) = c \operatorname*{fine\,lim}_{y \to x_0} G^U(x,y) = c \operatorname*{fine\,lim}_{y \to x_0} G^\Omega(x,y) ,$$

noting that $G^U(x,y) = G^\Omega(x,y)$ for $x, y \in \Omega$.

References

[1] BOBOC, N. & Gh. BUCUR : Green potentials on standard H-cones. INCREST, Preprint N$^{\underline{o}}$. 35/1985. To appear in Revue roumaine math. pures appl.

[2] BOBOC, N. & Gh. BUCUR : Potentials and supermedian functions on fine open sets in standard H-cones. Revue roumaine math. pures appl. 31 (1986), 745-775

[3] FUGLEDE, B. : Finely Harmonic Functions. Springer LNM 289 , Berlin-Heidelberg-New York 1972

[4] FUGLEDE, B. : Sur la fonction de Green pour un domaine fin. Ann. Inst. Fourier 25:3-4 (1975), 201-206

[5] FUGLEDE, B. : Integral representation of fine potentials. Math. Ann 262 (1983), 191-214

[6] FUGLEDE, B. : Représentation intégrale des potentiels fins. C.R. Acad. Sci. Paris, 300 (1985), 129-132.

NONLINEAR ELLIPTIC MEASURES

Juha HEINONEN, Jyväskylä, Finland

Suppose that $A : R^n \times R^n \to R^n$ is an elliptic operator given as

$$A(x,h) = \left(\Theta(x)h\cdot h\right)^{\frac{p}{2}-1} \Theta(x)h \ , \quad 1 < p \leqq n \ ,$$

where $\Theta(x)$ is a symmetric, measurable matrix function such that, for some $\lambda > 0$,

$$\lambda^{-1} |h|^2 \leqq \Theta(x)h\cdot h \leqq \lambda |h|^2$$

for all x , h in R^n . If $\Omega \subset R^n$ is an open set and E is a subset of $\partial\Omega$, then one can use the classical Perron method to define the A-*harmonic measure* of E in Ω ,

$$\omega_A(E) = \omega_A(E;\Omega) \ ,$$

which is a continuous weak solution of the equation

$$\operatorname{div} A(x,\nabla u) = 0$$

in Ω . It is clear that, for fixed $x \in \Omega$, the set function

$$E \to \omega_A^x(E) = \omega_A(E;\Omega)(x)$$

does not define an outer measure unless $p = 2$. However, the possibility that $\omega_A^x(E) = 1$ or $= 0$ is independent of x by Harnack's inequality. If $\omega_A(E) = 1$, then E *supports* the A-harmonic measure; if $\omega_A(E) = 0$, then E is of A-*harmonic measure zero*.

Elliptic measures in the linear case $p = 2$ have been studied e.g. in [3], while nonlinear "measures" are important in the theory of quasiregular mappings [4], [6].

The method of J. Bourgain [1] yields the following result, see [5] :

Theorem A. If $p = 2$, then the A-harmonic measure has always a support set $E \subset \partial\Omega$ such that the Hausdorff dimension of E is at most $n - \tau$, where $\tau > 0$ depends only on n and λ.

Question 1. Is Theorem A true also if $p \neq 2$?

Question 2. Is it true that the number $\tau > 0$ in Theorem A cannot depend only on n ?

Next, suppose that $\Omega = B^n = \{x \in R^n ; |x| < 1\}$ is the unit ball of R^n. It was shown in [3] that if $p = 2$, then the *doubling property* holds, that is, there is a constant $C = C(\lambda)$ so that

$$\omega_A^0\left(B^n(x,2r) \cap \partial B^n\right) < C\omega_A^0\left(B^n(x,r) \cap \partial B^n\right)$$

for all $x \in \partial B^n$ and $r > 0$. Here $B^n(x,r) = \{y \in R^n; |y - x| < r\}$.

Question 3. Is the doubling property true also if $p \neq 2$?

The A-harmonic measure in the unit ball may be completely singular with respect to the (n-1)-dimensional Hausdorff measure [2] (see also [6]).

Problem 4. Given a compact set $E \subset \partial B^n$, what metrical conditions on E would imply that $\omega_A(E) = 0$ or $\omega_A(E) > 0$?

References :

[1] BOURGAIN, J. : On the Hausdorff dimension of harmonic measure in higher dimension. - Invent. math. 87, 477-483 (1987)

[2] CAFFARELLI, L. A., E. B. FABES, C. E. KENIG : Completely singular elliptic-harmonic measures. - Indiana Univ. Math. J. 30 No. 6 (1981), 917-924

[3] CAFFARELLI, L., E. FABES, S. MORTOLA and S. SALSA : Boundary behavior of nonnegative solutions of elliptic operators in divergence form. - Indiana Univ. Math. J. 30, No. 4 (1981), 621--640

[4] GRANLUND, S., P. LINDQVIST, O. MARTIO : Phragmén-Lindelöf's and Lindelöf's theorems. - Ark. Mat. 23, 1 (1985), 103-128

[5] HEINONEN, J. : Boundary accessibility and elliptic harmonic measures. (to appear)

[6] HEINONEN, J., O. MARTIO : Estimates for F-harmonic measures and Øksendal's theorem for quasiconformal mappings. - to appear in Indiana Univ. Math. J.

PROBLEMS ON A RELATION BETWEEN MEASURES AND CORRESPONDING POTENTIALS

Mamoru KANDA, Tsukuba, Japan

Let $G(x,y)$ be a kernel on R^d and set $G\mu(x) = \int G(x,y)\mu(dy)$ for a measure μ if defined. In the sequel $G\mu$ is always assumed to be bounded on R^d for simplicity. For a kernel $G(x,y)$ which satisfies the maximum principle, it is well known that

$$G\mu_1 = G\mu_2 \text{ everywhere on } R^d \Longrightarrow \mu_1 = \mu_2 ,$$

where μ_1 and μ_2 are positive measures. For the Newton kernel there exist positive measures μ_1 and μ_2 such that

$$(1) \qquad G\mu_1 = G\mu_2 \text{ everywhere outside } \operatorname{supp}(\mu_1 + \mu_2) \text{ but } \mu_1 \neq \mu_2 .$$

However, for the Riesz kernel $G(x,y) = |x - y|^{\alpha-d}$, $0 < \alpha <$ $< \min(2,d)$, it can be proved that

$$(2) \qquad G\mu_1 = G\mu_2 \text{ everywhere on some open set in } \mathcal{C}\left(\operatorname{supp}(\mu_1 + \mu_2)\right) \Longrightarrow \mu_1 = \mu_2 .$$

If we look at only potentials, it would be difficult to see why this remarkable difference occurs between Newton potentials and Riesz potentials ($0 < \alpha < \min(2,d)$). As you know, the Newton kernel is the Green function for the Laplacian Δ and the Riesz kernel is the one for the isotropic stable generator A, for example, $Af(x) = \int \left(f(x + y) - f(x)\right) n(dy)$, $f \in C_0^\infty$ in case $0 < \alpha < 1$, where $n(dy) = |y|^{-d-\alpha} dy$. The Laplacian is a local operator but the stable generator is not. Indeed the statement (2) is a direct consequence of the antilocality of A, which was proved by Segal and Goodman for a special case [1] and by Murata [2] for a considerably wide class including isotropic stable generators.

So it would be natural to study at first the class of Hunt kernels of convolution type whose generator A has the Lévy-Khintchin representation with the Lévy measure $n(dy)$. Our first problem is

to characterize the class of A (using the Lévy measure) for which the statement (2) holds.

The Lévy measure $n(dy)$ of the isotropic stable generator is of the form $|y|^{-d-\alpha}dy$. But there are many interesting classes of biased stable generators. For example, $n(dy) = |y|^{-d-\alpha}dy$ on some cone and $n(dy) = 0$ outside the cone. For the Green function of the stable generator of the type above, the statement (2) does not hold, but some reasonable assertion (similar to (2) but depending on the cone) would hold. Indeed Ishikawa [3] got a result for $d = 1$. Our next problem is

to get a reasonably modified result similar to (2) for a class of biased potential kernels.

The problems proposed above are connected with the problem on the support of harmonic measures, but they are considerably delicate, because the results we want to show are phenomenons similar to analyticity.

References :

[1] SEGAL, I. E. & GOODMAN, R. W. : Anti-locality of certain Lorentz-invariant operators, J. Math. and Mech., 14, No. 4 (1965), 629-638

[2] MURATA, M. : Antilocality of certain functions of the Laplace operator, J. Math. Soc. Japan., vol. 25, No. 4 (1973), 556-563

[3] ISHIKAWA, Y. : Antilocality and one-sided antilocality for stable generators on the line, Tsukuba J. Math. Vol. 10, No.1 (1986) 1-9

[4] KANDA, M. : Notes on zero sets of convolution potentials on the line, Report at the Symposium on Potential Theory at Kyoto (1986)

OPEN PROBLEMS CONNECTED WITH LEVEL SETS OF HARMONIC FUNCTIONS

Bernhard KAWOHL, Heidelberg, Germany

Many problems in partial differential equations are relatively easy to solve in plane domains, but considerably harder in higher dimensions. Here are two of them :

Problem 1 : A free boundary problem

Let $\Omega \subset \mathbb{R}^2$ be a bounded convex domain with smooth boundary $\partial\Omega$ and consider the problem of finding a function u and a smooth closed arc Γ , being the boundary of a domain $D \supset \Omega$, and such that

(1.1) $$\Delta u = 0 \quad \text{in} \quad G = D \setminus \overline{\Omega} ,$$

(1.2) $$u = 1 \quad \text{on} \quad \partial\Omega ,$$

(1.3) $$u = 0 , \quad \frac{\partial u}{\partial n} = \text{const.} \quad \text{on} \quad \partial D .$$

This problem occurs in the modelling of potential flow. For its physical background one can consult e.g. [H]. If Γ is sufficiently smooth then one can infer the uniqueness of (u,Γ) and the convexity of D , see [T,H]. In [K] I could prove the convexity of D without using conformal maps. The initial hope to *generalize the convexity result to the analogous problem in higher dimensions* (D and Ω are subsets of $\mathbb{R}^n$ with $n \geq 3$) is dampened by several obstacles. First of all the existence question can (at present) only be answered by a variational approach which admits (and sometimes produces) singular free boundaries. This was shown by Alt and Caffarelli [AC]. If the free boundary is smooth (which is an open problem), one can show its uniqueness as in the paper of Tepper [T]. But even if the C^2-regularity of the free boundary were known, there are problems with proving the convexity of D .

My proof for $n = 2$ uses the fact that $v(x) = |\nabla u(x)|^2$ attains its minimum on the boundary of D, because $v \neq 0$ in G and

(1.4) $$\Delta v - 2\left(\frac{\nabla v}{v}\right) \nabla v = 0 \quad \text{in} \quad G$$

holds, as can be seen e.g. from [Sp,p.69]. For $n = 3$ one can not expect such a behaviour in general. The following function f is harmonic in a zero-neighborhood, but $w = |\nabla f|^2$ has a strict local minimum at zero, cf. [P]:

(1.5) $$f(x,y,z) = z + x^2 - y^2 + \left(\frac{2}{3} z^2 - x^2 - y^2\right)z \ .$$

If, however, D is already known to be convex, then the function $v(x) = |\nabla u(x)|^2$ attains its minimum on the boundary of D. This follows from the fact that all level sets of u are convex due to [L] and thus the level surfaces $\{x \in D \mid u(x) = t\}$ have positive Gaussian curvature $g(x)$. Therefore one can use the remarkable identity [We]

(1.6) $$\Delta\left(\log |\nabla u(x)|\right) = - g(x) \quad \text{in} \quad G$$

and the maximum principle to conclude that v is minimal on ∂G. Hopf's second lemma implies that v is minimal on ∂D. For $n = 2$ and nonconvex Ω there are other interesting relations between the fixed and free boundary, see [A].

<u>Problem 2 :</u> Existence of critical points

Let D be a simply connected, bounded domain in R^N, $n \geq 2$, with smooth boundary ∂D and let u be the solution to

(2.1) $$\Delta u = 0 \quad \text{in} \quad \mathbb{R}^n - D \ ;$$

(2.2) $$u = 1 \quad \text{on} \quad \partial D \ ,$$

(2.3a) $u(x)$ decays like $-\log|x|$ as $|x| \to \infty$, if $n = 2$

(2.3b) $u(x) \to 0$ uniformly as $|x| \to \infty$, if $n > 2$.

I consider the question whether u has critical points, i.e. points in which the gradient vanishes.

<u>Theorem 1</u>. If $n = 2$ there are no critical points.

The p r o o f is trivial. This theorem is false if D is not simply connected. It seems to be an *open problem* if this Theorem can be extended to $n > 2$. But the following partial results are known.

<u>Theorem 2</u>. If $n > 2$ and if D is starshaped with respect to y then u has no critical points. Moreover there is a constant c such that

$$|\nabla u(x)|^2 > cr^{-2}u^2(x) \quad \text{in } \mathbb{R}^n - \overline{D} \tag{2.4}$$

holds, where $r = |x - y|$.

For a p r o o f of Theorem 2 one has to notice that the auxiliary function $v(x) = u(x) + C(x - y)\nabla u(x)$ is harmonic whenever u is harmonic. Furthermore $v(x) \to 0$ as $|x| \to \infty$ and $v(x)$ is non-positive on D, provided C is sufficiently large. Theorem 2 is essentially due to Gergen [G], see also [W,St,Pf].

<u>Theorem 3</u>. If $n \geq 2$ and if D is convex, then the level sets $\{x \in \mathbb{R}^n - \overline{D} \mid u(x) > t\} \cup \overline{D}$ of u are convex.

This theorem is due to Lewis [L], for a simpler proof see [KO]. If D is not convex and $n = 2$ or 3, then for sufficiently small t the level sets are still convex, see [N or PS, p.166].

<u>References</u>:

[A] ACKER, A. : On the geometric form of free boundaries satisfying a Bernoulli boundary condition. I and II. Math.Methods Appl. Sci. 6 (1984) 449-456 and 8 (1986) 378-404

[AC] ALT, H. W. & L. A. CAFFARELLI : Existence and regularity for a minimum problem with free boundary. J. Reine Angew. Math. 325 (1981) 105-144

[G] GERGEN, J. J.: Note on the Green function of a starshaped threedimensional region. Amer. J. Math. 53 (1931) 746-752

[H] HAMILTON, R. S. : The inverse function theorem of Nash and Moser. Bull. Amer. Math. Soc. 7 (1982) 65-222

[KO] KAWOHL, B. : Rearrangements and convexity of level sets in PDE. Springer Lecture Notes in Math. 1150 (1985)

[K] KAWOHL, B. : Some qualitative properties of nonlinear partial differential equations. Proceedings of the MSRI microprogram on nonlinear diffusion equations and their equilibrium states, Berkeley, 1986, Eds. W. M. Ni, J. Serrin & L. Peletier, to appear

[L] LEWIS, J.: Capacitary functions in convex rings. Arch. Ration. Mech. Anal. 66 (1977) 201-224

[N] NIKLIBORC, W. : Über die Niveaukurven logarithmischer Flächenpotentiale. Math. Zeitschr. 36 (1933) 641-646

[Pf] PFALTZGRAFF, J. A. : Radial symmetrization and capacities in space. Duke Math. J. 34 (1967) 747-756

[P] POLYA, G.: Liegt die Stelle der größten Beanspruchung an der Oberfläche? Zeitschr. Angew. Math. Mech. 10 (1930) 353-360

[PS] POLYA, G. & G. SZEGÖ : Isoperimetric inequalities in mathematical physics. Annals of Math. Studies 27 (1951) Princeton Univ. Press

[Sp] SPERB, R.: Maximum principles and their applications. Acad. Press, New York, 1981

[St] STODDART, A. W. J.: The shape of level surfaces of harmonic functions in three dimensions. Michigan Math. J. 11 (1964) 225-229

[T] TEPPER, D. F.: Free boundary problem. SIAM J. Math. Anal. 5 (1974) 841-846

[W] WARSCHAWSKI, S. E.: On the Green function of a starshaped threedimensional region. Amer. Math. Monthly 57 (1950) 471--473

[We] WEATHERBURN, C. E.: On families of surfaces. Math. Annalen 99 (1928) 473-478

ON THE EXTREMAL BOUNDARY OF CONVEX COMPACT MEASURES WHICH REPRESENT A NON-REGULAR POINT IN CHOQUET SIMPLEX

D. G. KESELMAN, Astrakhan, USSR

Let w be a relatively compact set of R^n , $n > 1$, $\Gamma_w = \bar{w} \setminus w$, $C(\bar{w})$ - the space of all real continuous functions on a compact space $\bar{w}$, $\mathbf{1}$ - the function equal to one on the whole of $\bar{w}$. Let's consider in $C(\bar{w})$ the subspace :

$$H(\bar{w}) = \{f \in C(\bar{w}) : \Delta f = 0 \text{ on } w\} ,$$

where $\Delta f = 0$ is the Laplace equation. $H(\bar{w})$ contains $\mathbf{1}$ and separates points of $\bar{w}$, i.e. for any two different points $x, y \in \in \bar{w}$ there is a function $f \in H(\bar{w})$ such that $f(x) \neq f(y)$. Such subspaces, containing constants and separating points of the compact set $\bar{w}$, are called functional systems on $\bar{w}$.

Let z be a non-regular boundary point of w . Let's introduce the following notations :

ε_x = Dirac measure of the point $x \in \bar{w}$;

w_x = harmonic measure representing the point $x \in \bar{w}$, i.e. for any function $f \in H(\bar{w})$ the equality $w_x(f) = f(x)$ is valid;

$M_z^+(\Gamma_w)$ = the set of all probability Borel measures ν , which satisfy the following equalities $\operatorname{supp} \nu \subseteq \Gamma_w$ and $\nu(f) = f(z)$ $(\forall f \in H(\bar{w}))$.

As it was shown in [1], p. 293, $M_z^+(\Gamma_w) = [\varepsilon_z, w_z]$, i.e. if $\{x_n\}$ w and $z = \lim_{n\to\infty} x_n$, then the limit point of the sequence w_{x_n} must be situated on the segment $[\varepsilon_z, w_z]$ only. (The equality $\lambda = = \lim_{n\to\infty} w_{x_n}$ means that the equality $\lambda(f) = \lim_{n\to\infty} w_{x_n}(f)$ is valid for all $f \in C(\bar{w})$.)

Let S be a Choquet simplex; we shall introduce the following notations :

$C(S)$ = the space of all real continuous functions on a compact space S ;

$A(S)$ = the set of all real continuous affine functions on S ;

ε_x = the Dirac measure of point $x \in S$;

$E(S)$ = the extremal boundary of S ;

$Sh_s = \overline{E(S)} \setminus E(S)$ = the set of all non-regular points of S ;

$M_x^+(\overline{E(S)})$ = the set of all probability Borel measures ν which satisfy the following equalities :
$\operatorname{supp} \nu \subseteq \overline{E(S)}$ and $\nu(a) = a(x)$ ($\forall$ $a \in A(S)$).

If $z \in Sh_s$, then it is evident that the measures ε_z and μ_z are the extremal points of the set $M_z^+(\overline{E(S)})$. It is also evident that if $Sh_s = \{z\}$, then

(1) $$M_z^+(\overline{E(S)}) = [\varepsilon_z, \mu_z] .$$

As we have seen above, for the simplex

$$S_H = \{s \in H^*(\bar{w}) : s(\mathbf{1}) = 1 = \|s\|\}$$

the equality (1) is also valid. In this case we consider that $\bar{w} \subset \subset S_H$ ([2], pp. 71-75); then the harmonic measure representing the point z coincides with the maximal measure μ_z .

In connection with this the problem to describe the extremal boundary of the set $M_z^+(\overline{E(S)})$, where $z \in Sh_s$, arises. In particular, for what simplexes does the extremal boundary of $M_z^+(\overline{E(S)})$ consist of the finite number of points ? For what simplexes

$$M_z^+(\overline{E(S)}) = [\varepsilon_z, \mu_z] ?$$

How many extremal points has the set $M_z^+(\overline{E(S)})$ if S is the state space on the functional system

$$H_1(\bar{w}) = \{f \in C(\bar{w}) : \Pi f = 0 \text{ on } w\} ,$$

where $\Pi f = 0$ is the heat equation (i.e.

$$S = \{s \in H_1^*(\bar{w}) : s(\mathbf{1}) = 1 = \|s\|) ?$$

One step in this direction has been made. In [3] it has been proved that if z is a non-regular point of a metrizable simplex S and $\nu \in M_z^+\left(\overline{E(S)}\right)$, then in any convex dense subset $D \subset S$ one can choose a sequence $x_n \to z$ such that $\lim_{n\to\infty} \mu_{x_n} = \nu$.

The problem posed above is connected with the research of the behavior of the solution of the linear partial differential equation in the neighbourhood of a non-regular point.

References

[1] LANDKOF, N. S.: Foundations of modern potential theory. Moskva, Nauka 1966 (in Russian)

[2] ALFSEN, E. M.: Compact convex sets and boundary integrals. Berlin, Springer-Verlag, 1971.

[3] KESELMAN, D. G.: On Shilov boundary in Choquet simplex. - The theory of functions and functional analysis and their applications. 1984, № 42, 62-67 (in Russian).

THE PROBLEM OF CONSTRUCTION OF THE HARMONIC SPACE BASED ON CHOQUET SIMPLEX

D. G. KESELMAN, Astrakhan, USSR

Let w be open relatively compact subset of metrizable $\mathcal{B}$-harmonic space by Constantinescu-Cornea [1]. $C(\overline{w})$ is the space of all real continuous functions on a compact space $\overline{w}$. Let's consider the system of functions

$$H(\overline{w}) = \{f \in C(\overline{w}) : f \text{ is harmonic on } w\} .$$

In 1974 - 1976 Bliedtner and Hansen investigated simpliciality of the system of a similar type and, in particular, proved that if $H(\overline{w})$ contains constant functions and separates points of $\overline{w}$ (i.e. $H(\overline{w})$ contains $\mathbb{1}$ = the function equal to one on the whole of $\overline{w}$, and for any two different points $x, y \in \overline{w}$ there is a function $f \in H(\overline{w})$, such that $f(x) \neq f(y)$), i.e. if it is a functional system, then it is simplicial, i.e. the state space

$$S_H = \{s \in H^*(\overline{w}) : s(\mathbb{1}) = 1 = \|s\|\}$$

on $H(\overline{w})$ is a Choquet simplex.

In connection with this a converse problem appears to distinguish from the set of Choquet simplexes the subset of those simplexes, which are state spaces on harmonic systems of $H(\overline{w})$ type.

Let's consider the Dirichlet problem for the linear partial differential equation $D[f] = 0$ in a bounded open set $w \subset R^n$, $n > 1$, such that the set $H_D(\overline{w}) = \{f \in C(\overline{w}) : D[f] = 0 \text{ in } w\}$ is a functional system.

The problem : 1) to distinguish the set of those Choquet simplexes, which are the state spaces on the functional systems of $H_D(\overline{w})$ type;

2) to distinguish the set of Choquet simplexes, which are state spaces on functional systems of $H_D(\overline{w})$ type if D is an

elliptic differential operator or if D is a parabolic differential operator or if D is a differential operator of special type, associated with some special practical problems.

The solution of such a problem will give the opportunity to find out new properties of the solutions of the Dirichlet problem on $\overline{w} \subset R^n$ and will clarify the picture of behavior of the solutions in the neighbourhood of a non-regular point and on w .

The first step in this direction was made in [4], where on locally compact parts of a convex compact set the harmonic sheaf, possessing the Brelot convergence property, was constructed.

At the second step the faces, possessing the analogue of Harnack parabolic inequality, were distinguished and described in a Choquet simplex (cf. [5]). One may try to build the Bauer harmonic space on these faces.

Prof. N. S. Landkof from Rostov-on-Don was also interested in this problem.

References

[1] CONSTANTINESCU, C. & A. CORNEA : Potential theory on harmonic spaces. - Berlin, Springer-Verlag, 1972

[2] BLIEDTNER, I. & W. HANSEN : Simplicial cones in potential theory. - Invent. math., 1975, № 29, 83-110

[3] BLIEDTNER, I. & W. HANSEN : Simplicial Characterization of Elliptic Harmonic Spaces. - Math. Ann., 1976, v. 222, № 3, 261-274

[4] KESELMAN, D. G. : Convex sets and Harnack inequality. - Commentationes Math. Univ. Carolinae, 1986, v. 27, № 2, 359-370

[5] KESELMAN, D. G. : Choquet simplexes and Harnack inequalities. - Proceedings of the 14th Winter school on abstract analysis, Srní, January 1986, section of analysis (to appear).

THE PROBLEM ON QUASI-INTERIOR IN CHOQUET SIMPLEXES

D. G. KESELMAN, Astrakhan, USSR

Let S be a metrizable Choquet simplex and let's introduce the following notations : $E(S)$ is the extremal boundary of S ; $SH_s = \overline{E(S)} \setminus E(S)$ is the set of non-regular points of S ; $P(S)$ is the convex cone of all continuous and convex functions on S ;

$A(S) = P(S) \cap \{-P(S)\}$;

face (x) is the smallest face of S containing point x ;

μ_x is the maximal measure on S representing point x ;

$M_x^+(\overline{E(S)})$ is the set of all probability measures which represent the point x and are centred on $\overline{E(S)}$.

Let's define the Dirichlet operator $f \to U_f$ on the set B of all real bounded Borel functions on $E(S)$, where

$$U_f(x) = \int f d\mu_x , \quad x \in S .$$

Let's for $f \in B$ define the set

$$A_f = \{x \in S : U_f \text{ is continuous at } x\} .$$

Let $b : S \to R$ be the affine function of the first Baire class. As it was shown in [1], if the restriction $b|_{\overline{E(S)}}$ is continuous, then $b \in A(S)$.

Let w be an open relatively compact subset of a $\mathfrak{P}$-harmonic space (cf. [2]).

Let's define : $C(\overline{w})$ = the space of all real continuous functions on a compact space $\overline{w}$; $\mathbf{1}$ = the function equal to one on the whole of $\overline{w}$.

Let's consider the linear space :

$$H(\overline{w}) = \{f \in C(\overline{w}) : f \text{ is harmonic on } w\} .$$

If $H(\overline{w})$ contains $\mathbb{1}$ and separates points of $\overline{w}$ (i.e. for any two different points $x, y \in \overline{w}$ there is a function $f \in H(\overline{w})$ such that $f(x) \neq f(y)$), then by [3] the state space

$$S_H = \{s \in H^*(\overline{w}) : s(\mathbb{1}) = 1 = \|s\|\}$$

on $H(\overline{w})$ is a Choquet simplex.

Let's suppose that $\overline{w} \subset S_H$ ([4], pp. 71-75). In view of this injection the following statements are valid : 1) $\forall f \in H(\overline{w})$ there exists a unique continuous function $a \in A(S)$ such that $a|_{\overline{w}} = f$ and $\|a\| = \|f\|$;

2) the Choquet boundary for $H(\overline{w})$ coincides with the extremal boundary of S_H ;

3) the harmonic measure w_x with a barycentre $x \in \overline{w}$ coincides in S_H with the maximal measure μ_x .

Let g be a continuous real function on $\overline{E(S_H)}$; then U_t , where $t = g|_{E(S_H)}$, is an affine function of the first Baire class. If $U_t|_{\overline{E(S_H)}} = g$, i.e. the equality $\mu_z(g) = g(z)$ is valid ($\forall z \in Sh_{S_H}$), then U_t is continuous on S and $U_t|_{\overline{w}} \in H(\overline{w})$. Thus to verify that U_t is a harmonic function the interior of w is not needed, as it is enough to verify a barycentric equality at every non-regular boundary point of w . On the other hand, in solving the classical Dirichlet problem it's traditional to investigate the behavior of the solution both at any interior point and in the neighbourhoods of non-regular points. Here the interior is very important as it is connected with a differential structure of the operator which generates this harmonic space. However, after embedding w in an infinitely dimensional simplex, w is no longer its interior for S_H , because in the latter the interior is empty (cf. [5]).

A natural question arises whether interior of w is a trace on $\overline{w}$ of some subset from the state space S_H with special properties closely connected with non-regular points of w . The first step in this direction was the introduction of the following sets in a metrizable simplex :

$$A = \bigcap_{f \in C(\overline{E(S)})|_{E(S)}} A_f \quad \text{and} \quad A_0 = \bigcap_{f \in B} A_f \quad \text{(cf. [6])},$$

which have the following properties :

Theorem 1. 1) $A = \{x \in S : M_x^+(\overline{E(S)}) = \{\mu_x\}\}$;

2) $\exists\ p \in P(S)$ such that $A = A_t$, where $t = p|_{E(S)}$;

3) A is a G_δ-subset and is dense in S , i.e. it is of the second category.

If $Sh_s \neq \emptyset$ then the set $I = S \setminus A$ is convex and of the first category and has the following properties :

1) $x \in I \Leftrightarrow \text{face}\ (x) \cap I \neq \emptyset$;

2) for any two points $y \in S$ and $x \in I$ we have $]y;x] \subset I$.

Theorem 2. 1) For all functions f from B the following formula is valid :

$$A_f = \{x \in S : \hat{f}(x) = \check{f}(x)\},$$

where $\hat{f} = \inf\{a \in A(S) : a|_{E(S)} \geq f\}$, $\check{f} = \sup\{a \in A(S) : a|_{E(S)} \leq f\}$;

2) $x \in A_0 \Leftrightarrow \text{face}\ (x) \subset A_0$;

if $I_0 = S \setminus A_0 \neq \emptyset$, then

3) I_0 is convex and for any two points $y \in S$ and $x \in I_0$ the set $[x;y[\ \subset I_0$;

4) $\overline{I}_0 = S$.

However, $A \supset E(S)$, i.e. measures μ_x for the points $x \in A$ may have atomic components in contrary to the points $x \in w$. If there are no isolated points in the extremal boundary of the simplex S , then for all $x \in A_0$ the maximal measure μ_x has no atomic components. If we assume that $S = S_H$ = a state space on $H(\overline{w})$, then $A \cap \overline{w} = w$. However, it is unknown what is $A \setminus A_0$. For what simplexes S is the equality $\overline{A}_0 = S$ valid ? Shall we have in this case the set A_0 of the second category ? Does A_0 contain a locally compact subset dense in S on which a harmonic sheaf or a harmonic space may be constructed ?

Let's consider the cone $H = \{h = \hat{p}\ ,\ \ p \in P(S)\}$. Let's introduce on S a thin topology, i.e. the smallest topology on S in which all the functions from H are continuous. As it was shown in [1], if A_0 contains the subset w satisfying the following two properties

1) $\overline{w} = w \cup E(S)$,

2) a thin closure of w contains Sh_s (the analogue of Köhn and Sieveking lemma),

then simplex S is a state space on the functional system

$$\{f \in C(\overline{w}) : \mu_x(f) = f(x) \quad (\forall x \in w)\} .$$

The problem arises : what properties must a Choquet simplex possess for the set w to exist ?

References

[1] KESELMAN, D. G. : On the affine homeomorphism of the Choquet simplexes. Sib. Mat. J. 1981, 22, N° 4, 114-117

[2] CONSTANTINESCU, C. & A. CORNEA : Potential Theory on harmonic spaces. Berlin, Springer-Verlag, 1972.

[3] BLIEDTNER, J. & W. HANSEN : Simplicial cones in Potential Theory. Invent. math., 1975, N° 29, 83-110

[4] ALFSEN, E. M. : Compact convex sets and boundary integrals. Berlin, Springer-Verlag, 1971

[5] SIMONS, S. : Non-compact simplexes. Trans. Amer. Math. Soc., 1970, v. 149, N° 1, 155-161

[6] KESELMAN, D. G. : On some problems of Choquet Theory connected with Potential Theory . Proceedings of the 12-th Winter school on abstract analysis, Srní, January 1984, Section of analysis. Supplemento ai Rendiconti del Circolo Matematico di Palermo, serie 11, N° 5 (1984), 73-81.

BOUNDARY REGULARITY AND POTENTIAL-THEORETIC OPERATORS

Josef KRÁL, Praha, Czechoslovakia

Let $\mathcal{H}_k$ denote the k-dimensional Hausdorff measure (with the usual normalization, so that $\mathcal{H}_k(\langle 0,1\rangle^k) = 1$), $\Gamma = \{\theta \in \mathbb{R}^m;\ |\theta| = 1\}$, $\sigma_m = \mathcal{H}_{m-1}(\Gamma)$, $B_r(z) = \{x \in \mathbb{R}^m;\ |x - z| < r\}$. Consider an open set $G \subset \mathbb{R}^m$ with a compact boundary B and complement $C = \mathbb{R}^m \setminus G$ such that

$$(1) \qquad \mathcal{H}_m\big(B_r(z) \cap C\big) > 0 \quad \text{whenever} \quad z \in B\ , \quad r > 0\ .$$

Denote by $\mathcal{C}(B)$ the Banach space of all continuous real-valued functions on B with the supremum norm $\|\ldots\|$ and by $\mathcal{C}'(B)$ the space of all signed finite Borel measures with support in B . With each $\mu \in \mathcal{C}'(B)$ we associate its potential $\mathcal{U}\mu$ on G derived from the classical convolution kernel $x \mapsto \frac{1}{(m-2)\sigma_m}|x|^{2-m}$ if $m > 2$, $x \mapsto \frac{1}{2\pi}\ln\frac{1}{|x|}$ if $m = 2$. If $\mathcal{D}$ denotes the space of all infinitely differentiable functions with compact support in $\mathbb{R}^m$, then simple geometric conditions on G can be formulated guaranteeing that, for each $\mu \in \mathcal{C}'(B)$, the weak normal derivative of $\mathcal{U}\mu$ can be represented by another $\nu \in \mathcal{C}'(B)$ in the sense that

$$(2) \qquad \int_G \operatorname{grad}\phi(x) \cdot \operatorname{grad}\mathcal{U}\mu(x)\,dx = \int_B \phi\,d\nu\ , \quad \phi \in \mathcal{D}\ .$$

For this purpose we denote, for fixed $z \in \mathbb{R}^m$ and $r > 0$, by $n_r^G(\theta,z)$ the total number $(0 \leq n_r^G(\theta,z) \leq +\infty)$ of points x in $S_r(\theta,z) = \{z + \rho\theta;\ 0 < \rho < r\}$ enjoying the following property :

$$\mathcal{H}_1\big(S_r(\theta,z) \cap B_\rho(x) \cap G\big) > 0 \quad \text{and}$$
$$\mathcal{H}_1\big(S_r(\theta,z) \cap B_\rho(x) \cap C\big) > 0\ , \quad \forall \rho > 0\ .$$

Then $\theta \mapsto n_r^G(\theta,z)$ is Borel measurable and we may put

$$v_r^G(z) = \frac{1}{\sigma_m} \int_T n_r^G(\Theta,z)\, d\mathcal{H}_{m-1}(\Theta)\ , \quad V_r^G = \sup_{z\in B} v_r^G(z)\ ,$$

$$V_0^G = \lim_{r\downarrow 0} V_r^G\ .$$

Then $V_0^G < \infty$ is a necessary and sufficient condition implying that, for each $\mu \in \mathcal{C}'(B)$, there is a (uniquely determined) $\nu \in \mathcal{C}'(B)$ fulfilling (2); it turns out that the operator $\mu \mapsto \nu$ is adjoint to a bounded linear operator W^G acting on $\mathcal{C}(B)$. For each $f \in \mathcal{C}(B)$, $W^G f$ is naturally interpreted as the double layer potential of momentum density f (cf. [2]). If Q denotes the space of all compact linear operators acting on $\mathcal{C}(B)$ and I is the identity operator, then the formula

$$V_0^G = \inf\left\{\left\|W^G - \frac{1}{2} I - K\right\|;\ K \in Q\right\}$$

explains the importance of the condition

$$V_0^G < \frac{1}{2} \tag{3}$$

in connection with treatment of boundary value problems by the Fredholm-Radon method : (3) guarantees that W^G can be decomposed in the form $\frac{1}{2} I + K + Z$, where $K \in Q$ and $\|Z\| < \frac{1}{2}$. These analytical applications of V_0^G justify investigation of geometrical implications of (3) on the structure of B. While $\mathcal{H}_m(B) > 0$ is possible for $V_0^G \geq \frac{1}{2}$, (3) implies the existence of a closed set $F \subset B$ with $\mathcal{H}_{m-1}(F) = 0$ such that each $z \in B \setminus F$ has a neighbourhood in B isometric with a non-parametric $(m-1)$-dimensional Lipschitz surface (cf. [1]). In general, one cannot assume $F = \emptyset$. Nevertheless, in the plane case $m = 2$ the condition (3) implies that each $z \in B$ has a neighbourhood in B formed by an open arc, i.e. homeomorphic with $\mathbb{R}^1$.

<u>Problem</u>. If $m > 2$ and (3) holds, is it true that each $z \in B$ has a neighbourhood in B homeomorphic with $\mathbb{R}^{m-1}$?

References

[1] KRÁL, J. : The Fredholm method in potential theory, Trans. Amer. Math. Soc. 125 (1966), 511-547

[2] KRÁL, J. : Integral operators in potential theory, Lecture Notes in Mathematics vol. 823, Springer-Verlag, 1980

[3] KRÁL, J. : Boundary regularity and normal derivatives of logarithmic potentials, Proceedings of the Royal Society of Edinburgh 1987, to appear.

CONTRACTIVITY OF THE OPERATOR OF THE ARITHMETICAL MEAN

Josef KRÁL, Dagmar MEDKOVÁ, Praha, Czechoslovakia

Let $G \subset \mathbb{R}^m$ be an open set complementary to a compact set $C = \mathbb{R}^m \setminus G$ and denote by B the common boundary of G, C. Using the notation from the preceding note we assume that $V_0^G < \infty$ and $\mathcal{H}_m(B_r(z) \cap C) > 0$ for each $z \in B$ and $r > 0$. Then, on the space $\mathcal{C}(B)$ of all continuous real-valued functions on B, the double layer potential operator $W^G : f \to W^G f$ is naturally defined. The operator T^C of the arithmetical mean can be introduced by the equation

$$W^G = \frac{1}{2}(I + T^C) ,$$

where I is the identity operator on $\mathcal{C}(B)$ (cf. §3 in [4]). Writing $\operatorname{osc} f(B) = \max f(B) - \min f(B)$ for $f \in \mathcal{C}(B)$, we have for suitable constant $q_C \in \langle 0,\infty)$

$$\text{(1)} \qquad \operatorname{osc} T^C f(B) \leq q_C \operatorname{osc} f(B) , \quad f \in \mathcal{C}(B) ,$$

which shows that T^C can also be considered on the factor-space $\mathcal{C}_0(B) = \mathcal{C}(B)$ modulo the subspace $\tilde{\mathcal{C}}(B)$ of constant functions in $\mathcal{C}(B)$. T^C is said to be contractive if (1) holds with $q_C < 1$. For convex C, the following results are known: (1) always holds with $q_C = 1$, T^C is contractive if and only if C cannot be represented as an intersection of two cones with vertices in B, and the second iterate $(T^C)^2$ is always contractive. Historically, investigation of contractivity of T^C began in connection with attemts to justify convergence of the Neumann series (cf. [3], [4] for the corresponding comments). The least q_C fulfilling (1) has been called the structure constant of C by C. Neumann. If $V_0^G = 0$ then, even if C is not convex, the series

$$f + \sum_{n=1}^{\infty} [(T^C)^{2n} f - (T^C)^{2n-1} f]$$

converges uniformly for each $f \in \mathcal{C}(B)$ with $\lim_{n\to\infty} \operatorname{osc} (T^C)^n f = 0$ to a function $g \in \mathcal{C}(B)$ satisfying $(I + T^C)g = f$ modulo $\tilde{\mathcal{C}}(B)$, because T^C is compact (cf. [7], [4]). Nevertheless, the question of contractivity of T^C or of its modifications (guaranteeing the convergence of $\sum_{k=0}^{\infty} (\pm T^C)^k$ in the operator norm in $\mathcal{C}_0(B)$) is of continuing interest (cf. [1], [2], [5], [8]).

Problem. Find simple geometric conditions on C guaranteeing contractivity of T^C or, more generally, of the k-th iterate $(T^C)^k$, $k \geq 1$.

One can expect $(T^C)^2$ to be always contractive if C is in a suitable sense close to a convex set. Put for $z \in \mathbb{R}^m$:

$$C(z) = C \cup \{x + z;\ x \in C\} .$$

Improving results obtained in [6] one can show for $m = 2$ and any disc C of diameter $\operatorname{diam} C$ that $T^{C(z)}$ is contractive for $|z| < \operatorname{diam} C$ and fails to be contractive for $|z| \geq \operatorname{diam} C$; besides that, simple estimates can be derived for the structure constant of $C(z)$ provided $|z| < \operatorname{diam} C$.

Conjecture. If $C \subset \mathbb{R}^m$ is a smoothly bounded convex body, then for each unit vector $\Theta \in \mathbb{R}^m$ there is a positive constant $k_C(\Theta)$ such that $T^{C(\rho\Theta)}$ is contractive for $0 < \rho < k_C(\Theta)$ and fails to be contractive for $\rho > k_C(\Theta)$.

If this is true, it would be interesting to obtain estimates of the structure constant of $C(\rho\Theta)$ for $0 \leq \rho < k_C(\Theta)$.

References

[1] KLEINMAN, R. E. : Iterative solutions of boundary value problems, Lecture Notes in Math. vol. 561 (1976), 298-313 (Proc. of the Internat. Symp. "Function Theoretic Methods for Partial Differential Equations", editors: E. Meister, N. Weck, W. Wendland)

[2] KLEINMAN, R. E. & W. L. WENDLAND : On Neumann's method for the exterior Neumann problem for the Helmholtz equation, J. Math. Analysis and Applications vol. 57 (1977), 170-202

[3] KRÁL, J. & I. Netuka : Contractivity of C. Neumann's operator in potential theory, J. Math. Analysis and Applications 61 (1977), 607-619

[4] KRÁL, J. : Integral operators in potential theory, Lecture Notes in Mathematics vol. 823, Springer-Verlag 1980

[5] KRESS, R. & G. F. ROACH : On the convergence of successive approximations for an integral equation in a Green's function approach to the Dirichlet problem, J. Math. Analysis and Applications vol. 55 (1976), 102-111

[6] MEDKOVÁ, D. : Notice on contractivity of the Neumann operator, Časopis pro pěstování matematiky 112 (1987), 197-208

[7] SUZUKI, N. : On the convergence of the Neumann series in Banach space, Math. Ann. 220 (1976), 143-146

[8] WINZELL, B. : Estimates for the gradient of solutions of the Neumann problem, Report LITH MAT-76-2, Linköping University, Sweden.

FINE MAXIMA

Josef KRÁL, Ivan NETUKA, Praha, Czechoslovakia

In what follows we assume, for the sake of simplicity, that $f : \mathbb{R}^m \to \mathbb{R}$ is a function on the Euclidean space of dimension $m > 2$.

There is a variety of results in real analysis showing that the set of points where f has a sort of local maximum must be small in the corresponding sense.

It is easy to see that the set of points where f has a strict local maximum is at most countable (cf. [13]; see also [3], [16], [17] and the references given there).

More delicate questions concern approximate maxima (cf. [6], [2], [11], [12], [7]). Put

$$A_f(x) = \{y \in \mathbb{R}^m;\ f(y) \geq f(x)\}$$

and denote by $B(x,r)$ the ball of radius $r > 0$ centered at $x \in \mathbb{R}^m$. If $|\cdots|$ stands for the outer Lebesgue measure, then the upper symmetric density of $A \subset \mathbb{R}^m$ at $x \in \mathbb{R}^m$ is defined by

$$\overline{D}_x(A) = \limsup_{x \downarrow 0} |A \cap B(x,r)| \ / \ |B(x,r)| \ .$$

Let

$$E_a(f) = \{x \in \mathbb{R}^m;\ \overline{D}_x(A_f) < a\} \ .$$

J. Foran [5] showed that $|E_a(f)| = 0$ for $a = 2^{-m}$ and L. Zajíček [15] and J. Tišer [14] extended this result to $a = 1/2$, which is clearly the limit case.

The set of qualitative maxima has been investigated by M. J. Evans [4]. Let $Q(f)$ be the set of all $x \in \mathbb{R}^m$ for which there is an $r = r(x) > 0$ such that $A_f(x) \cap B(x,r)$ is of the first category. Then $Q(f)$ is of the first category.

Fine maxima have been examined in [9]. Denote by $e(A)$ the

set of all points at which $A \subset \mathbb{R}^m$ is thin (cf. [8], [10]), and put

$$M(f) = \{x \in \mathbb{R}^m ;\ x \in e(A_f(x))\} .$$

Ancona's theorem on compacta of positive capacity (cf. [1]) permits to prove that $M(f)$ is inner polar. If f is Borel measurable, then $M(f)$ is a Borel set and, by the Choquet capacitability theorem, $M(f)$ is polar.

Problem. Is it true that $M(f)$ is polar for general (not necessarily Borel measurable) $f : \mathbb{R}^m \to \mathbb{R}$?

For an arbitrary f, polarity has been established for the set of all strong fine maxima defined as follows. We say that a function $\phi : \mathbb{R}^m \to \mathbb{R}$ peaks at $x \in \mathbb{R}^m$ provided $\phi(y) < \phi(x)$ whenever $y \in \mathbb{R}^m \setminus \{x\}$; f is said to have strong fine maximum at x if there is a finely upper semicontinuous function $\phi_x : \mathbb{R}^m \to \mathbb{R}$ peaking at x such that $\phi_x(x) = f(x)$ and $\{y \in \mathbb{R}^m ;\ f(y) > \phi_x(y)\}$ is thin at x .

If $M_S(f)$ consists of all such $x \in \mathbb{R}^m$, then clearly $M_S(f) \subset M(f)$. For an arbitrary polar set $N \subset \mathbb{R}^m$ one can construct a function $f : \mathbb{R}^m \to \mathbb{R}$ such that $M(f) = N$ and $M_S(f) = \emptyset$, so that $M_S(f)$ may be considerably smaller than $M(f)$. The problem is, whether $M(f) \setminus M_S(f)$ is polar in general.

References

[1] ANCONA, A. : Sur une conjecture concernant la capacité et l'effilement, Théorie du Potentiel, Proceedings, Orsay 1983, Lecture Notes in Mathematics 1096, Springer-Verlag 1984, 34-68

[2] CZÁSZÁR, Á. : Sur la structure des ensembles de niveau des fonctions réelles à deux variables, Acta Sci. Math. Szeged 15 (1954), 183-202

[3] DROBOT, V. & M. MORAYNE : Continuous functions with a dense set of proper local maxima, Amer. Math. Monthly 92 (1985), 209-211

[4] EVANS, M. J. : Qualitative maxima of a funciton, Rev. Roumaine Math. Pures Appl. 30 (1985), 745-748

[5] FORAN, J. : On the density maxima of a function, Colloq. Math. 37 (1977), 245-254

[6] GOOD, I. J. : The approximate local monotony of measurable functions, Proc. Cambridge Philos. Soc. 36 (1940), 9-13

[7] GRANDE, Z. : Sur le maximum approximatif, Fund. Math. 123 (1984), 109-115

[8] HELMS, L. L. : Introduction to potential theory, Wiley-Interscience, New York 1969

[9] KRÁL, J. & I. NETUKA : Fine topology in potential theory and strict maxima of funcitons, Expositiones Mathematicae 5 (1987), 185-191

[10] LUKEŠ, J. & J. MALÝ & L. ZAJÍČEK : Fine topology methods in real analysis and potential theory, Lecture notes in Math. 1189, Springer-Verlag, Berlin, 1986

[11] O'MALLEY, R. J. : Strict essential maxima, Proc. Amer. Math. Soc. 33 (1972), 501-504

[12] PU, H. W. & H. H. PU : On the approximate maxima of a function, Rev. Roumaine Math. Pures Appl. 24 (1979), 281-284

[13] SAKS, S. : Theory of the integral, Hafner, New York, 1947

[14] TIŠER, J. : On strict preponderant maxima, Comment. Math. Univ. Carolinae 22 (1981), 561-567

[15] ZAJÍČEK, L. : On preponderant maxima, Colloq. Math. 46 (1982), 289-291

[16] POSEY, E. E. & J. E. VAUGHAN : Functions with a proper local maximum in each interval, Amer. Math. Monthly 90 (1983), 281--282

[17] POSEY, E. E. & J. E. VAUGHAN : Extrema of nowhere differential functions, Rocky Mountain J. of Mathematics 16 (1986), 661-668.

REPEATED SINGULAR INTEGRALS

E. R. LOVE, Melbourne, Australia

The Poincaré-Bertrand Formula for changing the order of integration in repeated Cauchy-principal-value integrals is

$$(1)\qquad \frac{1}{\pi^2} \,{}^{*}\!\!\int_{-\infty}^{\infty} \frac{ds}{s - x} \,{}^{*}\!\!\int_{-\infty}^{\infty} \frac{h(s,t)}{t - s}\, dt = \frac{1}{\pi^2} \,{}^{*}\!\!\int_{-\infty}^{\infty} dt \,{}^{**}\!\!\int_{-\infty}^{\infty} \frac{h(s,t)}{(t - s)(s - x)}\, ds - h(x,x) \; .$$

Early proofs of it go back to Hardy and Poincaré in 1909; the latter author used it in Celestial Mechanics.

Bertrand [1] used it in 1923 in connection with tides.

Muskhelishvili [2:p.56-60] proved it in 1946 assuming that h is Hölder continuous, a much lighter condition than earlier workers had required.

Tricomi [3:p.169-172] considered the case in which $h(s,t) = f(s)g(t)$, and proved the Formula for almost all x when $f \in L^p$, $g \in L^q$ and $p^{-1} + q^{-1} < 1$. His approach was strongly related to Hilbert transforms.

Love [4] extended Tricomi's result to include the case $p^{-1} + q^{-1} = 1$.

It is well known that any $h \in L^2(R^2)$ is expressible by

$$(2)\qquad h(s,t) = \sum_{n=1}^{\infty} f_n(s) g_n(t)$$

where f_n and g_n are in $L^2(R^1)$. Now [4] gives that (1) holds with $h(s,t)$ replaced by $f_n(s)g_n(t)$, and hence by a *finite* sum of such terms.

My problem is to prove or disprove the Poincaré-Bertrand Formula (1) when $h(s,t)$ is an *infinite* sum of such terms, as in (2); that is, whenever $h \in L^2(R^2)$. More generally, the same question arises when h is given by (2) with $f_n \in L^p$, $g_n \in L^q$ and $p^{-1} + q^{-1} = 1$.

References

[1] BERTRAND, G. : "La théorie des marées et les équations integrales", p. 220 in Ann. Sci. École Norm. Sup. (3) 40 (1923) 151-258

[2] MUSKHELISHVILI, N. I. : Singular Integral Equations (trans. J. R. M. Radok) (Wolters-Noordhoff 1946)

[3] TRICOMI, F. G. : Integral Equations (Interscience 1957)

[4] LOVE, E. R. : "Repeated Singular Integrals", J. London Math. Soc. (2) 15 (1977) 99-102.

COFINE POTENTIAL THEORY

Jaroslav LUKEŠ, Jan MALÝ, Praha, Czechoslovakia

Let X be a harmonic space with a countable base and a Green function G on $X \times X$. We associate with each point $z \in X$ the potential $g_z = G(\cdot,z)$ whose harmonic support is $\{z\}$. A set A is said to be a *cofine* neighborhood of z if $\hat{R}^{X \setminus A}_{g_z} \neq g_z$.

Let U be a Borel measurable relatively compact finely open subset of X. A point $z \in \partial_{cof} U$ (= the cofine boundary of U) is called *cofinely regular* if

$$\text{cofine} - \lim_{x \to z} \varepsilon_x^{X \setminus U}(f) = f(z)$$

for every $f \in C(\partial U)$.

Problems. (a) Characterize the set of all cofinely regular points.

(b) (The minimum principle.) Let U be a lower bounded lower semicontinuous function on U. Assume that

$$\varepsilon_x^{X \setminus V}(u) \leq u(x)$$

whenever $x \in V \subset \overline{V} \subset U$, and

(*) $$\text{cofine} - \liminf_{x \to z} u(x) \geq 0$$

for every $z \in \overline{U}^{cof} \setminus U$. Does it follow $u \geq 0$?

(c) The same question as in (b), but (*) is supposed to hold for all cofinely regular points only.

References

[1] JANSSEN, K. : A co-fine domination principle for harmonic spaces, Math. Z. 141 (1975), 185-191

[2] Le JAN, Y. : Quasi-continuous functions associated with a Hunt process, Proc. Amer. Math. Soc. 86 (1982), 133-137.

ESSENTIAL AND PRINCIPAL BALAYAGES

Jaroslav LUKEŠ, Jan MALÝ, Praha, Czechoslovakia

Let U be an open subset of a $\mathfrak{P}$-harmonic space X , and let $\mathcal{P}$ [$\mathcal{P}^c$] denote the convex cones of all [continuous] potentials on X . If $p \in \mathcal{P}$, define the *balayage* of p on CU by

$$\hat{R}_p^{CU} = \bigwedge \{q \in \mathcal{P} : q \geq p \text{ on } CU\}$$

where $\bigwedge M$ means the infimum of a set M with respect to the natural order in $\mathcal{P}$. The importance of this notion is supported by the following theorem.

Theorem. Let $p \in \mathcal{P}^c$, and let f be the restriction of p to the boundary of U . Then the function $\hat{R}_p^{CU}$ is harmonic on U and represents a solution of the generalized Dirichlet problem on U corresponding to the boundary condition f .

By analogy we can define the following operators,

$$T_p^{CU} = \sup \{q \in \mathcal{P}^c : q \leq p,\ q = \hat{R}_q^{CU}\},$$

$$P_p^{CU} = \sup \{q \in \mathcal{P} : q \leq p,\ q = \hat{R}_q^{CU}\}.$$

In [1], T^{CU} is called the *essential balayage* while P^{CU} is termed the *principal balayage* in [2] and [3]. These operators are closely related to solutions of the Dirichlet problem in a natural way and lead to important examples of Keldych operators (cf. C8 of this Collection).

The importance of T^{CU} was discovered by J. Bliedtner and W. Hansen in connection with the simpliciality of certain potential cones and their Choquet boundaries, while an analogue of principal balayage in more general setting of H-cones was exploited in [4], p. 427. It was shown in [1] that, in general, T^{CU} and P^{CU} can differ. So we can formulate our question as follows.

<u>Problem</u>. Characterize those open sets U on which the essential balayage of any $p \in \mathcal{P}^c$ coincides with the principal balayage.

It is known (cf. [3], [4], Ex. 13.B.2) that $T^{CU} = \hat{R}^{CU}$ on $\mathcal{P}^c$ if and only if $P^{CU} = \hat{R}^{CU}$ on $\mathcal{P}^c$, which is the case exactly when the set of all irregular points of U is negligible.

References

[1] BLIEDTNER, J. & W. HANSEN : Simplicial cones in potential theory, Inventiones Math. 29 (1975), 83-110

[2] LUKEŠ, J. : Théorème de Keldych dans la théorie axiomatique de Bauer des fonctions harmoniques. Czech. Math. J. 24(1974), 114-125

[3] LUKEŠ, J. : Principal solution of the Dirichlet problem in potential theory. Comment. Math. Univ. Carolinae 14 (1973), 773-778

[4] LUKEŠ, J. & J. MALÝ & L. ZAJÍČEK : Fine topology methods in real analysis and potential theory. Lecture Notes in Mathematics 1189, Springer, 1986.

LOCAL CONNECTEDNESS OF THE FINE TOPOLOGY

Jaroslav LUKEŠ, Jan MALÝ, Praha, Czechoslovakia

One of remarkable and fruitful results in fine potential theory was given by B. Fuglede in [3]: The fine topology of a harmonic space is locally connected provided the domination axiom D is assumed.

This result is no longer true if we replace the axiom D by the axiom of polarity. We would like also to remember the open *problem* quoted already in [2], Ex. 9.2.4 : Is the fine topology of any elliptic harmonic space locally connected ?

If the answer to this query is positive, then the following property could be close upon the characterization of harmonic spaces whose fine topology is locally connected.

Axiom of broken ellipticity. For any open subset U (of a harmonic space) and any $x \in U$ there is an open neighborhood V of x, $V \subset U$, such that for any absorbent set A in U, either $x \in A$ or $A \cap V = \emptyset$ (cf. [4], Ex. 11.D.7c).

(It may be said that in broken-elliptic spaces the presence of proper absorbent sets is not excluded but they appear very rarely, in fact. A typical example of a non-elliptic space satisfying the axiom of broken ellipticity is indicated in Ex. 11.D.6 of [4]; cf. also Ex. 3.1.7 of [2].)

If the fine topology of a harmonic space X is locally connected, then X satisfies the axiom of broken ellipticity (Ex. 11.D.6c of [4]). On the other hand, the fine topology of X is locally connected provided the following *Brelot property* (BP) holds :

(BP) For every superharmonic function s on X and $z \in X$ there is a fine neighborhood U of z such that $s \wedge U$

is continuous

(cf. [4], Ex. 11.D.6b).

It was shown by M. Brelot in [1] that in the presence of the domination axiom D the condition (BP) holds.

It would be interesting to have some results of the type

(BP) + (?) $\Rightarrow$ axiom D ,

local connectedness of the fine topology + (?) $\Rightarrow$ (BP), or even axiom D .

References

[1] BRELOT, M. : Recherches axiomatiques sur un théorème de Choquet concernant l'effilement, Nagoya Math. J. 30 (1967), 33-46

[2] CONSTANTINESCU, C. & A. CORNEA : Potential theory on harmonic spaces. Berlin, Heidelberg, New York, Springer, 1972

[3] FUGLEDE, B. : Connexion en topologie fine et balayage des mesures. Ann. Inst. Fourier 21 (1971), 227-244

[4] LUKEŠ, J. & J. MALÝ & L. ZAJÍČEK : Fine topology methods in real analysis and potential theory. Lecture Notes in Mathematics 1189, Springer, 1986.

ON THE LUSIN-MENCHOFF PROPERTY

Jaroslav LUKEŠ, Jan MALÝ, Praha, Czechoslovakia

A point x on the real line is a *point of density* of a Lebesgue measurable set M if

$$\lim_{h \to 0_+} \frac{1}{2h} \lambda\left(M \cap (x - h, x + h)\right) = 1 .$$

The classical Lebesgue density theorem states that almost every point of M is its point of density. Another important theorem bearing today a name of Lusin and Menchoff says that given a measurable set M and its closed subset F such that each point of F is a density point of M , there is a closed set P so that $F \subset P \subset M$ and each point of F is again a density point of P .

Around 1950, a development of topological methods in some parts of real analysis started with introducing the *density topology* d on the real line. The topology d is formed by measurable sets having each point as a point of density.

The density topology is substantially finer than the Euclidean one (notice that only finite sets are density compact) and it is not normal. Nevertheless, the mentioned Lusin-Menchoff theorem implies the complete regularity of d and further nice separation properties (which can be applied e.g. to constructions of pathological derivatives).

Also in potential theory on harmonic spaces, the fine topology possesses the following Lusin-Menchoff property : Given a finely open set M and its closed subset F , there is a closed set P such that $P \subset M$ and F is contained in the fine interior of P . This property serves as a useful tool in fine potential theory.

Having a more general structure of a standard H-cone (cf. [1]), the problem arises whether or not its fine topology possesses always the Lusin-Menchoff property. The answer is positive if the whole space is σ-compact (see Corollary 10.26 of [2]).

For further information and more details on the remarkable Lusin-Menchoff property, its history and some applications in topology, real analysis and fine potential theory we refer to [2].

References

[1] BOBOC, N. & Gh. BUCUR & A. CORNEA : Order and convexity in potential theory: H-cones. Lecture Notes in Mathematics 853, Springer, 1981

[2] LUKEŠ, J. & J. MALÝ & L. ZAJÍČEK : Fine topology methods in real analysis and potential theory. Lecture Notes in Mathematics 1189, Springer, 1986.

RELATIONS BETWEEN PARABOLIC CAPACITIES

Fumi-Yuki MAEDA, Hiroshima, Japan

Let X be a domain in R^{d+1} $(d \geq 1)$ on which we consider the heat equation $Lu \equiv \partial u/\partial t - \Delta_x u = 0$ and its adjoint equation $L^*u \equiv -\partial u/\partial t - \Delta_x u = 0$. The sheaf of solutions of $Lu = 0$ (resp. $L^*u = 0$) defines a structure $\mathcal{H}$ (resp. $\mathcal{H}^*$) of P-harmonic spaces on X. Let $\mathcal{P}$ (resp. $\mathcal{P}^*$) be the set of all continuous potentials for $\mathcal{H}$ (resp. $\mathcal{H}^*$). For $p \in \mathcal{P}$ (resp. $\mathcal{P}^*$), Lp (resp. L^*p) in the distribution sense is a non-negative measure, and $\partial p/\partial x_j$ $(j = 1,\dots,d)$ are L^2_{loc}-functions on X (see [M]). Given a compact set K in X, let

$$C_I(K) = \inf \left\{\int p\,d(Lp) \,\middle|\, p \in \mathcal{P}_K\right\}; \quad C_D(K) = \inf \left\{D[p] \,\middle|\, p \in \mathcal{P}_K\right\}$$

$$C_I^*(K) = \inf \left\{\int p\,d(L^*p) \,\middle|\, p \in \mathcal{P}_K^*\right\}; \quad C_D^*(K) = \inf \left\{D[p] \,\middle|\, p \in \mathcal{P}_K^*\right\},$$

where $\mathcal{P}_K = \{p \in \mathcal{P} \mid p \geq 1 \text{ on } K\}$, $\mathcal{P}_K^* = \{p \in \mathcal{P}^* \mid p \geq 1 \text{ on } K\}$

and $D[p] = \iint \sum_j (\partial p/\partial x_j)^2(x,t)\,dx\,dt$.

Problem 1. Are the capacities C_I and C_I^* different ?

It is known that $4^{-1}C_I(K) \leq C_I^*(K) \leq 4C_I(K)$ for any compact set K, so that C_I and C_I^* are equivalent to each other. On the other hand, it can be shown that if $X = R^d \times (-\infty,b)$ with $b < +\infty$, then neither C_I nor C_D^* is equivalent to C_D. Thus we may ask

Problem 2. Are C_I and C_D^* equivalent to C_D if X is bounded?

For related problems, see [G-Z].

References

[G-Z] GARIEPY, R. & W. P. ZIEMER : Thermal capacity and boundary regularity. J. Diff. Eq. 45 (1982), 377-388

[M] MAEDA, F-Y. : Dirichlet integral and energy of potentials on harmonic spaces with adjoint structure. Hiroshima Math. J. 17 (1987), to appear.

ISOVOLUMETRIC INEQUALITIES FOR THE LEAST HARMONIC MAJORANT OF $|x|^p$

Makoto SAKAI, Tokyo, Japan

Motivation . Let $w = f(z)$ be a holomorphic function defined in the unit disk U in the complex plane satisfying $f(0) = 0$. Then

$$\sup_{0<r<1} \frac{1}{2\pi} \int_0^{2\pi} |f(re^{is})|^2 \, ds \leq (1/\pi)\left(\text{area } f(U)\right) .$$

This is called the Alexander-Taylor-Ullman inequality [1] and the left-hand side is equal to the square of the Hardy norm $\|f\|_{H^2(U)}$ of f . Kobayashi [2] gave a new proof of the inequality. Let $D = f(U)$, let h be the least harmonic majorant of $|w|^2$ in D and consider the composite $h\circ f$ of f and h . Then $|f|^2 \leq h\circ f$ in U and $h\circ f$ is harmonic in U . Hence

$$\frac{1}{2\pi} \int_0^{2\pi} |f(re^{is})|^2 \, ds \leq \frac{1}{2\pi} \int_0^{2\pi} (h\circ f)(re^{is}) \, ds$$

and, by the mean-value property of harmonic functions, the right--hand side is equal to $h\left(f(0)\right) = h(0)$. It is known that $h(0) \leq (1/\pi)(\text{area } D)$. Combining this with the above estimation of $h(0)$, we obtain the Alexander-Taylor-Ullman inequality. Thus our motivation of the problem arose from the inequality $h(0) \leq (1/\pi)(\text{area } D)$. This was proved by Pólya and Szegö for simply connected domains and by Payne for multiply connected domains. How can we give its simple proof ? How can we generalize it ?

Definitions and known results . Let $h^{(p)}(x,D)$ be the least harmonic majorant of $|x|^p$ in a domain D of $\mathbf{R}^d$ which contains the origin 0 , where $0 < p < \infty$ if $d \geq 2$ and $1 \leq p \leq \infty$ if $d = 1$. Let $r(D)$ denotes the volume radius of D , namely, a ball with radius $r(D)$ has the same volume as D has. We fix the dimension d and set

$$c(p) = \sup \{h^{(p)}(0,D)^{1/p};\ 0 \in D \text{ and } r(D) = 1\}.$$

Since $h^{(p)}(0,\lambda D)^{1/p} = \lambda h^{(p)}(0,D)^{1/p}$ and $r(\lambda D) = \lambda r(D)$ for every $\lambda > 0$,

$$h^{(p)}(0,D)^{1/p} \leq c(p) r(D)$$

for every domain containing 0. By taking the unit ball, we see that $c(p) \geq 1$ for every p. It follows that

(1) $\quad c(p) = 1$ if $p \leq d + 2^{1-d}$

(2) $\quad$ there are positive constants C_1 and C_2 such that

$$C_1 p^{(d-1)/d} \leq c(p) \leq C_2 p^{(d-1)/d} \quad \text{for } p \geq 1,$$

see [3].

Formulation of the problem and a conjecture. What is the greatest value of p satisfying $c(p) = 1$? Our conjecture is "The greatest value of p satisfying $c(p) = 1$ is equal to $d + 2$ ".

A consequence of the solution of the problem. If the above conjecture is true, we get more sharp estimate than the Alexander--Taylor-Ullman inequality, namely,

$$\|f\|_{H^4(U)} \leq \{(1/\pi)(\text{area } f(U))\}^{1/2}$$

for holomorphic functions f in U satisfying $f(0) = 0$.

References

[1] ALEXANDER, H. & B. A. TAYLOR & J. L. ULLMAN : Areas of projections of analytic sets. Invent. Math. 16 (1972), 335-341

[2] KOBAYASHI, S. : Image areas and H_2 norms of analytic functions. Proc. Amer. Math. Soc. 91 (1984), 257-261

[3] SAKAI, M. : Isoperimetric inequalities for the least harmonic majorant of $|x|^p$. Trans. Amer. Math. Soc. 299 (1987), 431-472.

THE COPENHAGEN PROBLEMS

The following problems C 1 - C 15 are reproduced from

Potential Theory, Copenhagen 1979, Proceedings, Lecture Notes in Mathematics 787, Springer-Verlag, Berlin, 1980; pp. 316--319.

The proposers were asked for commentaries on the problems. Their answers are attached to the corresponding problem.

Problems C 3, C 6, C 12 have been solved.

C 1. We consider $U_\mu^{(\alpha)}(x) = \int |x-y|^{\alpha-n}\, d\mu(y)$, $x, y \in \mathbb{R}^n$, for $0 < \alpha < n$, $\mu \geq 0$, supp μ compact. For $\alpha \leq 2$ and supp $\mu \subset S \subset \mathbb{R}^n$ we have the following maximum principle : If $U_\mu^{(\alpha)} \leq M$, on supp μ , then $U_\mu^{(\alpha)} \leq M$ on S . If $\alpha > 2$ this result is no longer true in general. On the other hand, the result still holds for $2 < \alpha < 3$ if S an $(n-1)$-dimensional hyperplane in $\mathbb{R}^n$.

Generalize this fact to more general manifolds S than hyperplanes. Should the kernel $|x-y|^{\alpha-n}$ be changed ?

Literature : O. FROSTMAN, Sur un principe du maximum, in Complex analysis and its applications, dedicated to I. Vekua, "Nauka", Moscow, 1978, pp. 574-576.

H. Wallin

C 2. Let $f|F$ denote the restriction of f to $F \subset \mathbb{R}^n$. For various classes A of functions defined in $\mathbb{R}^n$ the trace $\{f|F : f \in A\}$ has been characterized for quite general sets F . If F is compact, K a quite general kernel, and A the class of continuous K-potentials of real measures on $\mathbb{R}^n$ with compact support, then the trace is the class of continuous functions on F if and only if the K-capacity of F is zero. If A is the class of Bessel potentials of order α of L^p-functions, the trace has also been characterized.

Characterize the trace of other classes of potentials.

Literature : H. WALLIN, Ark. Mat. 5, 1963; A. JONSSON, Ark. Mat. 17, 1979 and Studia Math. 67, 1980.

Added in 1987 : A. JONSSON, H. WALLIN : Function spaces on subsets of $\mathbb{R}^n$, Math. Reports, Vol. 2, Part 1, Harwood Acad. Publ., London, 1984.

H. Wallin

C 3. For $0 < \alpha < \min(n,2)$ and $D \subset \mathbb{R}^n$ a relatively compact open set denote by D^{α}_{reg} the set of regular points for the Dirichlet problem for α-harmonic functions (see e.g. N.S. LANDKOF, Foundations of Modern Potential Theory, Springer--Verlag, Berlin 1971).

Does there exist D such that for all α , α' we have $D^{\alpha}_{reg} \neq D^{\alpha'}_{reg}$ provided $\alpha \neq \alpha'$?

J. Veselý

Comments by the proposer : The problem has been affirmatively solved by T. Baba and M. Kanda. The solution can be found in T. Baba, Regular points for α-harmonic functions, Hiroshima Math. J. 13 (1983), 523-527. Another solution was previously suggested by M. Kanda in 1980 (private communication).

A more detailed study of the Riesz potentials, α-harmonicity and related topics in the framework of theory of balayage spaces appeared in J. Bliedtner, W. Hansen, Potential Theory (An Analytic and Probabilistic Approach to Balayage), Springer-Verlag, Berlin-Heidelberg, 1986.

C 4. Let $D \subset \mathbb{C}^n$ be a bounded strictly pseudo-convex doamin and let Δ be the Bergmann Laplacian of D.

Is the associated Martin compactification the topological closure of D? Are all the boundary points minimal?

J. C. Taylor

C 5. Let $M \subset \mathbb{R}^n$ be a smooth submanifold of codimension $k > 1$. Let $Lu(x) = \Delta u(x) - \phi(x)u(x)$, $x \notin M$. Assume that in a tubular neighbourhood of M $\phi(x)$ depends only on $\bar{x}$, the projection of x onto M.

When does the Martin boundary of $\mathbb{R}^n \setminus M$ relative to $Lu = 0$ ramify M, with each point being replaced by a sphere S^{n-k-1}?

J. C. Taylor

C 6. Let u be finely harmonic in a fine domain $U \subset \mathbb{R}^n$, and suppose that $u = 0$ everywhere in some fine neighbourhood of a point $x \in U$.

Does it follow that $u = 0$ everywhere in U?

B. Fuglede

Comments by the proposer: Problem 6 was solved in the negative by Terry J. Lyons in 1982. Clearly it suffices to construct, in the open unit ball B of $\mathbb{R}^n$, a sequence of points $x_j \in B \setminus \{0\}$ and a sequence of numbers $\alpha_j > 0$ such that

$$h(0) = \sum_{j \in \mathbb{N}} \alpha_j h(x_j)$$

for every harmonic polynomial h on $\mathbb{R}^n$, equivalently such that

$$|x|^{2-n} = \sum_{j \in \mathbb{N}} \alpha_j |x - x_j|^{2-n}$$

for every $x \in \mathbb{R}^n \setminus \bar{B}$. (If $n = 2$, replace $|x|^{2-n}$ by

$-\log|x|$.) Lyons' original construction of such sequences (x_j) , (α_j) for $n = 2$ (communicated in a letter of May 28, 1982 to Bent Fuglede) was replaced in [L] by an extremely simple construction by means of Vitali's covering lemma, as suggested orally by Makoto Sakai. A still further construction has since been given by Didier Pinchon [P]. Like in Lyons' original example, the sequence (x_j) in [P] accumulates solely at the boundary of B . - An application of Lyons' example in [L] has been made in [F].

References

[F] FUGLEDE, B. : Complements to Havin's theorem on L^2-approximation by analytic functions. Ann. Acad. Sci. Fennicae 10 (1985), 187-201

[L] LYONS, T. : Finely harmonic functions need not be quasi analytic. Bull. London Math. Soc. 16 (1984), 413-415

[P] PINCHON, D. : À propos d'un résultat de Terry Lyons. (Manuscript, 1986.)

C 7. It was shown by Brelot and Choquet, Espaces et lignes de Green, Ann. Inst. Fourier 3, 199-263 (1951), that if U is open in $\mathbb{R}^n$ and x is in U then harmonic measure on ∂U for x is supported by the set of points of ∂U accessible from x by rectifiable arcs in U . In view of the fact that, in this setting, fine domains are polygonal path connected, it would be interesting to know if the stated result generalizes.

In particular, if K is compact in $\mathbb{R}^n$ and x is in the fine interior K' of K , is then the Keldych measure for x supported by the points of the fine boundary of K accessible from x by rectifiable arcs in K' ? (In this context rectifiable is to mean of finite total length.)

T. Lyons and B. Øksendal

<u>C 8.</u> Determine the extreme points of the convex set of K-operators and Keldyš operators.

I. Netuka

<u>Comments by the proposer:</u> Let U be a relatively compact open subset of a $\mathcal{B}$-harmonic space X. Denote by $\mathcal{H}(U)$ the space of harmonic functions on U and put $H(U) = \{h \in C(\overline{U}); h_{|U} \in \mathcal{H}(U)\}$. We shall suppose that $H(U)$ contains a strictly positive function and linearly separates the points of $\overline{U}$.

Recall that an operator $A : C(\partial U) \to \mathcal{H}(U)$ is said to be a <u>Keldyš operator</u> (on U), if A is a positive linear operator such that

(1) $$A(h_{|\partial U}) = h_{|U}, \quad h \in H(U).$$

An increasing operator $A : C(\partial U) \to \mathcal{H}(U)$ satisfying (1) is termed a <u>K-operator</u>.

Let us denote by $\mathcal{K}$ the convex set of all Keldyš operators on U. Then the PWB-solution H^U of the generalized Dirichlet problem belongs, of course, to $\mathcal{K}$. For $x \in U$, let δ_x^U be the balayage of ε_x to the Choquet boundary $\mathrm{Ch}_{H(U)}\overline{U}$ of $\overline{U}$ w.r.t. $H(U)$. Then D^U defined by $D^U f(x) = \delta_x^U(f)$, $f \in C(\partial U)$, $x \in U$, is a Keldyš operator.

<u>Proposition.</u> The operators H^U and D^U are extreme points of $\mathcal{K}$.

<u>Sketch of the proof.</u>: Given a Keldyš operator A, define, for $x \in U$, the positive Radon measure α_x by $\alpha_x : f \to Af(x)$, $f \in C(\partial U)$. (We shall simply write $A \sim \{\alpha_x\}$.) Of course, $H^U \sim \{\mu_x^U\}$ (harmonic measures) and $D^U \sim \{\delta_x^U\}$.

It is known that, for every Keldyš operator $A \sim \{\alpha_x\}$ and every function w of the form $\min(h_1,\dots,h_n)$ with $h_j \in H(U)$, the following inequalities hold :

(2) $$\delta_x^U(w) \leq \alpha_x(w) \leq \mu_x^U(w)$$

(see J. Lukeš, I. Netuka : The Wiener type solution of the Dirichlet problem in potential theory, Math. Ann. 244 (1976), 173-178).

Let $D^U = \frac{1}{2} A_1 + \frac{1}{2} A_2$ with $A_1, A_2 \in \mathcal{K}$, $A_j \sim \{\alpha_x^j\}$ and $H^U = \frac{1}{2} B_1 + \frac{1}{2} B_2$ with $B_1, B_2 \in \mathcal{K}$, $B_j \sim \{\beta_x^j\}$. It follows immediately from (2) that $\alpha_x^1 = \alpha_x^2 = \delta_x^U$, $\beta_x^1 = \beta_x^2 = \mu_x$ on the smallest min-stable cone W containing $H(U)$. Since $(W - W)_{|\partial U}$ is uniformly dense in $C(\partial U)$, we have $A_1 = A_2 = D^U$ and $B_1 = B_2 = H^U$.

Notice that the statement concerning H^U follows also from the observation that $A \in \mathcal{K}$ is a Riesz homomorphism iff $A = H^U$ and from the following

<u>Lemma :</u> Let P, P_1, P_2 be positive linear mappings between vector lattices L_1, L_2. If $P = \frac{1}{2} P_1 + \frac{1}{2} P_2$, $x, y \in L_1$ and $P(x \wedge y) = P(x) \wedge P(y)$, then $P_1(x \wedge y) = P_1(x) \wedge P_1(y)$.

Indeed, $P_1(x \wedge y) = 2P(x \wedge y) - P_2(x \wedge y) = 2P(x) \wedge 2P(y) - P_2(x \wedge y) = \left(P_1(x) + P_2(x)\right) \wedge \left(P_1(y) + P_2(y)\right) - P_2(x \wedge y) \geq P_1(x) \wedge P_1(y) + \left[P_2(x) \wedge P_2(y) - P_2(x \wedge y)\right] \geq P_1(x) \wedge P_1(y) + 0 \geq P_1(x \wedge y)$.

The result of the above proposition was mentioned on the occasion of meetings on potential theory in Erlangen and Eichstätt, 1982.

An analogous result is stated without proof (for Keldyš operators on a compactification) also in the final remark of the paper by T. Ikegami : On the simplical cone of superharmonic functions in a resolutive compactification of a harmonic space, Osaka J. Math. 20 (1983), 881-898.

One should ask whether, for a set B with $Ch_{H(U)}\overline{U} \subset B \subset \partial U$, the operator $A \sim \{\varepsilon_x^B\}$ is an extreme point of $\mathcal{K}$. The answer is negative as shown by an example due to W. Hansen (private communication in a letter of May 31, 1982). The example is constructed for the heat equation in $\mathbb{R}^2$.

Methods of functional analysis can be used to show that $A \in \mathcal{K}$ is an extreme point of $\mathcal{K}$ iff 0 is the greatest harmonic minorant of

$$\{A(|f - h_{|\partial U}|);\ h \in H(U)\}$$

for every $f \in C(\partial U)$. A potential theoretic interpretation of this condition does not seem to be quite clear.

A list of references concerning the Keldyš operators is contained in the paper I. Netuka : The classical Dirichlet problem and its generalizations, Potential Theory, Copenhagen, 1979, Proceedings, Lecture Notes in Mathematics 787, Springer-Verlag, Berlin, 1980, 235-266.

We add a list of recent related references :

BAUER, H. : Simplices in potential theory. Aspects of Positivity in Functional Analysis. Tübingen 1985, R. Nagel, V. Schlottenbeck, M.P.H. Wolff (editors), North-Holland Math. Stud., 122, North-Holland, Amsterdam, 1986, 27-39

BLIEDTNER, J. & W. HANSEN : Potential Theory. An analytic and probabilistic approach to balayage. Springer-Verlag, Berlin, 1986

DEMBINSKI, V. : Regular points, transversal sets, and Dirichlet problem for standard processes. Z. Warsch. Verw. Gebiete 66 (1984), 507-527

HANSEN, W. : On the identity of Keldych solutions. Czechoslovak Math. J. 35 (1985), 632-638

IKEGAMI, T. : On a generalization of Lukeš' theorem. Osaka J. Math. 18 (1981), 699-702

LUKEŠ, J. & J. MALÝ & L. ZAJÍČEK : Fine topology methods in real analysis and potential theory. Lecture Notes in Mathematics 1189, Springer-Verlag, Berlin, 1986

NETUKA, I. : The Dirichlet problem for harmonic functions, Amer. Math. Monthly 87 (1980), 621-628

NETUKA, I. : La représentation de la solution généralisée à l'aide des solutions classiques du problème de Dirichlet, Séminaire de Théorie du Potentiel (Brelot, Choquet, Deny), Paris, N° 6, Lecture Notes in Mathematics 906, Springer-Verlag, Berlin, 1982, 261-268

NETUKA, I. : L'unicité du problème de Dirichlet généralisé pour un compact, ibid. 269-281

NETUKA, I. : Monotone extensions of operators and the first boundary value problem. Equadiff 5, Bratislava 1981, Proceedings, Teubner-Texte zur Mathematik 47, Teubner, Leipzig, 1982, 268-271

NETUKA, I. : Extensions of operators and the Dirichlet problem in potential theory. Rend. Circ. Mat. Palermo, Supplemento, Ser. II, N° 10 (1985), 143-163

NETUKA, I. : The Ninomiya operators and the generalized Dirichlet problem in potential theory. Osaka J. Math. 23 (1986), 741-750

C 9. Let X be a harmonic space, $V \subset X$ an open relatively compact set, and $H(V)$ the system of functions continuous on the closure of V and harmonic on V. For an $x \in V$ denote by $\mathcal{M}_x$ the set of measures μ, $\operatorname{supp} \mu \subset \bar{V}$, such that $\mu(f) = f(x)$ for all $f \in H(V)$.

Is it possible to make a choice of $\mu_x \in \mathcal{M}_x$ in such a way that the operator $Af(x) = \mu_x(f)$, $x \in V$, has the following properties :

(i) $A : C(\bar{V}) \to C(\bar{V})$,

(ii) $Af = f$ if and only if f is harmonic on V .

(For the classical case partial results were obtained by W.A. Veech : A converse to the mean value theorem for harmonic functions, Amer. J. of Math. 97 (1976), 1007-1027.)

J. Veselý

Comments by the proposer : An operator of a similar type was in the classical case introduced by H. Lebesgue (1912). Even in the classical case a characterization of the harmonicity described by (ii) requires some additional assumptions on f (e.g. "good" boundary behaviour) or V ("nice" boundary) or both. For references see the survey article : I. Netuka, Harmonic functions and mean value theorems (Czech), Časopis Pěst. Mat. 100 (1975), 391-409. Advanced notions including ergodicity, Martin boundary, etc. are related to the problem.

In the framework of harmonic spaces, the Lebesgue type operators and their use to solve the Dirichlet problem was studied in J. Veselý, Sequence solutions of the Dirichlet problem, Časopis Pěst. Mat. 106 (1981), 84-93. A partial solution of the above formulated problem can be found in J. Veselý, Restricted mean value property in axiomatic potential theory, Comment. Math. Univ. Carolinae 23 (1982), 613-628.

C 10. Characterize all harmonic spaces on which the initial and the fine topologies coincide.

J. Lukeš and I. Netuka

C 11. Characterize all sets $F \subset \mathbb{R}^n$ having the following property : There is a sheaf $\mathcal{H}$ such that $(X, \mathcal{H})$ is a harmonic space and F is an absorbent set in this space.

J. Lukeš and I. Netuka

C 12. Let X be a harmonic space with a countable base. (See C. Constantinescu and A. Cornea : Potential Theory on Harmonic Spaces. Springer-Verlag, Berlin 1972.) Let a point $x \in X$ and a semipolar set $S \subset X$ be given.

Does every fine neighbourhood U of x contain a fine neighbourhood V of x such that $(\varepsilon_x^{CV})^*(S) = 0$?

B. Fuglede

Comments by the proposer : An affirmative answer to this problem was soon afterwards obtained by W. Hansen, Semipolar sets are almost negligible, J. reine angew. Math. 314 (1980), 217-220. Let $(B_i)_{i\in I}$ be a family of pairwise disjoint Borel subsets of X . First an analytic proof is given for the fact that if $\bigcup_{i\in I} B_i$ is semi-polar then all B_i are polar except at most countably many of them. Let S be any semi-polar subset of X . If every B_i , $i \in I$, is finely closed it is next shown that for every $x \in X$ the set of all $i \in I$ such that $(\varepsilon_x^{B_i})^*(S) > 0$ is at most countable. This implies that $(\varepsilon_x^{CK})^*(S) = 0$ for many compact fine neighbourhoods K of x . In particular, such fine neighbourhoods form a neighbourhood base of x in the fine topology.

C 13. Let X be a harmonic space with the sheaf $\mathcal{H}$ of harmonic functions. Let $V \subset X$ be an open set and $h \in \mathcal{H}(V)$. A point $x \in \partial V$ is called a *point of continuability* of h if there is an open set V_1 containing x , and a function $h_1 \in \mathcal{H}(V_1)$ such that $h = h_1$ on $V_1 \cap V$. The set of all points of continuability of the function h will be denoted by C(h) .

Problem 1 : Characterize the set $Q = \bigcap\{C(h);\ h \in \mathcal{H}(V)\}$.

Problem 2 : How "large" is the system of all $h \in \mathcal{H}(V)$ with $C(h) \cap (\partial V \setminus Q) = \emptyset$?

I. Netuka and J. Veselý

Comments by the proposers : If V is a relatively compact open set and the space $\mathcal{H}(V)$ is replaced by the subspace $\mathcal{H}^c(V) := \{h \in \mathcal{H}(V);\ h$ continuously extendable to $\overline{V}\}$ or $\mathcal{H}^b(V) := \{h \in \mathcal{H}(V);\ h$ bounded$\}$, then the corresponding problems are solved; see : I. Netuka and J. Veselý, Harmonic continuation and removable singularities in the axiomatic potential theory, Math. Ann 234 (1978), 117-123 for the case of $\mathcal{H}^c(V)$ and

J. Hyvönen, On the harmonic continuation of bounded harmonic functions, Math. Ann. 245 (1979), 151-157 for $\mathcal{H}^b(V)$.

In contrast to the classical potential theory, the set Q is, in general, non-empty. To see this let us consider the heat equation in $\mathbb{R}^2$ and put $V =]0,5[\times]0,4[\setminus [2,3] \times [2,4]$.

We claim that $Q =]2,3[\times \{2\}$. It follows easily from the first of the above quoted articles that

$$Q \subset \big(]2,3[\times \{2\}\big) \cup \big(]0,2[\cup]3,5[\big) \times \{4\} .$$

Fix $z \in]2,3[\times \{2\}$ and an $h \in \mathcal{H}(V)$. Put $U =]1,4[\times]1,3[$ and choose a function g continuous on ∂U such that $g = h$ on $\partial U \setminus \big(]1,4[\times \{3\}\big)$. If $V_1 =]2,3[\times]1,3[$ and h_1 is the restriction of H_g^U to V_1 , then $h_1 \in \mathcal{H}(V_1)$ and $h = h_1$ on $V_1 \cap V$. We see that $z \in Q$ and $]2,3[\times \{2\} \subset Q$.

Denote by W the fundamental solution of the heat equation. Then $a_n = \inf W\big([0,5] \times \{1/n\}\big) > 0$. Define

$$g(x,t) = \sum_{n=1}^{\infty} a_n^{-1} W(x,\, t - 4 + 1/n) \;, \quad (x,t) \in V .$$

Then $f \in \mathcal{H}(V)$ and for any $y \in]0,2[\cup]3,5[$ we have $f(x,t) \to \to \infty$ as $(x,t) \to (y,4)$, $(x,t) \in V$. Consequently, no point of $\big(]0,2[\cup]3,5[\big) \times \{4\}$ belongs to Q .

Notice that to be a point of continuability is not a local property. It is sufficient to consider points on horizontal parts of ∂V .

<u>C 14.</u> Let S be an H-cone such that its dual S^* separates S . It is known that S may be identified with a convex subcone of S^{**} ; see N. Boboc, Gh. Bucur, and A. Cornea, H-cones and potential theory, Ann. Inst. Fourier <u>25</u> (1975), 71-108. In some cases S is a solid part of S^{**} .

Is this property true generally ?

N. Boboc, Gh. Bucur

<u>C 15.</u> Let S a standard H-cone such that the axiom C is fulfilled (see the reference in the preceding problem).

Does there exist a Dirichlet space such that its convex cone of potentials is a solid part of S with respect to the natural order, and dense in order from below in S ?

N. Boboc, Gh. Bucur

SELECTED PROBLEMS FROM THE COLLECTION

"RESEARCH PROBLEMS IN COMPLEX ANALYSIS"

Problems R 1 - R 14 are reproduced from

A. HAYMAN, W. K. : Research problems in function theory, Athlone Press, University of London, 1967

C. ANDERSON, J. M. & K. F. BARTH & D. A. BRANNAN & W. K. HAYMAN : Research problems in complex analysis, Bull. London Math. Soc. 9 (1977), 129-162

H. CAMPBELL, D. M. & J. G. CLUNNIE & W. K. HAYMAN : Research problems in complex analysis, Aspects of contemporary complex analysis (ed. D. A. Brannan and J. G. Clunnie), Academic Press, London, 1980

I. BARTH, K. F. & D. A. BRANNAN & W. K. HAYMAN : Research problems in complex analysis, Bull. London Math. Soc. 16 (1984), 490-517

as follows :

Problem	Number	From	Page
R 1	3.10	**A**	23
R 2	3.16	**C**	132
R 3	3.18	**C**	132
R 4	7.26	**C**	146
R 5	7.30	**C**	147
R 6	3.19	**H**	551
R 7	3.20	**H**	551
R 8	7.48	**H**	566
R 9	9.10	**H**	568
R 10	3.24	**I**	494
R 11	3.25	**I**	494
R 12	3.27	**I**	494
R 13	3.29	**I**	494
R 14	3.30	**I**	494

(We wish to thank Professor Hayman for his assistance in preparation of this part.)

R 1. Suppose that $u(X)$ is harmonic in the unit ball $|X| < 1$ and remains continuous with partial derivatives of all orders on $|X| = 1$, where X is a point (x_1, x_2, x_3) in space and

$$|X|^2 = x_1^2 + x_2^2 + x_3^2 .$$

Suppose further that there is a set E of positive area on $|X| = 1$ such that both u and its normal derivative vanish on E. Is it true that $u \equiv 0$?

L. Bers

Comment by W. K. Hayman (1974) : The result has been proved by Mergelyan provided the complement of E is sufficiently thin at some point of E. However, the complete question remains open. Mergelyan lectured on the topic at Nice, but I have been unable to trace a published result.

R 2. A compact set E in $\mathbb{R}^n$ $(n \geq 3)$ is said to be thin at P_0 if

$$\int_0^1 \frac{c(P_0,\rho)}{\rho^{n-1}}\, d\rho < \infty , \qquad (1)$$

where $c(P_0,\rho) = \operatorname{cap}\left[E \cap \{P : |P - P_0| \leq \rho\}\right]$; this is the integrated form of the Wiener criterion. It follows from Kellogg's theorem that the points of E where (1) holds form a polar set. Can one give a direct proof of this fact, which shows perhaps that (1) is best possible ?

P. J. Rippon

R 3. It is known that the set E of least capacity C and given volume is a ball. If E displays some measure of asymmetry (for instance, if every ball with the same volume as E in space contains a minimum proportion δ in the complement of E), can one obtain a lower bound for the capacity of E which exceeds C by some positive function of δ?

E. Fraenkel (communicated by W.K. Hayman)

R 4. Is there a homeomorphism of the open unit ball in $\mathbb{R}^3$ onto $\mathbb{R}^3$, whose coordinate functions are harmonic ? In other words, do there exist u_1, u_2, u_3 harmonic in $|x| < 1$, $x = (x_1,x_2,x_3)$, such that

$$(x_1,x_2,x_3) \to (u_1,u_2,u_3)$$

is a homeomorphism of $|x| < 1$ onto all of $\mathbb{R}^3$? The analogous problem in $\mathbb{R}^2$ is answered negatively; the result is due to T. Rado, and is an important lemma in the theory of minimal surfaces.

H. S. Shapiro

R 5. Let $u(z)$ be a real bounded continuous function on $U = \{|z| < 1\}$, and suppose that to each $z \in U$ there corresponds a real number $r(z)$ with $0 < r(z) < 1 - |z|$ such that

$$\frac{1}{2\pi}\int_0^{2\pi} u\left(z + r(z)\, e^{i\Theta}\right) d\Theta = u(z) . \qquad (2)$$

Must $u(z)$ be harmonic on U ? Volterra showed that this was true in the case that $u(z)$ is given to be continuous on $\overline{U}$; the case in which (2) is replaced by an areal-mean-value (and the continuity condition on $u(z)$ is relaxed) has been studied by Veech (Ann. of Math. 97 (1973), 189-216 and Bull. Amer. Math. Soc. 78 (1972), 444-446) and others.

L. Zalcman

R 6. Let C_0 be a tangential path in $\{|z| < 1\}$ which ends at $z = 1$ and let C_Θ be any rotation of C_0. Littlewood showed that there exists a function $u(z)$, harmonic and satisfying $0 < u(z) < 1$ in $\{|z| < 1\}$, such that

$$\lim_{\substack{|z|\to 1 \\ z \in C_\Theta}} u(z)$$

does *not* exist for almost all Θ, $0 \leq \Theta \leq 2\pi$. Surprisingly, it seems to be unknown whether there exists a $v(z)$, positive and harmonic in $\{|z| < 1\}$, such that

$$\lim_{\substack{|z|\to 1 \\ z \in C_\Theta}} v(z)$$

does *not* exist for *all* Θ, $0 \leq \Theta \leq 2\pi$.

The corresponding result is known for bounded holomorphic functions in $\{|z| < 1\}$. See, for example, Chapter 2 of Collingwood and Lohwater.

K. Barth

R 7. Suppose that you have a continuous real function $u(x)$ on $\mathbb{R}^n$ and you want to know whether a homeomorphism $\Phi : \mathbb{R}^n \to \mathbb{R}^n$ and a harmonic function v on $\mathbb{R}^n$ exist such that

$$v(x) = u(\Phi(x)) .$$

Is it necessary and sufficient that there should exist mappings $\mu_2, \mu_3, \ldots, \mu_n$ so that

$$U = (u, \mu_2, \mu_3, \ldots, \mu_n)$$

is a light open mapping of $\mathbb{R}^n$ into $\mathbb{R}^n$? The case $n = 2$ is a result of Stoilow which is in Whyburn's "Topological analysis", for example.

L.A. Rubel (communicated by D.A. Brannan)

R 8. A domain $D \subseteq \mathbb{R}^n$ is said to be linearly accessible, if each point in the complement of D can be joined to ∞ by a ray which does not meet D. Let $g(\cdot, x_0)$ be the Green's function for D with pole at x_0 in D. Is $\{x : g(x,x_0) > t\}$ linearly accessible for $0 < t < \infty$? This conclusion is valid in $\mathbb{R}^2$ (see Sheil-Small, J. Lond. Math. Soc. 6 (1973), 385-398).

J. Lewis

R 9. Let F be a closed subset of $\mathbb{R}^n$, $n \geq 2$. Call F a set of harmonic approximation if every function continuous on F and harmonic in the interior of F can be uniformly approximated there by a harmonic function in $\mathbb{R}^n$. Give necessary and sufficient conditions that F be a set of harmonic approximation. If F is nowhere dense, Šaginjan has done this. If F is the closure of its interior, Gauthier, Ow and I have given necessary conditions and sufficient conditions when $n = 2$ but not necessary and sufficient conditions.

M. Goldstein

Comment : For references and a relation to Rubel's problem see $\mathbb{H}$, pp. 568-570.

R 10. For which $p > 0$ does there exist a function u ($\not\equiv 0$) harmonic on $\mathbb{R}^3$ and vanishing on the cone $x_1^2 + x_2^2 = px_3^2$?

H.S. Shapiro

R 11. Is there a harmonic polynomial $P(x_1,x_2,x_3)$ ($\not\equiv 0$) that is divisible by $x_1^4 + x_2^4 + x_3^4$?

H.S. Shapiro

R 12. Let D be an unbounded doamin in $\mathbb{R}^n$, $n \geq 2$. Is there a positive continuous function $\varepsilon(|x|)$ such that, if u is harmonic in D and $|u(x)| < \varepsilon(|x|)$, then $u \equiv 0$?

For $n = 2$, the answer is yes. The answer is also yes if we restrict our attention to positive harmonic functions. For fine domains and finely harmonic functions, it follows from an example of Lyons (J. Funct. Anal. 37 (1980), 1-18, 19-26) that the answer is no.

P.M. Gauthier and W. Hengartner

R 13. It is known that the Newtonian potential of a uniform mass distribution spread over an ellipsoid K in $\mathbb{R}^n$ $(n \geq 2)$ is a quadratic function of the coordinates of $x = (x_1, \ldots, x_n)$ for $x \in K$.

Nikliborc and Dive independently proved that for $n = 2$ and $n = 3$, the ellipsoid is the only body with this property. Prove this converse assertion for $n > 3$ (preferably by a new method, since Nikliborc and Dive both use methods involving highly nontrivial calculations). (For references see **I**, p. 516.)

H.S. Shapiro

R 14. Let $K(z,z')$ denote the kernel of the double layer potential occurring in Fredholm's theory where $z, z' \in \Gamma$, Γ being a smooth Jordan curve. (Recall that $K(z,z') = \frac{\cos \Phi}{|z - z'|}$, where Φ is the angle between the inward normal to Γ at z and the line (z,z').)

When Γ is a circle, the function $z \to K(z,z')$ is a constant (i.e. the same for each choice of z'); and consequently the integral operator

$$T_\Gamma : f \to \int_\Gamma f(z)\, K(z,z')\, ds_z$$

is of rank one (as an operator from $C(\Gamma)$ to $C(\Gamma)$). Are there any other Γ for which the rank of T_Γ is finite ?

H.S. Shapiro

COMMENTS ON PROBLEMS (added in October 1987)

P 2. Addendum (provided by the proposer in August 1987)

In dimension $N = 2$ it is known that only one Martin boundary point lies above an irregular boundary point for a domain in $\mathbb{R}^N$. This is due to

M. BRELOT : Sur le principe des singularités positives et la topologie de R.S. Martin. Ann. Univ. Grenoble 23 (1948), 113-138.

See also Satz 13.8 in

C. CONSTANTINESCU und A. CORNEA : Ideale Ränder Riemannscher Flächen. Erg. d. Math., N.F., Bd.32. Springer, Berlin-Göttingen-Heidelberg 1963.

R 9. Partial solution (communicated by P.M. Gauthier in August 1987)

The problem is solved in $\mathbb{R}^2$ (also on Riemann surfaces) by Theorem 2.5.1 in the paper

T. BAGBY, P.M. GAUTHIER : Approximation by harmonic functions on closed subsets of Riemann surfaces (preprint).

R 12. Solution (communicated by P.M. Gauthier in August 1987)

The problem is answered affirmatively in the paper

D.H. ARMITAGE, T. BAGBY, P.M. GAUTHIER : Note on the decay of solutions of elliptic equations, Bull. Lond. Math. Soc. 17 (1985), 554-556.

R 13. Solution (communicated by the M. Sakai in September 1987)

The problem is solved in the paper

E. Di BENEDETTO, A. FRIEDMAN : Bubble growth in porous media, Indiana Univ. Math. J. 35 (1986), 573-606

(cf. the proof of Theorem 5.1).

SCIENTIFIC PROGRAMME

Monday, July 20

Survey lectures

9.00-9.55 W. HANSEN: Balayage spaces - a natural setting for potential theory

10.10-11.05 G. WILDENHAIN: Potential theory methods for higher order elliptic equations

11.20-12.15 A.ANCONA: Positive harmonic functions and hyperbolicity

Section A

14.20-14.45 P. LOEB: Compactifications and integral representations for bounded harmonic functions

14.55-15.20 J. BLIEDTNER: Approximation of (finely) superharmonic functions by continuous potentials

15.30-15.55 M. MEYER: Balayage spaces on topological sums

16.15-16.40 H.-H. MÜLLER: Subordination for balayage spaces

16.50-17.15 T. IKEGAMI: The Dirichlet problem on ends

17.25-17.50 A. IBRAGIMOV: Zaremba s problem and related problems of potential theory

Section B

14.20-14.45 J. WERMER: Polynomial hulls and envelopes of holomorphy

14.55-15.20 V. AZARIN: Cluster sets of entire functions and their applications

15.30-15.55 J.L. FERNÁNDEZ: Internal distortion under conformal mapping

16.15-16.40 M. SAKAI: Simply connected quadrature domains

16.50-17.15 Ü. KURAN: A sharper mean-value inequality for subharmonic functions

17.25-17.50 S. GARDINER: Integrals of subharmonic functions over affine sets

Section C

14.20-14.45 O. KOUNCHEV: Extremal problems for the inverse potential problem

14.55-15.20 S. SALSA: On a classical inverse problem for the Newtonian potential

15.30-15.55 N. JACOB: On Dirichlet´s boundary value problem for certain anisotropic differential and pseudodifferential operators (with K.DOPPEL)

16.15-16.40 M. KANDA: On the maximum principle for potential kernels of second order differential operators

16.50-17.15 G. LUMER: Potential-theoretic methods in time-dependent reaction-diffusion equations of Kolmogorov-Petrovskii-Piskounov type

17.25-17.50 A. SHIDFAR: On the integration of the linear elastostatic displacement equation

Tuesday, July 21

Survey lectures

9.00 - 9.55 E.M. LANDIS: Potential theory and partial differential equations

10.10-11.05 M. OHTSUKA: Weighted extremal length and Beppo Levi functions

11.20-12.15 T. LYONS: Dirichlet processes and stochastic calculus

Section A

14.20-14.45 U. SCHIRMEIER: Continuity of measure representation

14.55-15.20 L. CSINK: Stochastic characterization of harmonic morphisms

15.30-15.55 S.L. ERIKSSON-BIQUE: Hyperharmonic cones

16.15-16.40 E. POPA: Morphisms of Stonian cones

16.50-17.15 A. de la PRADELLE: Sur la perturbation des résolvantes

17.25-17.50 H. MAAGLI: Perturbation and comparison of semigroups

Section B

14.20-14.45 E. LOVE: On a method for potential problems using the Poisson transformation

14.55-15.20 H. LEUTWILER: On the Appell transformation

15.30-15.55 H. HUEBER: On the computation of Green functions for sublaplacians and the Martin compactification of $\Delta_K - 2$

16.15-16.40 T. STURM: Gauge theorems for Schrödinger operators

16.50-17.15 A. BOUKRICHA: The Picard principle and the behaviour of continuous solutions of the Schrödinger equation at an isolated singularity

17.25-17.50 D.G. KESELMAN: Harnack inequality and some characteristic of non-regular points of simplexes

Section C

14.20-14.45 T.J. SJÖDIN: Non linear potential theory in Lebesgue spaces with mixed norm

14.55-15.20 R. KRESS: A Neumann boundary value problem for force-free magnetic fields

15.30-15.55 G. HSIAO: On the unified characterization of capacity (with R.KLEINMAN)

16.15-16.40 P. MATTILA: Integralgeometric properties of capacities

16.50-17.15 O. MARTIO: Counterexamples for unique continuation

17.25-17.50 T. KILPELÄINEN: Polar sets in a nonlinear potential theory

Wednesday, July 22

Survey lectures

9.00 - 9.55 E. FABES: Potential theory methods in boundary value problems

10.10-11.05 K.-L. CHUNG: Probability methods in potential theory

11.20-12.15 G.F. ROACH: Iteration methods for potential problems (with T.S.ANGELL and R. KLEINMAN)

Section A

14.20-14.45 E. BERTIN: Non-linear potential theory

14.55-15.20 A. CORNEA: A characterization of the fine sheaf property

15.30-15.55 G. LEHA: On diffusion semigroups generated by semi-elliptic differential operator in Hilbert space

16.15-16.40 I. PAVLOV: On the Poisson equation on the infinite dimensional torus

16.50-17.15 M. ITO: Diffusion kernels of logarithmic type

17.25-17.50 M. HMISSI: Caractérisation des semigroupes deterministes par leurs cônes de fonction excessives

Section B

14.20-14.45 M. BIROLI: Wiener estimates for solutions of elliptic equations in nonvariational form

14.55-15.20 G.M. LIEBERMAN: Regularity of boundary points for some classes of elliptic equations

15.30-15.55 J. HEINONEN: Boundary accessibility and elliptic harmonic measures

16.15-16.40 B. MAIR: Generalized approach regions

16.50-17.15 R. WITTMANN: Green functions and uniform Harnack inequality

17.25-17.50 R. SERAPIONI: Pointwise regularity of solutions of non-uniformly elliptic equations

Section C

14.20-14.45 T.S. ANGELL: On iterative methods in potential problems (with R.E. KLEINMAN and G.F. ROACH)

14.55-15.20 H. KRETSCHMAR: On mathematical modelling of non-Newtonian flow by perturbation methods

15.30-15.55 J. NEDOMA: On the new potential problem in gravity field theory

16.15-16.40 A. PISKOREK: On thermoelastic potential

16.50-17.15 A. PRILEPKO: The inverse problems in potential theory

17.25-17.50 P. LEHTOLA: The obstacle problem in non-linear potential theory

Thursday, July 23

Section A

9.00 - 9.25 NGUYEN X.-L.: A convergence theorem for finely hyperharmonic function with applications in fine holomorphy

9.35 -10.00 K. GOWRISANKARAN: Fine limits of multiply superharmonic functions and applications

10.25-10.45 Ö. AKIN: The integral representation of the positive solutions of the generalized Weinstein equation

10.55-11.20 B. KORENBLUM: A sharper form of a theorem of Kolmogorov

11.30-11.55 F.-Y. MAEDA: Energy and capacities on harmonic spaces with adjoint structure

Section B

9.00 - 9.25 B. KAWOHL: On the convexity of level sets for elliptic and parabolic exterior boundary value problems

9.35 -10.00 P. BLANCHET: Fusion of the solutions of an elliptic partial differential equation

10.20-10.45 D. FEYEL: The equation $\frac{1}{2}\Delta u - u\mu = 0$ where μ is a positive measure

10.55-11.20 H. AIKAWA: On weighted Beppo Levi functions

11.30-11.55 H. BEN SAAD: Discrete fine topology

Section C

9.00 - 9.25 M. ROECKNER: Traces of harmonic functions and a new support for the free field

9.35-10.00 A. HORNBERG: Jump relations for surface potentials in case of continuous densities

10.20-10.45 S. ZARGARYAN: On the asymptotic behavior of solutions of a system of boundary equations of elasticity in a neighborhood of corner points of the contour

10.55-11.20 M. COSTABEL: Crack singularities in three-dimensional elasticity

11.30-11.55 N. SHOPOLOV: On the class of linear integral operators of the type potential

Friday, July 24

Survey lectures

10.10-11.05 N. BOBOC, Gh. BUCUR: Order and convexity in potential theory

11.20-12.15 I. LAINE: Axiomatic non-linear potential theories

14.20-15.15 W. WENDLAND: Boundary potentials and boundary element methods

15.30-16.25 B. FUGLEDE: Fine potential theory

PARTICIPANTS OF THE CONFERENCE

AIKAWA Hiroaki
Department of Mathematics, Faculty of Science, Gakushuin University, 1-5-1 Mejiro,Toshima-ku, Tokyo 171, Japan

AKIN Omer
Department of Mathematics, Faculty of Science and Arts, Firat University, Elazig Turkey

ANCONA Alano
Universite Paris XI, Campus d'Orsay, Bat.425 (Mathematiques), 91405 Orsay, France

ANGEL Thomas S.
Department of Mathematical Sciences, University of Delaware, Newark, 501 Ewing Hall Delaware 19716, USA

AZARIN Vladimir
Kharkov Instit. Railway Transport, Feyerbacha Square 7, 31050 Kharkov, USSR

BARTH Thomas
Fachbereich Mathematik, Universitat Essen, P.O. Box 103764, D-4300 Essen, FRG

BAUER Heinz
Mathematisches Institut, Universitat Erlangen-Nurnberg, Bismarckstrasse 1 1/2 D-8520 Erlangen, FRG

BEN SAAD Hedi
Department de Mathematiques, Faculte des Sciences de Tunis, Campus Universitaire 1060 Tunis, Tunis

BERTIN Emile
Mathematisch Instituut, Budapestlaan 6, Postbus 80.010, 3508 TA Utrecht, Netherlands

BIROLI Marco
Dipartimento di Matematica, Politecnico di Milano, Via Bonardi, 9-20133 Milano, Italy

BLANCHET Pierre
Universite de Montreal, Department de Mathematiques et de Statistique, C.P.6128, Succ.A H3C 3J7 Montreal, Quebec, Canada

BLIEDTNER Jurgen
Fachbereich Mathematik, Johan Wolfgang Goethe Universitat, Robert-Mayer-Strasse 6-10 D-6000 Frankfurt am Main, FRG

BOBOC Nicu
Institut de Mat. INCREST, Str. Academici 14, Bucuresti 1, Romania

BOUKRICHA Abderrahman
Departement de Mathematiques, Faculte des Sciences, Universite de Tunis, 1060 Tunis Tunis

BYRNE Catrione
Springer-Verlag Heidelberg, Tiergartenstrasse 17, D-6900 Heidelberg, FRG

CHLEBIK Miroslav
Matematicky Ustav CSAV, Czechoslovak Academy of Sciences, Zitna 25, 11567 Praha 1, CSSR

CHUNG Kai Lai
Department of Mathematics, Stanford University, Stanford, CA 94305, USA

CORNEA Aurel
Mathematisch.-Geogr. Fakultat, Universitat Eichstatt, Ostenstrasse 26-28 D-8078 Eichstatt, FRG

COSTABEL Martin
Mathematisches Institut, Schlossgartenstr. 7, D-6100 Darmstadt, FRG

CSINK Laszlo
Eotvos University, Faculty of Arts, P.O. Box 107, H-1364 Budapest, Hungary
DONT Miroslav
Department of Mathematics, FEL CVUT, Šuchbatarova 2, 16627 Praha 6, CSSR
DONTOVA Eva
Department of Mathematics, FJFI CVUT, Husova 5, 11000 Praha 1, CSSR
EDWARDS David A.
Mathematical Institute, Oxford University, 24-29 St. Giles, 0X1 3LB Oxford
Great Britain
ERIKSSON-BIQUE Sirka-Liisa
Department of Mathematics, University of Joensuu, P.O. Box 111, SF 80101 Joensuu
Finland
FABES Eugene
School of Mathematics, University of Minnesota, 127 Vincent Hall, 206 Church Street
S.E., Minneapolis, Minnesota 55455, USA
FERNANDEZ Jose L.T.
Department of Mathematics, University of Maryland, College Park, MD 20742, USA
FEYEL Denis
Equipe d'Analyse, Tour 46, Universite Paris VI, 4, Place Jussieu, F-75230 Paris
VI-Cedex 05, France
FUGLEDE Bent
Matematiske Institut, University of Copenhagen, Universitetsparken 5
DK-2100 Copenhagen, Denmark
GARDINER Stephen
Department of Mathematics, University College, Belfield, Dublin 4, Ireland
GOWRISANKARAN Kohur
Department of Mathematics and Statistics, Mc Gill University, 805 Sherbrooke
H3A 2K6 Montreal West, Canada
HANSEN Wolfhard
Universitat Bielefeld, Facultat fur Mathematik, Postfach 8640, 4800 Bielefeld 1, FRG
HEINONEN Juha
Department of Mathematics, University of Jyvaskyla, Seminaarinkatu 15, 40100 Jyvaskyla
Finland
HLADNIK Milan
Department of Mathematics, Physics and Mechanics, E.K. University of Ljubljana
Jadranska 19, 61000 Ljubljana, Yugoslavia
HMISSI Mohamed
Departement de Mathematiques, Faculte des Sciences de Tunis, 1060 Belvedere Tunis
Tunis
HOLOTA Petr
Vyzkumny ustav geodeticky a topogr., a kartograficky, Zdiby 98, 250 66 Praha, CSSR
HORNBERG Alexander
Mathematisches Institut II, Universitat Karlsruhe, Englerstrasse 2
D-7500 Karlsruhe 1, FRG
HSIAO George
Department of Mathematical Sciences, University of Delaware, Newark, 501 Ewing Hall
Delaware 19716, USA
HUEBER Hermann
Mathematisches Fakultat, University of Bielefeld, Universitatsstrasse 1, Postfach 8640
D-4800 Bielefeld 1, FRG
IBRAGIMOV A.I.
Institute of Mathematics, Academy of Sciences of Azerbeidzan SSR, USSR
IKEGAMI Teruo
Department of Mathematics, Osaka City Univ., Sugimoto 3-3 138, Sumiyoshi-ku, Osaka 558
Japan
ITO Masayuki
Department of Mathematics, Nagoya University, Furocho, Chikusa-ku, Nagoya 464, Japan

JACOB Niels
Mathematisches Institut, Universitat Erlangen-Nurnberg, Bismarckstrasse 1 1/2
D-8520 Erlangen, FRG
JANSSEN Klaus
Institut fur Statistik und Dokument., Universitat Dusseldorf, Universitatsstr.1
D-4000 Dusseldorf, FRG
KANDA Mamoru
Institute of Mathematics, University of Tsukuba, Sakura-mura, Niihari-gun, Ibaraki 305
Japan
KAWOHL Bernhard
Universitat Heidelberg, Im Neuenhermes Feld 294, SFB 123, D-6900 Heidelberg, FRG
KELLY George
Mathematical Physics Department, University College, Cork, Ireland
KESELMAN Dimitrii Gidarevich
Department of Mathematics, Astrybvuz, Tatischeva street 16, 414025 Astrakhan, USSR
KIESER Ralf
Mathematisches Institut A, Universitat Stuttgart, Pfaffenwaldring 57
D-7000 Stuttgart 80, FRG
KILPELAINEN Tero
Department of Mathematics, University of Jyvaskyla, Seminaarinkatu 15
SF-40100 Jyvaskyla, Finland
KLEINMAN Ralph
Department of Mathematical Sciences, University of Delaware, Newark, 501 Ewing Hall
Delaware 19716, USA
KOHLENBACH Ulrich
Department of Mathematics, Universitat Frankfurt/Main, Robert Mayer Strasse 6-8
D-6000 Frankfurt am Main 1, FRG
KORENBLUM Boris
Department of Mathematics and Statistics, SUNY, Albany, New York 1222, USA
KOUNCHEV Ognyan
Department of Mathematics with Computer Centre, Bulgarian Academy of Sciences
ul. G. Bonchev, bl.8, 1113 Sofia, Bulgaria
KRAL Josef
Matematicky Ustav CSAV, Czechoslovak Academy of Sciences, Zitna 25, 11567 Praha 1, CSSR
KRESS Reiner
Institut fur Numerische und Angewandte Math., Universitat Gottingen, Lotzestrasse 16-18
D-3400 Gottingen, FRG
KRETZSCHMAR Horst
Technische Universitat Karl-Marx-Stadt, Sektion Mathematik, Postfach 964, 9010
Karl-Marx-Stadt, GDR
KURAN Ulku
Department of Pure Mathematics, University of Liverpool, L69 3BX Liverpool, England
LAINE Ilpo
Department of Mathematics, University of Joensuu, P.O. Box 111, SF-80101 Joensuu
Finland
LANDIS E. M.
Department of Mathematics, Moscow State University, 117234 Moscow, USSR
LEHA Gottlieb
Fakultat fur Mathematik und Informatik, Universitat Passau, Innstrasse 27
D-8390 Passau, FRG
LEHTOLA Pasi
Department of Mathematics, University of Jyvaskyla, Seminaarinkatu 15
SF-40100 Jyvaskyla, Finland
LEUTWILER Heinz
Mathematisches Institut, Universitat Erlangen-Nurnberg, Bismarckstrasse 1 1/2
D-8520 Erlangen, FRG
LEVENBERG Norman
Wellesley College, Wellesley, Massachussetts 02181, USA

LIEBERMAN Gary N.
Department of Mathematics, Iowa State University, 400 Carver Hall,Ames, Iowa 50011, USA
LOEB Peter
Department of Mathematics, University of Illions, 1409 West Green Street
61801 Illions, Urbana, USA
LOVE Eric
Department of Mathematics, University of Melbourne, Parkville, Victoria 3052, Australia
LUKES Jaroslav
Department of Mathematical Analysis, Matematicko-fyzikalni fakulta UK, Sokolovska 83
18600 Praha 8, CSSR
LUMER Gunter
Departement de Mathematiques, Faculte des Sciences, Universite de l'Etat
Avenue Maistriau 15, B-7000 Mons, Belgium
LUMER-NAIM Linda
6 Rue Jules Lejeune, B-1060 Bruxelles, Belgium
LYONS Terence
Department of Mathematics, University of Edinburgh, J. C. Maxwell Building
Edinburgh EH9 3JZ, Mayfield Road, Scotland
MAAGLI Habib
Faculte des Sciences de Tunis, Campus Universitaire, 1060 Tunis, Tunis
MAEDA Fumi-Yuki
Department of Mathematics, Faculty of Science, Hiroshima University
1-1-89 Higashi-senda-machi, 730 Hiroshima, Japan
MAIR Bernard
Department of Mathematics, Texas Tech University, Lubbock,Texas 79409, USA
MALY Jan
Department of Mathematical Analysis, Matematicko-fyzikalni fakulta UK, Sokolovska 83
18600 Praha 8, CSSR
MARTIO Olli
Department of Mathematics, University of Jyvaskyla, Seminaarink 15, 40100 Jyvaskyla
Finland
MATTILA Pertti
Department of Mathematics, University of Helsinky, Hallituskatu 15, 00100 Helsinki
Finland
MEDKOVA Dagmar
Statni vyzkumny ustav pro stavbu stroju, 19000 Praha 9 - Bechovice, CSSR
MEYER Michael
Institut fur Statistik, Universitat Dusseldorf, Universitatstrasse 1
D-4000 Dusseldorf 1, FRG
MULLER Hans-Helge
Institut fur Statistik, Universitat Dusseldorf, Universitatstrasse 1
D-4000 Dusseldorf 1, FRG
NECAS Jindrich
Mathematical Institute of Charles University, Matematicko-fyzikalni fakulta UK
Sokolovska 83, 18600 Praha 8, CSSR
NEDOMA Jiri
Stredisko vypocetni techniky CSAV, Pod vodarenskou vezi 2, 18207 Praha 8-Liben, CSSR
NETUKA Ivan
Mathematical Institute of Charles University, Matematicko-fyzikalni fakulta UK
Sokolovska 83, 18600 Praha 8, CSSR
NGUYEN Xuan-Loc
Institute of Computer Science and Cybernetics, Vien KHTT-DK, Lieu Giai. Ba-Dinh., Hanoi
Vietnam
OHTSUKA Makoto
Department of Mathematics, Faculty of Science, Gakushuin University, 1-5-1 Mejiro,
Toshima-ku, Tokyo 171, Japan

PAVLOV Igor
Rostov Civil Engineering Institute, 344022 Sotsialisticheskaya street 162
Rostov-on-Don, USSR
PESONEN Martti E.
Department of Mathematics, University of Joensuu, P.O. Box 111, 80101 Joensuu, Finland
PIRO-GRIMALDI Anna
Dipartimento di Matematica, Universita di Cagliari, Viale Merello 92, 09100 Cagliari
Italy
PIRO-VERNIER Stella
Dipartimento di Matematica, Universita di Cagliari, Viale Merello 92, 09100 Cagliari
Italy
PISKOREK Adam
Faculty of Mathematics, Computer Sciences and Mechanics, Institut of Mechanics
University of Warsaw, Palac Kultury i Nauki IX p.937, 00-901 Warszawa, Poland
POLASEK Jan
Katedra vypocetni techniky a inf., Strojni fakulta CVUT, Suchbatorova 4, 166 07 Praha 6
CSSR
POPA Eugen
Department of Mathematics, University "Al.I.Cuza", Calea 23 August Nr.11, 6600 Iasi
Romania
POSTI Eero
Department of Mathematics, University of Joensuu, P.O.Box 111, 80101 Joensuu, Finland
PRADELLE Arnaud,de la
Equipe d'Analyse,Tour 46, Universite Paris VI, 4, place Jussieu, 75230 Paris-Cedex 05
France
PRILEPKO Aleksey Ivanovich
Department of Mathematics, Moscow Institute of Ingeneering and Physics
Kaschirskoye Shosse 31, 115409 Moscow, USSR
PYRIH Pavel
UDZ Praha, Krizova 25, 15000 Praha 5, CSSR
RAGNEDDA Francesco
Dipartimento di Matematica, Universita di Cagliari, Viale Merello 92, 09100 Cagliari
Italy
RAKOSNIK Jiri
Matematicky ustav CSAV, Czechoslovak Academy of Sciences, Zitna 25, 11567 Praha 1, CSSR
ROACH G. F.
Department of Mathematics, University of Strathclyde, Livingstone Tower
26 Richmond Street, Glasgow G1 1XH, Great Britain
ROCKNER Michael
Department of Mathematics, University of Edinburgh, J.C.Maxwell Building
Edinburgh EH9 3JZ, Mayfield Road, Great Britain
SAKAI Makoto
Department of Mathematics, Tokyo Metropolitan University, Fukasawa, Setagaya-ku
Tokyo 158, Japan
SALSA Sandro
Dipartmeno di Matematica, Politecnico di Milano, Via Bonardi 7, 20133 Milano, Italy
SCHIRMEIER Ursula
Mathematisch.- Geogr. Fakultat, Universitat Eichstatt, Ostenstrassen 26-28
D-8078 Eichstatt, FRG
SERAPIONI Raul
Department of Mathematics, Universita degli studi di Trento, 38050 Povo (Trento), Italy
SHIDFAR Abdullah
Department of Mathematics, Iran University of Sciences Technology, Narmak, Teheran - 16
Iran
SHOPOLOV Nikolaj
26 Galitchiza street, Sofia 1126, Bulgaria
SJODIN Tord Jonas
Department of Mathematics, University of Umea, S-90187 Umea, Sweden

STEINER Uwe
Fachbereich Mathematik, J. W. Goethe Universitat, Robert-Mayer-Strasse 6-8
D-6000 Frankfurt am Main, FRG
STURM Theo
Mathematisches Institut, Universitat Erlangen-Nurnberg, Bismarckstrasse 1 1/2
D-8520 Erlangen, FRG
THOMAS Robin
Department of Mathematical Analysis, Matematicko-fyzikalni fakulta UK, Sokolovska 83
18600 Praha 8, CSSR
VESELY Jiri
Mathematical Institute of Charles University, Matematicko-fyzikalni fakulta UK
Sokolovska 83, 18600 Praha 8, CSSR
WENDLAND Wolfgang
Mathematisches Institut A, Universitat Stuttgart, Pfaffenwaldring 57
D-7000 Stuttgart 80, FRG
WERMER John
Department of Mathematics, Brown University, Providence, R.I. 02912, USA
WILDENHAIN Gunther
Sektion Mathematik, Wilhelm-Pieck-Universitat, Universitatsplatz 1, 2500 Rostock, GDR
WIRSCHING Gunter
Mathematisch.-Geogr. Fakultat, Universitat Eichstatt, Ostenstrasse 26-28
D-8078 Eichstatt, FRG
WITTMAN Rainer
Mathematisch.-Geogr. Fakultat, Universitat Eichstatt, Ostenstrasse 26-28
D-8078 Eichstatt, FRG
ZARGARYAN Stepan
Yerevan Polytechnical Institute, 105 Teryan street, 375009 Yerevan, USSR

LECTURE NOTES IN MATHEMATICS

Edited by A. Dold and B. Eckmann

Some general remarks on the publication of proceedings of congresses and symposia

Lecture Notes aim to report new developments - quickly, informally and at a high level. The following describes criteria and procedures which apply to proceedings volumes. The editors of a volume are strongly advised to inform contributors about these points at an early stage.

§1. One (or more) expert participant(s) of the meeting should act as the responsible editor(s) of the proceedings. They select the papers which are suitable (cf. §§ 2, 3) for inclusion in the proceedings, and have them individually refereed (as for a journal). It should not be assumed that the published proceedings must reflect conference events faithfully and in their entirety. Contributions to the meeting which are not included in the proceedings can be listed by title. The series editors will normally not interfere with the editing of a particular proceedings volume - except in fairly obvious cases, or on technical matters, such as described in §§ 2, 3. The names of the responsible editors appear on the title page of the volume.

§2. The proceedings should be reasonably homogeneous (concerned with a limited area). For instance, the proceedings of a congress on "Analysis" or "Mathematics in Wonderland" would normally not be sufficiently homogeneous.

One or two longer survey articles on recent developments in the field are often very useful additions to such proceedings - even if they do not correspond to actual lectures at the congress. An extensive introduction on the subject of the congress would be desirable.

§3. The contributions should be of a high mathematical standard and of current interest. Research articles should present new material and not duplicate other papers already published or due to be published. They should contain sufficient information and motivation and they should present proofs, or at least outlines of such, in sufficient detail to enable an expert to complete them. Thus resumes and mere announcements of papers appearing elsewhere cannot be included, although more detailed versions of a contribution may well be published in other places later.

Surveys, if included, should cover a sufficiently broad topic, and should in general not simply review the author's own recent research. In the case of surveys, exceptionally, proofs of results may not be necessary.

"Mathematical Reviews" and "Zentralblatt für Mathematik" require that papers in proceedings volumes carry an explicit statement that they are in final form and that no similar paper has been or is being submitted elsewhere, if these papers are to be considered for a review. Normally, papers that satisfy the criteria of the Lecture Notes in Mathematics series also satisfy this

.../...

requirement, but we would strongly recommend that the contributing authors be asked to give this guarantee explicitly at the beginning or end of their paper. There will occasionally be cases where this does not apply but where, for special reasons, the paper is still acceptable for LNM.

§4. Proceedings should appear soon after the meeeting. The publisher should, therefore, receive the complete manuscript within nine months of the date of the meeting at the latest.

§5. Plans or proposals for proceedings volumes should be sent to one of the editors of the series or to Springer-Verlag Heidelberg. They should give sufficient information on the conference or symposium, and on the proposed proceedings. In particular, they should contain a list of the expected contributions with their prospective length. Abstracts or early versions (drafts) of some of the contributions are very helpful.

§6. Lecture Notes are printed by photo-offset from camera-ready typed copy provided by the editors. For this purpose Springer-Verlag provides editors with technical instructions for the preparation of manuscripts and these should be distributed to all contributing authors. Springer-Verlag can also, on request, supply stationery on which the prescribed typing area is outlined. Some homogeneity in the presentation of the contributions is desirable.

Careful preparation of manuscripts will help keep production time short and ensure a satisfactory appearance of the finished book. The actual production of a Lecture Notes volume normally takes 6 -8 weeks.

Manuscripts should be at least 100 pages long. The final version should include a table of contents and as far as applicable a subject index.

§7. Editors receive a total of 50 free copies of their volume for distribution to the contributing authors, but no royalties. (Unfortunately, no reprints of individual contributions can be supplied.) They are entitled to purchase further copies of their book for their personal use at a discount of 33.3 %, other Springer mathematics books at a discount of 20 % directly from Springer-Verlag. Contributing authors may purchase the volume in which their article appears at a discount of 33.3 %.

Commitment to publish is made by letter of intent rather than by signing a formal contract. Springer-Verlag secures the copyright for each volume.

LECTURE NOTES

ESSENTIALS FOR THE PREPARATION OF CAMERA-READY MANUSCRIPTS

Springer
Springer-Verlag
Berlin Heidelberg New York
London Paris Tokyo Hong Kong

The preparation of manuscripts which are to be reproduced by photo-offset require special care. Manuscripts which are submitted in technically unsuitable form will be returned to the author for retyping. There is normally no possibility of carrying out further corrections after a manuscript is given to production. Hence it is crucial that the following instructions be adhered to closely. If in doubt, please send us 1 - 2 sample pages for examination.

General. The characters must be uniformly black both within a single character and down the page. Original manuscripts are required: photocopies are acceptable only if they are sharp and without smudges.

On request, Springer-Verlag will supply special paper with the text area outlined. The standard TEXT AREA (OUTPUT SIZE if you are using a 14 point font) is 18 x 26.5 cm (7.5 x 11 inches). This will be scale-reduced to 75% in the printing process. If you are using computer typesetting, please see also the following page.

Make sure the TEXT AREA IS COMPLETELY FILLED. Set the margins so that they precisely match the outline and type right from the top to the bottom line. (Note that the page number will lie outside this area). Lines of text should not end more than three spaces inside or outside the right margin (see example on page 4).

Type on one side of the paper only.

Spacing and Headings (Monographs). Use ONE-AND-A-HALF line spacing in the text. Please leave sufficient space for the title to stand out clearly and do NOT use a new page for the beginning of subdivisons of chapters. Leave THREE LINES blank above and TWO below headings of such subdivisions.

Spacing and Headings (Proceedings). Use ONE-AND-A-HALF line spacing in the text. Do not use a new page for the beginning of subdivisons of a single paper. Leave THREE LINES blank above and TWO below headings of such subdivisions. Make sure headings of equal importance are in the same form.

The first page of each contribution should be prepared in the same way. The title should stand out clearly. We therefore recommend that the editor prepare a sample page and pass it on to the authors together with these instructions. Please take the following as an example. Begin heading 2 cm below upper edge of text area.

MATHEMATICAL STRUCTURE IN QUANTUM FIELD THEORY

John E. Robert
Mathematisches Institut, Universität Heidelberg
Im Neuenheimer Feld 288, D-6900 Heidelberg

Please leave THREE LINES blank below heading and address of the author, then continue with the actual text on the same page.

Footnotes. These should preferable be avoided. If necessary, type them in SINGLE LINE SPACING to finish exactly on the outline, and separate them from the preceding main text by a line.

Symbols. Anything which cannot be typed may be entered by hand in BLACK AND ONLY BLACK ink. (A fine-tipped rapidograph is suitable for this purpose; a good black ball-point will do, but a pencil will not). Do not draw straight lines by hand without a ruler (not even in fractions).

Literature References. These should be placed at the end of each paper or chapter, or at the end of the work, as desired. Type them with single line spacing and start each reference on a new line. Follow "Zentralblatt für Mathematik"/"Mathematical Reviews" for abbreviated titles of mathematical journals and "Bibliographic Guide for Editors and Authors (BGEA)" for chemical, biological, and physics journals. Please ensure that all references are COMPLETE and ACCURATE.

IMPORTANT

Pagination. For typescript, number pages in the upper right-hand corner in LIGHT BLUE OR GREEN PENCIL ONLY. The printers will insert the final page numbers. For computer type, you may insert page numbers (1 cm above outer edge of text area).

It is safer to number pages AFTER the text has been typed and corrected. Page 1 (Arabic) should be THE FIRST PAGE OF THE ACTUAL TEXT. The Roman pagination (table of contents, preface, abstract, acknowledgements, brief introductions, etc.) will be done by Springer-Verlag.

If including running heads, these should be aligned with the inside edge of the text area while the page number is aligned with the outside edge noting that right-hand pages are odd-numbered. Running heads and page numbers appear on the same line. Normally, the running head on the left-hand page is the chapter heading and that on the right-hand page is the section heading. Running heads should not be included in proceedings contributions unless this is being done consistently by all authors.

Corrections. When corrections have to be made, cut the new text to fit and paste it over the old. White correction fluid may also be used.

Never make corrections or insertions in the text by hand.

If the typescript has to be marked for any reason, e.g. for provisional page numbers or to mark corrections for the typist, this can be done VERY FAINTLY with BLUE or GREEN PENCIL but NO OTHER COLOR: these colors do not appear after reproduction.

COMPUTER-TYPESETTING. Further, to the above instructions, please note with respect to your printout that
- the characters should be sharp and sufficiently black;
- it is not strictly necessary to use Springer's special typing paper. Any white paper of reasonable quality is acceptable.

If you are using a significantly different font size, you should modify the output size correspondingly, keeping length to breadth ratio 1 : 0.68, so that scaling down to 10 point font size, yields a text area of 13.5 x 20 cm (5 3/8 x 8 in), e.g.

Differential equations.: use output size 13.5 x 20 cm.

Differential equations.: use output size 16 x 23.5 cm.

Differential equations.: use output size 18 x 26.5 cm.

Interline spacing: 5.5 mm base-to-base for 14 point characters (standard format of 18 x 26.5 cm).
If in any doubt, please send us 1 - 2 sample pages for examination. We will be glad to give advice.

Vol. 1173: H. Delfs, M. Knebusch, Locally Semialgebraic Spaces. XVI, 329 pages. 1985.

Vol. 1174: Categories in Continuum Physics, Buffalo 1982. Seminar. Edited by F.W. Lawvere and S.H. Schanuel. V, 126 pages. 1986.

Vol. 1175: K. Mathiak, Valuations of Skew Fields and Projective Hjelmslev Spaces. VII, 116 pages. 1986.

Vol. 1176: R.R. Bruner, J.P. May, J.E. McClure, M. Steinberger, H_∞ Ring Spectra and their Applications. VII, 388 pages. 1986.

Vol. 1177: Representation Theory I. Finite Dimensional Algebras. Proceedings, 1984. Edited by V. Dlab, P. Gabriel and G. Michler. XV, 340 pages. 1986.

Vol. 1178: Representation Theory II. Groups and Orders. Proceedings, 1984. Edited by V. Dlab, P. Gabriel and G. Michler. XV, 370 pages. 1986.

Vol. 1179: Shi J.-Y. The Kazhdan-Lusztig Cells in Certain Affine Weyl Groups. X, 307 pages. 1986.

Vol. 1180: R. Carmona, H. Kesten, J.B. Walsh, École d'Été de Probabilités de Saint-Flour XIV – 1984. Édité par P.L. Hennequin. X, 438 pages. 1986.

Vol. 1181: Buildings and the Geometry of Diagrams, Como 1984. Seminar. Edited by L. Rosati. VII, 277 pages. 1986.

Vol. 1182: S. Shelah, Around Classification Theory of Models. VII, 279 pages. 1986.

Vol. 1183: Algebra, Algebraic Topology and their Interactions. Proceedings, 1983. Edited by J.-E. Roos. XI, 396 pages. 1986.

Vol. 1184: W. Arendt, A. Grabosch, G. Greiner, U. Groh, H.P. Lotz, U. Moustakas, R. Nagel, F. Neubrander, U. Schlotterbeck, One-parameter Semigroups of Positive Operators. Edited by R. Nagel. X, 460 pages. 1986.

Vol. 1185: Group Theory, Beijing 1984. Proceedings. Edited by Tuan H.F. V, 403 pages. 1986.

Vol. 1186: Lyapunov Exponents. Proceedings, 1984. Edited by L. Arnold and V. Wihstutz. VI, 374 pages. 1986.

Vol. 1187: Y. Diers, Categories of Boolean Sheaves of Simple Algebras. VI, 168 pages. 1986.

Vol. 1188: Fonctions de Plusieurs Variables Complexes V. Séminaire, 1979–85. Edité par François Norguet. VI, 306 pages. 1986.

Vol. 1189: J. Lukeš, J. Malý, L. Zajíček, Fine Topology Methods in Real Analysis and Potential Theory. X, 472 pages. 1986.

Vol. 1190: Optimization and Related Fields. Proceedings, 1984. Edited by R. Conti, E. De Giorgi and F. Giannessi. VIII, 419 pages. 1986.

Vol. 1191: A.R. Its, V.Yu. Novokshenov, The Isomonodromic Deformation Method in the Theory of Painlevé Equations. IV, 313 pages. 1986.

Vol. 1192: Equadiff 6. Proceedings, 1985. Edited by J. Vosmansky and M. Zlámal. XXIII, 404 pages. 1986.

Vol. 1193: Geometrical and Statistical Aspects of Probability in Banach Spaces. Proceedings, 1985. Edited by X. Fernique, B. Heinkel, M.B. Marcus and P.A. Meyer. IV, 128 pages. 1986.

Vol. 1194: Complex Analysis and Algebraic Geometry. Proceedings, 1985. Edited by H. Grauert. VI, 235 pages. 1986.

Vol.1195: J.M. Barbosa, A.G. Colares, Minimal Surfaces in $\mathbb{R}^3$. X, 124 pages. 1986.

Vol. 1196: E. Casas-Alvero, S. Xambó-Descamps, The Enumerative Theory of Conics after Halphen. IX, 130 pages. 1986.

Vol. 1197: Ring Theory. Proceedings, 1985. Edited by F.M.J. van Oystaeyen. V, 231 pages. 1986.

Vol. 1198: Séminaire d'Analyse, P. Lelong – P. Dolbeault – H. Skoda. Seminar 1983/84. X, 260 pages. 1986.

Vol. 1199: Analytic Theory of Continued Fractions II. Proceedings, 1985. Edited by W.J. Thron. VI, 299 pages. 1986.

Vol. 1200: V.D. Milman, G. Schechtman, Asymptotic Theory of Finite Dimensional Normed Spaces. With an Appendix by M. Gromov. VIII, 156 pages. 1986.

Vol. 1201: Curvature and Topology of Riemannian Manifolds. Proceedings, 1985. Edited by K. Shiohama, T. Sakai and T. Sunada. VII, 336 pages. 1986.

Vol. 1202: A. Dür, Möbius Functions, Incidence Algebras and Power Series Representations. XI, 134 pages. 1986.

Vol. 1203: Stochastic Processes and Their Applications. Proceedings, 1985. Edited by K. Itô and T. Hida. VI, 222 pages. 1986.

Vol. 1204: Séminaire de Probabilités XX, 1984/85. Proceedings. Edité par J. Azéma et M. Yor. V, 639 pages. 1986.

Vol. 1205: B.Z. Moroz, Analytic Arithmetic in Algebraic Number Fields. VII, 177 pages. 1986.

Vol. 1206: Probability and Analysis, Varenna (Como) 1985. Seminar. Edited by G. Letta and M. Pratelli. VIII, 280 pages. 1986.

Vol. 1207: P.H. Bérard, Spectral Geometry: Direct and Inverse Problems. With an Appendix by G. Besson. XIII, 272 pages. 1986.

Vol. 1208: S. Kaijser, J.W. Pelletier, Interpolation Functors and Duality. IV, 167 pages. 1986.

Vol. 1209: Differential Geometry, Peñíscola 1985. Proceedings. Edited by A.M. Naveira, A. Ferrández and F. Mascaró. VIII, 306 pages. 1986.

Vol. 1210: Probability Measures on Groups VIII. Proceedings, 1985. Edited by H. Heyer. X, 386 pages. 1986.

Vol. 1211: M.B. Sevryuk, Reversible Systems. V, 319 pages. 1986.

Vol. 1212: Stochastic Spatial Processes. Proceedings, 1984. Edited by P. Tautu. VIII, 311 pages. 1986.

Vol. 1213: L.G. Lewis, Jr., J.P. May, M. Steinberger, Equivariant Stable Homotopy Theory. IX, 538 pages. 1986.

Vol. 1214: Global Analysis – Studies and Applications II. Edited by Yu.G. Borisovich and Yu.E. Gliklikh. V, 275 pages. 1986.

Vol. 1215: Lectures in Probability and Statistics. Edited by G. del Pino and R. Rebolledo. V, 491 pages. 1986.

Vol. 1216: J. Kogan, Bifurcation of Extremals in Optimal Control. VIII, 106 pages. 1986.

Vol. 1217: Transformation Groups. Proceedings, 1985. Edited by S. Jackowski and K. Pawalowski. X, 396 pages. 1986.

Vol. 1218: Schrödinger Operators, Aarhus 1985. Seminar. Edited by E. Balslev. V, 222 pages. 1986.

Vol. 1219: R. Weissauer, Stabile Modulformen und Eisensteinreihen. III, 147 Seiten. 1986.

Vol. 1220: Séminaire d'Algèbre Paul Dubreil et Marie-Paule Malliavin. Proceedings, 1985. Édité par M.-P. Malliavin. IV, 200 pages. 1986.

Vol. 1221: Probability and Banach Spaces. Proceedings, 1985. Edited by J. Bastero and M. San Miguel. XI, 222 pages. 1986.

Vol. 1222: A. Katok, J.-M. Strelcyn, with the collaboration of F. Ledrappier and F. Przytycki, Invariant Manifolds, Entropy and Billiards; Smooth Maps with Singularities. VIII, 283 pages. 1986.

Vol. 1223: Differential Equations in Banach Spaces. Proceedings, 1985. Edited by A. Favini and E. Obrecht. VIII, 299 pages. 1986.

Vol. 1224: Nonlinear Diffusion Problems, Montecatini Terme 1985. Seminar. Edited by A. Fasano and M. Primicerio. VIII, 188 pages. 1986.

Vol. 1225: Inverse Problems, Montecatini Terme 1986. Seminar. Edited by G. Talenti. VIII, 204 pages. 1986.

Vol. 1226: A. Buium, Differential Function Fields and Moduli of Algebraic Varieties. IX, 146 pages. 1986.

Vol. 1227: H. Helson, The Spectral Theorem. VI, 104 pages. 1986.

Vol. 1228: Multigrid Methods II. Proceedings, 1985. Edited by W. Hackbusch and U. Trottenberg. VI, 336 pages. 1986.

Vol. 1229: O. Bratteli, Derivations, Dissipations and Group Actions on C*-algebras. IV, 277 pages. 1986.

Vol. 1230: Numerical Analysis. Proceedings, 1984. Edited by J.-P. Hennart. X, 234 pages. 1986.

Vol. 1231: E.-U. Gekeler, Drinfeld Modular Curves. XIV, 107 pages. 1986.

Vol. 1232: P.C. Schuur, Asymptotic Analysis of Soliton Problems. VIII, 180 pages. 1986.

Vol. 1233: Stability Problems for Stochastic Models. Proceedings, 1985. Edited by V.V. Kalashnikov, B. Penkov and V.M. Zolotarev. VI, 223 pages. 1986.

Vol. 1234: Combinatoire énumérative. Proceedings, 1985. Edité par G. Labelle et P. Leroux. XIV, 387 pages. 1986.

Vol. 1235: Séminaire de Théorie du Potentiel, Paris, No. 8. Directeurs: M. Brelot, G. Choquet et J. Deny. Rédacteurs: F. Hirsch et G. Mokobodzki. III, 209 pages. 1987.

Vol. 1236: Stochastic Partial Differential Equations and Applications. Proceedings, 1985. Edited by G. Da Prato and L. Tubaro. V, 257 pages. 1987.

Vol. 1237: Rational Approximation and its Applications in Mathematics and Physics. Proceedings, 1985. Edited by J. Gilewicz, M. Pindor and W. Siemaszko. XII, 350 pages. 1987.

Vol. 1238: M. Holz, K.-P. Podewski and K. Steffens, Injective Choice Functions. VI, 183 pages. 1987.

Vol. 1239: P. Vojta, Diophantine Approximations and Value Distribution Theory. X, 132 pages. 1987.

Vol. 1240: Number Theory, New York 1984–85. Seminar. Edited by D.V. Chudnovsky, G.V. Chudnovsky, H. Cohn and M.B. Nathanson. V, 324 pages. 1987.

Vol. 1241: L. Gårding, Singularities in Linear Wave Propagation. III, 125 pages. 1987.

Vol. 1242: Functional Analysis II, with Contributions by J. Hoffmann-Jørgensen et al. Edited by S. Kurepa, H. Kraljević and D. Butković. VII, 432 pages. 1987.

Vol. 1243: Non Commutative Harmonic Analysis and Lie Groups. Proceedings, 1985. Edited by J. Carmona, P. Delorme and M. Vergne. V, 309 pages. 1987.

Vol. 1244: W. Müller, Manifolds with Cusps of Rank One. XI, 158 pages. 1987.

Vol. 1245: S. Rallis, L-Functions and the Oscillator Representation. XVI, 239 pages. 1987.

Vol. 1246: Hodge Theory. Proceedings, 1985. Edited by E. Cattani, F. Guillén, A. Kaplan and F. Puerta. VII, 175 pages. 1987.

Vol. 1247: Séminaire de Probabilités XXI. Proceedings. Edité par J. Azéma, P.A. Meyer et M. Yor. IV, 579 pages. 1987.

Vol. 1248: Nonlinear Semigroups, Partial Differential Equations and Attractors. Proceedings, 1985. Edited by T.L. Gill and W.W. Zachary. IX, 185 pages. 1987.

Vol. 1249: I. van den Berg, Nonstandard Asymptotic Analysis. IX, 187 pages. 1987.

Vol. 1250: Stochastic Processes – Mathematics and Physics II. Proceedings 1985. Edited by S. Albeverio, Ph. Blanchard and L. Streit. VI, 359 pages. 1987.

Vol. 1251: Differential Geometric Methods in Mathematical Physics. Proceedings, 1985. Edited by P.L. García and A. Pérez-Rendón. VII, 300 pages. 1987.

Vol. 1252: T. Kaise, Représentations de Weil et GL_2 Algèbres de division et GL_n. VII, 203 pages. 1987.

Vol. 1253: J. Fischer, An Approach to the Selberg Trace Formula via the Selberg Zeta-Function. III, 184 pages. 1987.

Vol. 1254: S. Gelbart, I. Piatetski-Shapiro, S. Rallis. Explicit Constructions of Automorphic L-Functions. VI, 152 pages. 1987.

Vol. 1255: Differential Geometry and Differential Equations. Proceedings, 1985. Edited by C. Gu, M. Berger and R.L. Bryant. XII, 243 pages. 1987.

Vol. 1256: Pseudo-Differential Operators. Proceedings, 1986. Edited by H.O. Cordes, B. Gramsch and H. Widom. X, 479 pages. 1987.

Vol. 1257: X. Wang, On the C*-Algebras of Foliations in the Plane. V, 165 pages. 1987.

Vol. 1258: J. Weidmann, Spectral Theory of Ordinary Differential Operators. VI, 303 pages. 1987.

Vol. 1259: F. Cano Torres, Desingularization Strategies for Three-Dimensional Vector Fields. IX, 189 pages. 1987.

Vol. 1260: N.H. Pavel, Nonlinear Evolution Operators and Semigroups. VI, 285 pages. 1987.

Vol. 1261: H. Abels, Finite Presentability of S-Arithmetic Groups. Compact Presentability of Solvable Groups. VI, 178 pages. 1987.

Vol. 1262: E. Hlawka (Hrsg.), Zahlentheoretische Analysis II. Seminar, 1984–86. V, 158 Seiten. 1987.

Vol. 1263: V.L. Hansen (Ed.), Differential Geometry. Proceedings, 1985. XI, 288 pages. 1987.

Vol. 1264: Wu Wen-tsün, Rational Homotopy Type. VIII, 219 pages. 1987.

Vol. 1265: W. Van Assche, Asymptotics for Orthogonal Polynomials. VI, 201 pages. 1987.

Vol. 1266: F. Ghione, C. Peskine, E. Sernesi (Eds.), Space Curves. Proceedings, 1985. VI, 272 pages. 1987.

Vol. 1267: J. Lindenstrauss, V.D. Milman (Eds.), Geometrical Aspects of Functional Analysis. Seminar. VII, 212 pages. 1987.

Vol. 1268: S.G. Krantz (Ed.), Complex Analysis. Seminar, 1986. VII, 195 pages. 1987.

Vol. 1269: M. Shiota, Nash Manifolds. VI, 223 pages. 1987.

Vol. 1270: C. Carasso, P.-A. Raviart, D. Serre (Eds.), Nonlinear Hyperbolic Problems. Proceedings, 1986. XV, 341 pages. 1987.

Vol. 1271: A.M. Cohen, W.H. Hesselink, W.L.J. van der Kallen, J.R. Strooker (Eds.), Algebraic Groups Utrecht 1986. Proceedings. XII, 284 pages. 1987.

Vol. 1272: M.S. Livšic, L.L. Waksman, Commuting Nonselfadjoint Operators in Hilbert Space. III, 115 pages. 1987.

Vol. 1273: G.-M. Greuel, G. Trautmann (Eds.), Singularities, Representation of Algebras, and Vector Bundles. Proceedings, 1985. XIV, 383 pages. 1987.

Vol. 1274: N. C. Phillips, Equivariant K-Theory and Freeness of Group Actions on C*-Algebras. VIII, 371 pages. 1987.

Vol. 1275: C.A. Berenstein (Ed.), Complex Analysis I. Proceedings, 1985–86. XV, 331 pages. 1987.

Vol. 1276: C.A. Berenstein (Ed.), Complex Analysis II. Proceedings, 1985–86. IX, 320 pages. 1987.

Vol. 1277: C.A. Berenstein (Ed.), Complex Analysis III. Proceedings, 1985–86. X, 350 pages. 1987.

Vol. 1278: S.S. Koh (Ed.), Invariant Theory. Proceedings, 1985. V, 102 pages. 1987.

Vol. 1279: D. Ieşan, Saint-Venant's Problem. VIII, 162 Seiten. 1987.

Vol. 1280: E. Neher, Jordan Triple Systems by the Grid Approach. XII, 193 pages. 1987.

Vol. 1281: O.H. Kegel, F. Menegazzo, G. Zacher (Eds.), Group Theory. Proceedings, 1986. VII, 179 pages. 1987.

Vol. 1282: D.E. Handelman, Positive Polynomials, Convex Integral Polytopes, and a Random Walk Problem. XI, 136 pages. 1987.

Vol. 1283: S. Mardešić, J. Segal (Eds.), Geometric Topology and Shape Theory. Proceedings, 1986. V, 261 pages. 1987.

Vol. 1284: B.H. Matzat, Konstruktive Galoistheorie. X, 286 pages. 1987.

Vol. 1285: I.W. Knowles, Y. Saitō (Eds.), Differential Equations and Mathematical Physics. Proceedings, 1986. XVI, 499 pages. 1987.

Vol. 1286: H.R. Miller, D.C. Ravenel (Eds.), Algebraic Topology. Proceedings, 1986. VII, 341 pages. 1987.

Vol. 1287: E.B. Saff (Ed.), Approximation Theory, Tampa. Proceedings, 1985–1986. V, 228 pages. 1987.

Vol. 1288: Yu. L. Rodin, Generalized Analytic Functions on Riemann Surfaces. V, 128 pages, 1987.

Vol. 1289: Yu. I. Manin (Ed.), K-Theory, Arithmetic and Geometry. Seminar, 1984–1986. V, 399 pages. 1987.